THE LIBRARY OF SCIENTIFIC THOUGHT

GENERAL EDITOR:

Paul Edwards, NEW YORK UNIVERSITY

Milton K. Munitz

EDITOR

THEORIES

of the

UNIVERSE

FROM BABYLONIAN MYTH TO MODERN SCIENCE

THE FREE PRESS

New York   London   Toronto   Sydney   Singapore

THE FREE PRESS
A Division of Simon & Schuster Inc.
1230 Avenue of the Americas
New York, NY 10020

Copyright © 1957 by The Free Press

First Free Press Paperback Edition 1965

THE FREE PRESS and colophon are trademarks
of Simon & Schuster Inc.

Manufactured in the United States of America

10

Library of Congress Catalog Card Number: 57-6746

ISBN: 0-02-922270-2

To the Memory of my Parents

ACKNOWLEDGMENTS

Grateful acknowledgment is made to the following publishers who have kindly granted permission to reprint copyrighted material:

The University of Chicago Press: *The Intellectual Adventure of Ancient Man* by H. Frankfort, et al., Copyright, 1946, by the University of Chicago Press.

Harvard University Press: *Kosmos* by W. de Sitter, Copyright, 1932, by the President and Fellows of Harvard College.

E. P. Dutton and Company: Aristotle *Metaphysics,* translated by John Warrington, Everyman's Library, Copyright, 1956, by J. M. Dent and Sons, Ltd.

Encyclopedia Britannica, Inc.: *Great Books of the Western World,* Copyright, 1952, by Encyclopedia Britannica, Inc.

Yale University Press: *The Realm of the Nebula* by Edwin Hubble, Copyright, 1936, by the Yale University Press.

Abelard-Schuman, Inc.: *Astronomical Thought in Renaissance England* by Francis R. Johnson, Copyright, 1937, by the Johns Hopkins Press.

University of California Press: *Dialogue Concerning the Two Chief World Systems* by Galileo Galilei, translated by Stillman Drake, Copyright, 1953, by The Regents of the University of California; Newton's *Principia* (a revision of Motte's translation), Copyright, 1934, by the Regents of the University of California.

Peter Smith: *Relativity* by Albert Einstein, translated by R. W. Lawson, Copyright, 1920, by Henry Holt and Company.

D. Van Nostrand Company: *The Primeval Atom* by Canon Georges Lemaitre, Copyright, 1950, by Editions du Griffon.

The Library of Living Philosophers, Inc.: *Albert Einstein: Philosopher-Scientist,* edited by P. A. Schilpp, Copyright, 1949, by The Library of Living Philosophers, Inc.

The Scientific American, Inc.: from *The Scientific American,* March 1954, Copyright, 1954, by The Scientific American, Inc.

Harper and Brothers: *The Nature of the Universe* by Fred Hoyle, Copyright, 1950, by Fred Hoyle.

Preface

THE FOLLOWING volume of selections is devoted to the subject of cosmology, the study of the astronomical or physical universe as a whole. It is intended for the general reader. The material is drawn wherever possible from the non-technical writings of those who played an important part in the main lines of development of the subject.

The questions which cosmology seeks to answer are perennial ones. Whoever looks up into the starry heavens is prompted to speculate about the beginning and end of things, about space, time, and creation. The questions as formulated by modern scientific cosmologists, and the methods employed to solve them are refinements upon the same queries raised by every man in his reflective moments. However, if we would understand such progress as has been made by more advanced techniques and the variety of current trends in scientific cosmology which we shall survey in this volume, it is important to see these against the background of earlier efforts and achievements. For it was out of these earlier gropings and partial advances that the ground was prepared for the present-day surge forward in our knowledge of the subject. Beginning with one of the earliest ventures in the form of myth, I have, accordingly, selected those materials which would exhibit the main stages of progress in cosmological inquiry from antiquity up to the present time.

The student of philosophy and particularly of the philosophy of science will find the selections here included of special interest since they offer a wealth of material for the analysis of concepts and methods in an important area of intellectual concern. The study of the "logic" of cosmology, as distinguished from the pursuit of cosmological inquiry itself, is both a critical examination of the procedures involved in obtaining knowledge of the universe and an analysis of the meaning or meanings of the concept "universe." It cannot be more profitably conducted than by a close study of texts such as these.

It almost goes without saying that I have not attempted to provide a sourcebook of relevant materials for the study of the entire history of cosmology in all its aspects, including subsidiary ones or offshoots from the main lines of advance, whether in its earlier or contemporary phases. Nor is this intended as a collection of those papers, generally written in the technical language of mathematics as in the recent literature of the subject, that would be of interest primarily to the specialist. In view, however, of the general dearth of books in this area, it is hoped that the present volume at least will do something to satisfy a growing and already widespread interest in the subject.

I wish to thank Dr. Charles A. Muses of The Falcon's Wing Press and Mr. Jeremiah Kaplan of The Free Press for their splendid co-operation and thoughtful consideration of the many details that went into the production of this book as well as a companion volume *Space, Time and Creation*. I am most grateful also to Professor Paul Edwards as Editor of The Library of Scentific Thought for his many helpful suggestions, and to my wife, Dr. Lenore B. Munitz, for her valuable assistance in the preparation of this volume.

M. K. M.

Contents

General Introduction

THE PAST several decades have seen remarkable advances made in the scientific study of the physical universe on its most inclusive astronomical scale. Giant telescopes, such as those at Mount Wilson and Palomar Mountain have probed the heavens to depths hitherto beyond reach by observational means. The light coming from these far reaches of space travels over distances that needs to be reckoned in terms of billions of light-years. The same light, moreover, tells of events that occurred billions of years ago. The data gathered by these telescopes and various other auxiliary instruments inform us of a universe whose major constituents are now recognized to be galaxies and clusters of galaxies. It is to one such galaxy, a vast swarm of stars separated from other similar systems by enormous distances, that our own sun and its planetary family belongs.

In scientific cosmology, as in other branches of science, the work of the observer is supplemented by and co-ordinated with that of the theorist. The latter constructs various "models" of the universe as a whole that provide a pattern of intelligibility for what is already discovered by the astronomer. These models also help the astronomer in his search for fresh empirical materials. The models are grounded in the conceptual resources of mathematical physics and, since the path-breaking suggestions of Einstein in 1917, have generally exploited the leads provided by relativity theory.

The present scene in cosmology is one of great activity, indeed the most intense in the entire history of the subject. It is one in which constant additions and corrections are made in the storehouse of accumulated facts. It is one, too, in which, despite certain broad areas of agreement, wide-ranging differences on the level of theory have come to identify various "schools of thought." Through patient research and sifting of differences, a general pattern of growth in information and insight has already begun to characterize this most recent phase of interest in what are, after all, age-old questions. It is a pattern that may be expected to continue.

It will prove helpful to distinguish for the purposes of the present survey *four* main stages in the general development of our subject. The first stage is characterized by a changeover from the methods pursued by mythical cosmogonies to the more rational, embryonic scientific speculations of the pre-Socratic Ionian and Pythagorean schools. The second stage is marked by the emergence in the classic period of Greek philosophy and astronomy of the geocentric-finite cosmology as this is expounded by Plato, Aristotle and

1

Ptolemy. It was basically this cosmology which remained the orthodox point of view of enlightened thought until its gradual collapse and overthrow at the beginning of the modern era. The third stage stretches from the period of mounting attacks on the classic cosmology in the very late Middle Ages and, continuing on its constructive side with the contributions of such figures as Copernicus, Newton, Kant, Herschel and others, comes to an end at the turn of the present century. This third phase is marked by astronomic discoveries on an observational level and theoretical trials of alternative possibilities on the conceptual side that at once carry the subject forward and prepare the ground for the current phase. The fourth period may be said to begin in the second decade of the present century. Here the use of great telescopes effected a breakthrough into the realm of galaxies and an exploration of its properties, while the work of the theoretical physicist from Einstein's to Hoyle's, making use of the most advanced tools of mathematics, led to the construction of various improved models of the universe. It was out of such labors that current conceptions of "the expanding universe" were born.

First Steps in

Cosmological Speculation

Introduction

IN THE ATTEMPT to make the brute facts of experience intelligible, analogy plays a fundamental role. By means of it what is already familiar or understood is appealed to in order to make clear the unfamiliar and the unexplained. The use of analogy runs as a common thread from the earliest and crudest efforts of myth to the latest and most sophisticated reaches of science, though, of course, the types of analogies used in these contrasting areas will be widely different as will the sanctions for their use.

Cosmological speculation when it first makes its appearance at the dawn of intellectual history takes the form of myth. Here one finds the use of imagery borrowed from some familiar area of human experience as a basis for making intelligible the origin and structure of the universe as a whole. What distinguishes such myths from later philosophical and scientific efforts is the fact that the imagery selected is of a type which these more sophisticated accounts discard as too anthropomorphic and generally inappropriate. For the favorite analogies appealed to in myth consist basically of three types: craftsmanship as practiced by some artificer or creator, the process of biological generation from seeds or eggs, and the imposition of a social order by some powerful authority to yield a community living according to law. Most cosmogonic myths the world over employ the models furnished by art, biological reproduction, or the pattern of submission to focused authority (often commingled or superimposed on one another in a particular account) as the springboards from which they would account for the coming-into-existence of the world and the structure it is found to possess in its gross astronomical features.

Among such myths, that which goes by the name *Enuma Elish,* being the Babylonian and Assyrian account of how the world came into existence, is of central importance and great interest. Apart from being a well-developed illustration of cosmogonic myth among ancient peoples, it provides many significant parallels to the account to be found in the Biblical Book of Genesis. Thorkild Jacobsen's analysis of the contents of this myth, in what follows, lays bare not only its main themes but also the background of geographical and cultural facts in terms of which these themes become meaningfully related. The discerning reader will note the way in which the genesis of the universe is variously ascribed to the operation on a cosmic scale of processes of craftsmanship, biological generation, and the coming-into-existence of order based on authoritative command. The faint echoes of these modes of imagery

5

are still to be found in portions of the Bible, indicating an undoubted continuity in the traditions that provided the materials for the composition of both works.

The great achievement of the pre-Socratic philosophers was that of liberating the subject of cosmology from the use of myth. They substituted for its study purely physical ideas, or in addition, as in the case of the Pythagoreans, mathematical ones. The concepts were, to be sure, crude, as were the observational materials upon which they were directed. But the step forward was enormous. Yet even here, as F. M. Cornford shows in his discussion of the "Pattern of Ionian Cosmogony," the break with the past was neither as absolute nor as thoroughgoing as it is frequently assumed to be. The pattern set by cosmogonic myths continues to provide many of the presuppositions and indeed the over-all framework within which such more rational efforts and speculative trials were developed.

The type of thinking initiated by the Milesian school of pre-Socratic thinkers—Thales, Anaximander and Anaximenes—in the sixth century B.C. was carried forward in many directions. One of the most remarkable outcomes of such speculations, representing a culmination of their materialistic thought, was to be found in the Atomist school. Originally worked out in its main features by Leucippus and Democritus in the fifth century B.C., the teachings of atomism were later adopted as a basis for the primarily ethical philosophy of Epicureanism. The great work of the Roman poet Lucretius, *The Nature of the Universe (De Rerum Natura)*, of which selections are given below, belongs to this tradition. In it we find systematically presented the basic axioms on which the atomist philosophy is grounded and the consequences drawn from them for an understanding of the physical universe as a whole. It elaborates the conception of a universe whose order arises out of a blind interplay of atoms rather than as a product of deliberate design; of a universe boundless in spatial extent, infinite in its duration and containing innumerable worlds in various stages of development or decay. It was this conception of an infinite and, at bottom, irrational universe against which Plato, Aristotle, and the whole tradition of theologically oriented thought in Western culture set themselves in sharp and fundamental opposition. It was the same conception, however, which once more came into the foreground of attention at the dawn of modern thought and has remained up to the present time an inspiration for those modes of scientific thinking that renounce any appeal to teleology in the interpretation of physical phenomena.

The contributions of the Pythagorean school to the early development of cosmology, beginning in the sixth century B.C., were twofold: (1) it stressed the concept that the universe is indeed a *cosmos,* an orderly pattern whose formal structure can be grasped and expressed in the language of mathematics, the language of figure and number; (2) through the speculations of Philolaus, a member of the school in the fifth century B.C. (whose work is recounted by Theodor Gomperz below), it paved the way for later radical astronomical hypotheses. Among these were the views of Heracleides of Pontus (*c.* 388-310

B.C.) who defended the idea of the daily rotation of the earth on its axis, and that the sun, not the earth, was the center of the orbits of Venus and Mercury, and the non-geocentric cosmological views of Aristarchus of Samos, the "ancient Copernicus" (c. 310-230 B.C.) who taught both the daily rotation of the earth and its annual revolution about the sun.

The recognition of the value of mathematics as the means for making intelligible the orderly regularities observed in the motions of the heavenly bodies was, of course, of momentous importance for ancient astronomy and for the subsequent career of science generally. On its constructive side, the Pythagorean vision as mediated by Plato (who was steeped in their thought) led to the development of the first successful theory of planetary motions at the hands of Plato's pupil, Eudoxus of Cnidus, in the form of the theory of homocentric or concentric spheres. Even when this theory was abandoned later in favor of the use of eccentrics and epicycles by Hipparchus and Ptolemy, it was still the original Pythagorean confidence in the power of mathematics which remained, as it has up to the present, at the core of scientific theorizing. The other seminal idea in ancient astronomy due to the Pythagoreans, the conception of a non-geocentric cosmology, although it was tied up with the belief in a fictitious "central fire" (not to be confused with the sun), was an idea that did not gain any marked acceptance in ancient thought. The dominant cosmology in antiquity and the Middle Ages settled down to the elaboration of a world-view which bypassed this brief and brilliant venture in speculative astronomy. Once more it was the revival of science at the beginning of the modern era which was to bring to fruition and exploit the possibilities latent in these suggestions, beginning with the epoch-making work of Copernicus.

THORKILD JACOBSEN:

Enuma Elish—"The Babylonian Genesis"

A PROPER cosmogony treating of the fundamental problems of the cosmos as it appeared to the Mesopotamians—its origin and the origin of the order which it exhibits—does not appear until the earlier half of the second millennium B.C. Then it is given in a grandiose composition named *Enuma elish,* 'When above.'[1] *Enuma elish* has a long and complicated history. It is written in Akkadian,[2] seemingly Akkadian of approximately the middle of the second millennium B.C. At that period, then, the composition presumably received the form in which we now have it. Its central figure is Marduk, the god of Babylon, in keeping with the fact that Babylon was at that time the political and cultural centre of the Mesopotamian world. When later on, in the first millennium B.C., Assyria rose to become the dominant power in the Near East, Assyrian scribes apparently replaced Marduk with their own god Assur and made a few changes to make the story fit its new hero. This later version is known to us from copies of the myth found in Assyria.

The substitution of Assur for Marduk as the hero and central figure of the story seems to have been neither the only nor the first such substitution made. Behind our present version with Marduk as the hero undoubtedly lies a still earlier version wherein, not Marduk, but Enlil of Nippur played the central role. This more original form can be deduced from many indications in the myth itself. The most important of these is the fact that Enlil, although he was always at least the second most important Mesopotamian deity, seems to play no part whatever in the myth as we have it, while all the other important gods have appropriate roles. Again, the role which Marduk plays is not in keeping with the character of the god. Marduk was originally an agricultural or perhaps a solar deity, whereas the central role in *Enuma elish* is that of a

From H. and H. A. Frankfort, John A. Wilson, Thorkild Jacobsen, William A. Irwin, *The Intellectual Adventure of Ancient Man,* Chicago, 1946, pp. 168-83. Reprinted with the kind permission of The University of Chicago Press. Also published under the title *Before Philosophy* in Pelican Books.
1. Latest translation: A. Heidel, *The Babylonian Genesis,* University of Chicago Press, 1951. See literature there quoted.
2. A Semitic language which had long been spoken side by side with Sumerian in Mesopotamia and which by the end of the third millennium B.C. completely superseded its rival and became the only language spoken in the country.

god of the storm such as Enlil was. Indeed, a central feat ascribed to Marduk in the story—the separating of heaven and earth—is the very feat which other mythological material assigns to Enlil, and with right, for it is the wind which, placed between the sky and the earth, holds them apart like the two sides of an inflated leather bag. It seems, therefore, that Enlil was the original hero of the story and was replaced by Marduk when our earliest known version was composed around the middle of the second millennium B.C. How far the myth itself goes back, we cannot say with certainty. It contains material and reflects ideas which point backward through the third millennium B.C.

A. Fundamentals of Origin

We may now turn to the content of the myth. It falls roughly into two sections, one dealing with the origin of the basic features of the universe, the other telling how the present world order was established. There is, however, no rigid separation of these two themes. The actions of the second part of the myth are foreshadowed in, and interlock with, the events told in the first.

The poem begins with a description of the universe as it was in the beginning:

> When a sky above had not (yet even) been mentioned
> (And) the name of firm ground below had not (yet even) been
>   thought of;
> (When) only primeval Apsu, their begetter,
> And Mummu and Ti'amat—she who gave birth to them all—
> Were mingling their waters in one;
> When no bog had formed (and) no island could be found;
> When no god whosoever had appeared,
> Had been named by name, had been determined as to (his) lot,
> Then were gods formed within them.[3]

This description presents the earliest stage of the universe as one of watery chaos. The chaos consisted of three intermingled elements: Apsu, who represents the sweet waters; Ti'amat, who represents the sea; and Mummu, who cannot as yet be identified with certainty but may represent cloud banks and mist. These three types of water were mingled in a large undefined mass. There was not yet even the idea of a sky above or firm ground beneath; all was water; not even a swampy bog had been formed, still less an island; and there were yet no gods.

Then, in the midst of this watery chaos, two gods come into existence: Lahmu and Lahamu. The text clearly intends us to understand that they were begotten by Apsu, the sweet waters, and born of Ti'amat, the sea. They represent, it would seem, silt which had formed in the waters. From Lahmu and Lahamu derive the next divine pair: Anshar and Kishar, two aspects of

---

3. I.e., within Apsu, Mummu and Ti'amat.

'the horizon'. The myth-maker apparently viewed the horizon as both male and female, as a circle (male) which circumscribed the sky and as a circle (female) which circumscribed the earth.

Anshar and Kishar give birth to Anu, the god of the sky; and Anu engenders Nudimmut. Nudimmut is another name for Ea or Enki, the god of the sweet waters. Here, however, he is apparently to be viewed in his oldest aspect as representing the earth itself; he is *En-ki,* 'lord of the earth'. Anshar is said to have made Anu like himself, for the sky resembles the horizon in so far as it, too, is round. And Anu is said to have made Nudimmut, the earth, in his likeness; for the earth was, in the opinion of the Mesopotamians, shaped like a disc or even like a round bowl:

> Lahmu and Lahamu appeared and they were named;
> Increasing through the ages they grew tall.
> Anshar and Kishar (then) were formed, surpassing them;
> They lived for many days, adding year unto year.
> Their son was Anu, equal to his fathers.
> Anshar made his firstborn, Anu, to his own likeness,
> Anu, to his own likeness also, Nudimmut.
> Nudimmut excelled among the gods, his fathers;
> With ears wide open, wise, mighty in strength,
> Mightier than his father's father Anshar,
> He had no equal among his fellow-gods.

The speculations which here meet us, speculations by which the ancient Mesopotamians thought to penetrate the mystery concealing the origin of the universe, are obviously based upon observation of the way in which new land is actually formed in Mesopotamia. Mesopotamia is an alluvial country. It has been built through thousands of years by silt which has been brought down by the two great rivers, the Euphrates and the Tigris, and has been deposited at their mouths. This process still goes on; and day by day, year by year, the country slowly grows, extending farther out into the Persian Gulf. It is this scene—where the sweet waters of the rivers meet and blend with the salt waters of the sea, while cloud banks hang low over the waters—which has been projected back into the beginning of time. Here still is the primeval watery chaos in which Apsu, the sweet waters, mingles with Ti'amat, the salt waters of the sea; and here the silt—represented by the first of the gods, Lahmu and Lahamu—separates from the water, becomes noticeable, is deposited.

Lahmu and Lahamu gave birth to Anshar and Kishar; that is, the primeval silt, born of the salt and the sweet waters in the original watery chaos, was deposited along its circumference in a gigantic ring: the horizon. From Anshar, the upper side of this ring, and from Kishar, its lower side, grew up through days and years of deposits Anu, heaven, and Nudimmut-Enki, earth. As *Enuma elish* describes this, Anu, the sky, was formed first; and he engendered Nudimmut, the earth.

This presentation breaks the progression by pairs—Lahmu-Lahamu, Anshar-Kishar—after which we expect a third pair An-Ki, 'heaven and earth'; instead, we get Anu followed by Nudimmut. This irregularity suggests that we are here dealing with an alteration of the original story perhaps made by the redactor who introduced Marduk of Babylon as hero of the myth. He may have wanted to stress the male aspect of the earth, Ea-Enki, since the latter figured as father of Marduk in Babylonian theology. Originally, therefore, Anshar-Kishar may have been followed by An-Ki, 'heaven and earth'. This conjecture is supported by a variant of our story preserved in the great ancient Mesopotamian list of gods known as the An-Anum list. Here we find an earlier, more intact version of the speculation: from the horizon, from Anshar and Kishar as a united pair, grew the sky and the earth. Sky and earth are apparently to be viewed as two enormous discs formed from the silt which continued to be deposited along the inside of the ring of the horizon as the latter 'lived many days, added year unto year'. Later on, these discs were forced apart by the wind, who puffed them up into the great bag within which we live, its under side being the earth, its upper side the sky.

In speculating about the origin of the world, the Mesopotamians thus took as their point of departure things they knew and could observe in the geology of their own country. Their earth, Mesopotamia, is formed by silt deposited where fresh water meets salt water; the sky, seemingly formed of solid matter like the earth, must have been deposited in the same manner and must have been raised later to its present lofty position.

B. Fundamentals of World Order

Just as observed facts about the physical origin of his own country form the basis for the Mesopotamian's speculations about the origin of the basic features in the universe, so, it would seem, does a certain amount of knowledge about the origin of his own political organization govern his speculations as to the origin of the organization of the universe. The origin of the world order is seen in a prolonged conflict between two principles, the forces making for activity and the forces making for inactivity. In this conflict the first victory over inactivity is gained by authority alone; the second, the decisive victory, by authority combined with force. The transition mirrors, on the one hand, a historical development from primitive social organization, in which only custom and authority unbacked by force are available to ensure concerted action by the community, to the organization of a real state, in which the ruler commands both authority and force to ensure necessary concerted action. On the other hand, it reflects the normal procedure within the organized state, for here also authority alone is the means brought to bear first, while force, physical compulsion, is only resorted to if authority is not sufficient to produce the conduct desired.

To return to *Enuma elish*: With the birth of the gods from chaos, a new principle—movement, activity—has come into the world. The new beings contrast sharply with the forces of chaos that stand for rest and inactivity. In a typically mythopoeic manner this ideal conflict of activity and inactivity is given concrete form in a pregnant situation: the gods come together to dance.

> The divine companions thronged together
> and, restlessly surging back and forth, they disturbed Ti'amat,
> disturbed Ti'amat's belly,
> dancing within (her depth) where heaven is founded.
> Apsu could not subdue their clamour,
> and Ti'amat was silent . . .
> but their actions were abhorrent to her
> and their ways not good. . . .

The conflict is now manifest. The first power of chaos to come out openly against the gods and their new ways is Apsu.

> Then Apsu, the begetter of the great gods,
> called his servant Mummu, saying to him:
> 'Mummu, my servant, who dost gladden my heart,
> come let us go to Ti'amat.'
> They went; and seated before Ti'amat,
> about the gods their firstborn they took counsel.
>
> Apsu began to speak,
> saying to pure Ti'amat:
> 'Abhorrent have become their ways to me,
> I am allowed no rest by day, by night no sleep.
> I will abolish, yea, I will destroy their ways,
> that peace may reign (again) and we may sleep.'

This news causes consternation among the gods. They run around aimlessly; then they quiet down and sit in the silence of despair. Only one, the wise Ea-Enki, is equal to the situation.

> He of supreme intelligence, skilful, ingenious,
> Ea, who knows all things, saw through their scheme.
> He formed, yea, he set up against it
> the configuration of the universe,
> and skilfully made his overpowering sacred spell.
> Reciting it he cast it on the water (—on Apsu—),
> poured slumber over him, so that he soundly slept.

The waters to which Ea here recites his spell, his 'configuration of the universe', are Apsu. Apsu succumbs to the magic command and falls into a deep slumber. Then Ea takes from him his crown and drapes himself in Apsu's cloak of fiery rays. He kills Apsu and establishes his abode above him. Then he locks up Mummu, passes a string through his nose, and sits holding him by the end of this nose-rope.

What all this signifies is perhaps not immediately evident; yet it can be understood. The means which Ea employs to subdue Apsu is a spell, that is, a word of power, an authoritative command. For the Mesopotamians viewed authority as a power inherent in commands, a power which caused a command to be obeyed, caused it to realize itself, to come true. The authority, the power in Ea's command, was great enough to force into being the situation expressed in the command. And the nature of this situation is hinted at when it is called 'the configuration of the universe'; it is the design which now obtains. Ea commanded that things should be as they are, and so they became thus. Apsu, the sweet waters, sank into the sleep of death which now holds the sweet waters immobile underground. Directly above them was established the abode of Ea—earth resting upon Apsu. Ea holds in his hands the nose-rope of captive Mummu, perhaps—if our interpretation of this difficult figure is correct—the cloud banks which float low over the earth. But whatever the details of interpretation may be, it is significant that this first great victory of the gods over the powers of chaos, of the forces of activity over the forces opposing activity, was won through authority and not through physical force. It was gained through the authority implicit in a command, the magic in a spell. It is significant also that it was gained through the power of a single god acting on his own initiative, not by the concerted efforts of the whole community of the gods. The myth moves on a primitive level of social organization where dangers to the community are met by the separate action of one or more powerful individuals, not by co-operation of the community as a whole.

To return to the story: In the dwelling which Ea has thus established on Apsu is born Marduk, the real hero of the myth as we have it; but in more original versions it was undoubtedly Enlil's birth that was told at this juncture. The text describes him:

> Superb of stature, with lightning glance,
> and virile gait, he was a leader born.
> Ea his father, seeing him, rejoiced,
> and brightened and his heart filled with delight.
> He added, yea, he fastened on to him twofold divinity.
> Exceeding tall he was, surpassing in all things.
> Subtle beyond conceit his measure was,
> incomprehensible, terrible to behold.
> Four were his eyes and four his ears;
> fire blazed whenever he moved his lips.

But while Marduk grows up among the gods, new dangers threaten from the forces of chaos. They maliciously chide Ti'amat:

> When they killed Apsu, thy husband,
> thou didst not march at his side but sat quietly.

Finally they succeed in rousing her. Soon the gods hear that all the forces of chaos are making ready to do battle with them:

> Angry, scheming, restless day and night,
> they are bent on fighting, rage and prowl like lions.

> Gathered in council, they plan the attack.
> Mother Hubur—creator of all forms—
> adds irresistible weapons, has borne monster serpents,
> sharp toothed, with fang unsparing;
> has filled their bodies with poison for blood.
> Fierce dragons she has draped with terror,
> crowned with flame and made like gods,
> so that whoever looks upon them shall perish with fear,
> and they, with bodies raised, will not turn back their breast.

At the head of her formidable army Ti'amat has placed her second husband, Kingu. She has given him full authority and entrusted to him the 'tablets of destinies', which symbolize supreme power over the universe. Her forces are ranged in battle order ready to attack the gods.

The first intelligence of what is afoot reaches the always well-informed Ea. At first, a typical primitive reaction, he is completely stunned, and it takes some time before he can pull himself together and begin to act.

> Ea heard of these matters,
> lapsed into dark silence, wordlessly sat.
> Then, having deeply pondered and his inner turmoil quieted,
> arose and went to his father Anshar,
> went before Anshar, his father who begot him.
> All Ti'amat had plotted he recounted.

Anshar also is deeply disturbed and smites his thigh and bites his lip in his mental anguish. He can think of no better way out than to send Ea against Ti'amat. He reminds Ea of his victory over Apsu and Mummu and seems to advise him to use the same means he used then. But this time Ea's mission is unsuccessful. The word of an individual, even the powerful word of Ea, is no match for Ti'amat and her host.

Anshar then turns to Anu and bids him go. Anu is armed with authority even greater than that of Ea, for he is told:

> If she obey not thy command,
> speak unto her our command, that she may subside.

If Ti'amat cannot be overpowered by the authority of any one god, the command of all gods, having behind it their combined authority, must be used against her. But that, too, fails; Anu is unable to face Ti'amat, returns to Anshar, and asks to be relieved of the task. Unaided authority, even the highest which the gods command, is not enough. Now the gods face their hour of gravest peril. Anshar, who has thus far directed the proceedings, falls silent.

> Anshar grew silent, staring at the ground,
> he shook his head, nodded toward Ea.
> Ranged in assembly, all the Anunnaki
> lips covered, speechless sat.

Then, finally, rising in all his majesty, Anshar proposes that Ea's son, young

Marduk, 'whose strength is mighty', champion his fathers, the gods. Ea is willing to put the proposal to Marduk, who accepts readily enough but not without a condition:

> If I am to be your champion,
> vanquish Ti'amat, and save you,
> then assemble and proclaim my lot supreme.
> Sit down together joyfully in Ubshuukkinna;
> let me, like you, by word of mouth determine destiny,
> so that whatever I decide shall not be altered,
> and my spoken command shall not (come) back (to me),
>   shall not be changed.

Marduk is a young god. He has abundant strength, the full prowess of youth, and he looks ahead to the physical contest with complete confidence. But, as a young man, he lacks influence. It is for authority on a par with that of the powerful senior members of the community that he asks. A new and unheard-of union of powers is here envisaged: his demand foreshadows the coming state with its combination of force and authority in the person of the king.

And so the call goes out, and the gods foregather in Ubshuukkinna, the court of assembly in Nippur. As they arrive, they meet friends and relatives who have similarly come to participate in the assembly, and there is general embracing. In the sheltered court the gods sit down to a sumptuous meal; wine and strong drink soon put them in a happy and carefree mood, fears and worries vanish, and the meeting is ready to settle down to more serious affairs.

> They smacked their tongues and sat down to the feast;
> They ate and drank,
> Sweet drink dispelled their fears.
> They sang for joy, drinking strong wine.
> Carefree they grew, exceedingly, their hearts elated.
> Of Marduk, (of) their champion, they decreed the destiny.

The 'destiny' mentioned is full authority on a par with that of the highest gods. The assembly first gives Marduk a seat of honour and then proceeds to confer the new powers on him:

> They made a princely dais for him.
> And he sat down, facing his fathers, as a councillor.
> 'Thou are of consequence among the elder gods.
> Thy rank is unsurpassed and thy command is Anu('s).
> Marduk, thou are of consequence among the elder gods;
> Thy rank is unequalled and thy command is Anu('s).
> From this day onward shall thy orders not be altered;
> To elevate and to abase—this be within thy power.
> What thou hast spoken shall come true, thy word shall
>   not prove vain.
> Among the gods none shall encroach upon thy rights.'

What the assembly of the gods here confers upon Marduk is kingship: the

combination of authority with powers of compulsion; a leading voice in the
counsels of peace; leadership of the army in times of war; police powers to
penalize evildoers.

> We gave thee kingship, power over all things.
> Take thy seat in the council, may thy word prevail.
> May thy weapon not yield, may it smite thy foes.
> Grant breath of life to lord(s) who put (their) trust
>   in thee.
> But if a god embraces evil, shed his life.

Having conferred authority upon Marduk, the gods want to know that he
really has it, that his command now possesses that magic quality which makes
it come true. So they make a test:

> They placed a garment in their midst
> And said to Marduk their firstborn:
> 'O Lord, thy lot is truly highest among gods.
> Command annihilation and existence, and may both
>   come true.
> May thy spoken word destroy the garment,
> Then speak again and may it be intact.'
> He spoke—and at his word the garment was destroyed.
> He spoke again, the garment reappeared.
> The gods, his fathers, seeing (the power of) his word,
> Rejoiced, paid homage: 'Marduk is king.'

Then they give him the insignia of kingship—sceptre, throne, and royal robe(?)
—and arm him for the coming conflict. Marduk's weapons are the weapons
of a god of storm and thunder—a circumstance understandable when we
remember that the story was originally the story of the storm-god Enlil. He
carries the rainbow, the arrows of lightning, and a net held by four winds.

> He made a bow, designed it as his weapon,
> let the arrow ride firmly on the bowstring.
> Grasping his mace in his right hand, he lifted it;
> and fastened bow and quiver at his side.
> He bade lightning precede him,
> and made his body burn with searing flame.
> He made a net to encircle Ti'amat,
> bade the four winds hold on, that none of her escape.
> The south wind, north wind, east wind, west wind,
> Gifts from his father Anu, did he place along the edges
>   of the net.

In addition, he fashions seven terrible storms, lifts up his mace, which is the
flood, mounts his war chariot, 'the irresistible tempest', and rides to battle
against Ti'amat with his army, the gods milling around him.

At the approach of Marduk, Kingu and the enemy army lose heart and
are plunged into utter confusion; only Ti'amat stands her ground and
challenges the young god to battle. Marduk returns the challenge, and the
fight is on. Spreading his mighty net, Marduk envelops Ti'amat in its meshes.

As she opens her jaws to swallow him, he sends in the winds to hold them open. The winds swell her body, and through her open mouth Marduk shoots an arrow which pierces her heart and kills her. When her followers see Marduk treading on their dead champion, they turn and try to flee; but they are caught in the meshes of his net, and he breaks their weapons and takes them captive. Kingu also is bound, and Marduk takes from him the 'tablets of destinies'.

When complete victory has thus been achieved, Marduk returns to Ti'amat's body, crushes her skull with his mace, and cuts her arteries; and the winds carry her blood away. Then he proceeds to cut her body in two and to lift up half of it to form the sky. To make sure that the waters in it will not escape, he sets up locks and appoints guards. He carefully measures the sky which he has thus made; and, as Ea after his victory over Apsu had built his abode on the body of his dead opponent, so now Marduk builds his abode on that part of Ti'amat's body which he has made into the sky. By measuring he makes certain that it comes directly opposite Ea's dwelling to form a counterpart of it.

Here we may pause again for a moment to ask what all this means. At the root of the battle between Marduk or Enlil and Ti'amat, between wind and water, there probably lies an age-old interpretation of the spring floods. Every spring the waters flood the Mesopotamian plain and the world reverts to a— or rather to 'the'—primeval watery chaos until the winds fight the waters, dry them up, and bring back the dry land. Remnants of this concept may be seen in the detail that the winds carry away Ti'amat's blood. But such age-old concepts had early become vehicles for cosmological speculation. We have already mentioned the existence of a view that heaven and earth were two great discs deposited by silt in the watery chaos and forced apart by the wind, so that the present universe is a sort of inflated sack surrounded by waters above and below. This speculation has left clear traces in Sumerian myths and in the An-Anum list, and here in *Enuma elish* we have a variant of it: it is the primeval sea, Ti'amat, that is blown up and killed by the winds. Half of her—the present sea—is left down here; the other half is formed into the sky, and locks are affixed so that the water does not escape except once in a while when some of it falls down as rain.

Thus, through the use which it makes of its mythological material, *Enuma elish* accounts in two ways for the creation of the sky. First, the sky comes into being in the person of the god Anu, whose name means sky and who is the god of the sky; then, again, the sky is fashioned by the wind-god out of half of the body of the sea.

In a period, however, when emphasis had already shifted from the visual aspects of the great components of the universe to the powers felt as active in and through them, Anu, as the power behind the sky, would already be felt as sufficiently different from the sky itself to make this inherent contradiction less acute.

Quite as significant as the direct cosmological identification of the actors in these events, however, is the bearing which the events have on the establishing of the cosmic order. Under pressure of an acute crisis, a threatening war, a more or less primitively organized society has developed into a state.

Evaluating this achievement in modern, and admittedly subjective, terms, we might say that the powers of movement and activity, the gods, have won their final and decisive victory over the powers of rest and inertia. To accomplish this, they have had to exert themselves to the utmost, and they have found a method, a form of organization, which permits them to pull their full weight. As the active forces in a society become integrated in the form of the state and thus can overcome the ever threatening tendencies to chaos and inertia, so the active forces in the Mesopotamian universe through that same form, the state, overcome and defeat the powers of chaos, of inactivity and inertia. But, however that may be, this much is certain—that the crisis has imposed upon the gods a state of the type of Primitive Democracy. All major issues are dealt with in a general assembly, where decrees are confirmed, designs are formulated, and judgments are pronounced. To each god is assigned a station, the most important going to the fifty senior gods, among whom are the seven whose opinion is decisive. In addition to this legislative and judiciary assembly, however, there is now an executive, the young king, who is equal in authority to the most influential members of the assembly, is the leader of the army in war, the punisher of evildoers in peacetime, and generally active, with the assent of the assembly, in matters of internal organization.

It is to tasks of internal organization that Marduk turns after his victory. The first was organizing the calendar—ever a matter for the ruler of Mesopotamia. On the sky which he had fashioned he set up constellations of stars to determine, by their rising and setting, the year, the months, and the days. The 'station' of the planet Jupiter was established to make known the 'duties' of the days, when each had to appear:

> To make known their obligations,
> that none might do wrong or be remiss.

He also set on heaven two bands known as 'the ways' of Enlil and Ea. On both sides of the sky, where the sun comes out in the morning and leaves in the evening, Marduk made gates and secured them with strong locks. In the midst of the sky he fixed the zenith, and he made the moon shine forth and gave it its orders.

> He bade the moon come forth; entrusted night to her;
> Made her a creature of the dark, to measure time;
> And every month, unfailingly, adorned her with a crown.
> 'At the beginning of the month, when rising over the land,
> Thy shining horns six days shall measure;
> On the seventh day let half (thy) crown (appear).
> At full moon thou shalt face the sun.

. . . .
(But) when the sun starts gaining on thee in the depth
    of heaven,
Decrease thy radiance, reverse its growth.'

The text goes on with still more detailed orders.

Many further innovations introduced by the energetic young ruler are lost in a large lacuna which breaks the text at this point. When the text becomes readable again, Marduk—seemingly in response to a plea from them—is occupied with plans for relieving the gods of all toilsome menial tasks and for organizing them into two great groups:

> Arteries I will knot and bring bones into being.
> I will create Lullu, 'man' be his name,
> I will form Lullu, man.
> Let him be burdened with the toil of the gods,
>   that they may freely breathe.
> Next, I will dispose of the ways of gods;
> Verily— they are clustered like a ball,
>   I shall make them distinct.

Distinct, that is, in two groups. Following a suggestion of his father, Ea, Marduk then calls the gods to assembly; and in the assembly he asks them, now functioning as a court, to state who it was who was responsible for the attack, who stirred up Ti'amat. And the assembly indicts Kingu. So Kingu is bound and executed, and from his blood mankind is created under Ea's direction.

> They bound him, held him before Ea,
> Condemned him, severed his arteries.
> And from his blood they formed mankind.
> Ea then toil imposed on man, and set gods free.

The exceeding skill which went to fashion man commands the admiration of our poet.

> That work was not meet for (human) understanding.
> (Acting) on Marduk's ingenious suggestions Ea created.

Thereupon Marduk divided the gods and assigned them to Anu, to abide by Anu's instructions. Three hundred he stationed in heaven to do guard duty, and another three hundred were given tasks on earth. Thus the divine forces were organized and assigned to their appropriate tasks throughout the universe.

The gods are truly grateful for Marduk's efforts. To express their gratitude, they take pick in hand for the last time and build him a city and temple with throne daises for each of the gods to use when they meet there for assembly. The first assembly is held on the occasion of the dedication of the temple. As usual, the gods first sit down to a banquet. Thereupon matters of state are discussed and decided, and then, when the current business has been disposed of, Anu rises to confirm Marduk's position as king. He determines the eternal status of Marduk's weapon, the bow; he determines the status of his throne; and, finally, he calls upon the assembled gods to confirm and determine Marduk's own status, his functions in the universe, by recounting his fifty names, each expressing one aspect of his being, each defining one of his functions. With the catalogue of these names the poem comes to an end. The names summarize what Marduk is and what he signifies: the final victory over chaos and the establishing of the ordered, organized universe, the cosmic state of the Mesopotamians.

F. M. CORNFORD:

Pattern of Ionian Cosmogony

HISTORIES of philosophy and of natural science begin with the earliest Ionian system, initiated by Thales, rounded out by Anaximander, and somewhat simplified by Anaximenes. Every reader is struck by the rationalism which distinguishes it from mythical cosmogonies. This characteristic must certainly not be underrated. The Milesians brought into the world of common experience much that had previously lain beyond that world. It is difficult for us to recover the attitude of mind of a Hesiod towards his vision of the past. As he looked back in time from his own age and the life he dealt with every day, past the earlier ages—the heroic, the bronze, the silver, the golden—to the dominion of Cronos, to the elder gods and to the birth of these gods themselves from the mysterious marriage of Heaven and Earth, it must have seemed that the world became less and less like the familiar scene. The events—the marriage and birth of gods, the war of Olympians and Titans, the Prometheus legend—were not events of the same order as what happened in Boeotia in Hesiod's day. We may get a similar impression by thinking of the Book of Genesis. As we follow the story from the Creation, through the series of mythical events which the Hebrews took over from Babylon, down to the call of Abraham, we seem to emerge gradually into the world we know, peopled with men like ourselves. So the past must have looked to everybody before the appearance of Ionian rationalism. It was an extraordinary feat to dissipate the haze of myth from the origins of the world and of life. The Milesian system pushed back to the very beginning of things the operation of processes as familiar and ordinary as a shower of rain. It made the formation of the world no longer a supernatural, but a natural event. Thanks to the Ionians, and to no one else, this has become the universal premiss of all modern science.

But there is something to be added on the other side. If we give up the idea that philosophy or science is a motherless Athena, an entirely new discipline breaking in from nowhere upon a culture hitherto dominated by poetical and mystical theologians, we shall see that the process of rationalization

From F. M. Cornford, *Principium Sapientiae*, Cambridge, 1952, Chapter XI. Reprinted with the kind permission of Cambridge University Press.

had been at work for some considerable time before Thales was born. We shall also take note of the re-emergence in the later systems of figures which our own science would dismiss as mythical—the Love and Strife of Empedocles and the ghost of a creator in the Nous of Anaxagoras. And when we look more closely at the Milesian scheme, it presents a number of features which cannot be attributed to rational inference based on an open-minded observation of facts.

In the first place the Milesians proceed on certain tacit assumptions which it never occurs to them even to state, because they are taken over from poetical cosmogony. . . . The chief question they answer is: How did the present world-order, with the disposition of the great elemental masses and the heavenly bodies, come to exist as we now see it? Here at once it is assumed that the world had a beginning in time. The Ionians also asserted that it would some day come to an end, and be superseded by another world. Now there is nothing in the appearance of Nature to suggest that the world-order is not eternal, as we may see from the fact that Aristotle could declare that it was; not to mention Heracleitus and Parmenides, who, from their opposite standpoints, denied that any cosmogony was possible.

With this assumption goes the equally unfounded dogma that the order arose by differentiation out of a simple state of things, at first conceived as a single living substance, later, by the pluralists, as a primitive confusion in which 'all things', now separate, 'were together'.

Next, the differentiation is apparently attributed to the inherent hostility of certain primary 'opposites'—the Hot and the Cold, the Moist and the Dry—driving them apart. This hostility is personified by Heracleitus as War, the father of all things, and by Empedocles as the evil genius of Strife. In Anaximander the opposites prey upon one another and invade one another's provinces in 'unjust' aggression.

There is also a contrary principle of attraction between unlikes or opposites drawing them together into reconciliation and harmony—the Love of Empedocles. In Anaximander's scheme the hot and the cold, the moist and the dry, after they have been separated apart, interact and recombine. One of the consequences of this interaction is the birth of the first living creatures, when the heat of the sun warms the moist slime of earth.

If we now reduce these assumptions to a still more abstract scheme, we get the following:

(1) In the beginning there is a primal Unity, a state of indistinction or fusion in which factors that will later become distinct are merged together.

(2) Out of this Unity emerge, by separation, pairs of opposite things or 'powers'; the first being the hot and the cold, then the moist and the dry. This separating out finally leads to the disposition of the great elemental masses constituting the world-order, and the formation of the heavenly bodies.

(3) The Opposites interact or reunite, in meteoric phenomena and in the production of individual living things, plants and animals.

This formula, clothed in concrete terms, recurs in an Ionian system, evidently of the fifth century,[1] summarized by Diodorus (1, 7, 1). It opens with the words:

At the original formation of the universe heaven and earth had one form (μίαν ἔχειν Ιδέαν οὐρανόν τε καὶ γῆν), their nature being mingled. After that, when their bodies had taken up their stations apart from one another (διαστάντων τῶν σωμάτων ἀπ' ἀλλήλων), the world (κόσμος) embraced the whole order that is visible in it; the air was in continuous motion, and the fiery part of it ran together to the uppermost regions, its nature being buoyant because of its lightness. For this reason the sun and all the rest of the heavenly bodies were involved in the whole whirl; while the slimy and muddy part, together with the assembled moisture, established itself in one place by reason of its weight. The moisture was then collected to form the sea, and the more solid parts became soft muddy land.

The sun's heat then acted on the moisture and produced bubble-like membranes, such as may now be seen formed in marshy places. Life was generated in these, fed at first by the surrounding mist at night, and in the day time solidified by the heat. Out of these membranes, when they burst, all sorts of living creatures sprang: birds, the creeping things of earth, and the fishes. Later, when the earth had become more solid, it could no longer give birth to the larger creatures, but all living things were generated by the union of the sexes.

A sketch of the early history of mankind and the rise of civilization follows. Diodorus then points out that the formula is succinctly stated by Melanippe the Wise in Euripides:

The story is not mine—I had it from my mother—how (1) Heaven and Earth were once one form, and (2) when they were separated apart from one another, (3) they gave birth to all things and brought them to light, trees and winged creatures, fishes, and mortal men.[2]

In the group of closely related Orphic theogonies, Gruppe[3] saw

one central doctrine, which may best be summed up in the words in which it is ascribed to Orpheus' pupil Musaios (Diog. L. *prooem.* 3): ἐξ ἑνὸς τὰ πάντα γίνεσθαι, καὶ εἰς ταὐτὸν ἀναλύεσθαι.

'Everything comes to be out of One and is resolved into One.' At one time Phanes, at another Zeus contained the seeds of all being within his own body, and from this state of mixture in the One has emerged the whole of our manifold world, and all nature animate or inanimate. This central thought, that everything existed at first together in a confused mass, and that the process of creation was one of separation and division, with the corollary that the end of our era will be a

---

1. Attributed to Democritus' Μικρὸς Διάκοσμος in Diels-Kranz, *Vors.*[e] II, p. 134. It has been pointed out, however, that there is no trace of Atomism, and that consequently it is more likely to be a pre-Atomist system. See Bignone, *Empédocle*, p. 583.

2. Eurip. *Melanippe, frag.* 484:

οὐκ ἐμὸς ὁ μῦθος ἀλλ' ἐμῆς μητρὸς πάρα,
ὡς Οὐρανός τε Γαῖά τ' ἦν μορφὴ μία,
ἐπεὶ δ' ἐχωρίσθησαν ἀλλήλων δίχα
τίκτουσι πάντα κἀνέδωκαν εἰς φάος,
δένδρη, πετεινά, θῆρας οὕς θ' ἅλμη τρέφει
γένος τε θνητῶν.

3. I quote Mr. W. K. C. Guthrie's summary of Gruppe's views in his *Orpheus and Greek Religion*, p. 74.

return to the primitive confusion, has been repeated with varying degrees of mythological colouring in many religions and religious philosophies.

In Apollonius' *Argonautica* (1,496) Orpheus sings 'how (1) earth and heaven and sea were once joined together in one form, and (2) by deadly strife were separated each from the other'; how the heavenly bodies hold their fixed place in the sky, and the mountains and rivers were formed, and (3) 'all creeping things came into being.' Behind Apollonius is the tradition of the Orphic cosmogony parodied in the *Birds* of Aristophanes. There the primitive state of indistinction is called 'Chaos and Night, black Erebus and Tartarus,' before earth, air and sky existed. Night is the first principle of the Orphic cosmogony recorded by Eudemus; it is not mere absence of light, but dark, cold, moist air. Aristotle compares the Night from which the theologians generate the world with the 'all things together' of the physical philosophers and the Chaos of Hesiod.[4] In the *Birds* Night produces the wind-born World-Egg from which is hatched out the winged Eros. (In Athenagoras' version of the myth the upper half of the egg forms the Heaven, the lower the Earth.) The function of Eros, who appears between them, is to reunite the sundered parents in marriage. 'There was no race of immortals till Eros united all in marriage.'[5] Then Heaven, Ocean and Earth were born—the three great departments of the world—and all the generations of the blessed gods.

We may now turn back to Hesiod's cosmogony, the one complete document of its kind which we can be certain was familiar to Anaximander.

First a word as to the type of poem to which Hesiod's *Theogony* belongs. It announces itself in the prelude as, in the first place, a hymn to the Muses: 'Let us begin our song with the Heliconian Muses, who hold the high and bold mount of Helicon.' It was they who came to Hesiod, as he tended his sheep, and breathed into him the inspired song, that he might celebrate what has been and shall be and, before all else, the Muses themselves. But the song which they inspire, namely the theogony which follows the prelude, is itself a hymn, sung by the Muses in praise of Zeus, the Lord of the aegis, and the other Olympians, and the elder gods, and Dawn and Helios, Earth, Ocean and Night and all the sacred race of the immortals.[6]

The hymn is one of the oldest forms of poetry.[7] In Greece the traditional metre is the hexameter, also appropriate to the oracle and to the epic. The hymn is in essence an incantation, inviting the presence of a god at the sacrifice and enhancing the efficiency of the ritual. Its effectiveness is increased

---

4. Ar. *Met.* 1071 b 26: εἰ ὡς λέγουσιν οἱ θεολόγοι οἱ ἐκ Νυκτὸς γεννῶντες, ἢ ὡς οἱ φυσικοὶ ἦν ὁμοῦ πάντα χρήματά φασι. 1072 a 7: οὐκ ἦν ἄπειρον χρόνον χάος ἢ νύξ.

5. *Birds*, 700: πρὶν Ἔρως ξυνέμειξεν ἅπαντα. | ξυμμιγνυμένων δ' ἑτέρων ἑτέροις γένετ' Οὐρανὸς Ὠκεανός τε | καὶ Γῆ πάντων τε θεῶν μακάρων γένος ἄφθιτον.

6. *Theog.* 11: (Μοῦσαι) ὑμνεῦσαι Δία τ' αἰγίοχον, κτλ. 36: Μουσάων ἀρχώμεθα, ταὶ Διὶ πατρὶ | ὑμνεῦσαι τέρπουσι μέγαν νόον κτλ.

7. The following remarks on the hymn are suggested by C. Autran, *Homère et les origicce' de l'épopée grecque*, i, 40.

by a recital of the history of the god and his exploits; hence it becomes biographical. Later, the use of the form is extended to heroes and to men; the famous deeds of the men of old are sung by the minstrel in the epic.[8] At every stage genealogies form a more or less important part. They are a didactic element, preserving what is believed to be the pre-history of the race, and, in some cases, the actual ancestry of important families, which serves as a basis for legal claims to property.[9] In Hesiod, the genealogies are designed to fit together into one pantheon a number of divinities, of very diverse origin, round the dominating figure of the European sky-god Zeus. The *Theogony* can thus be regarded as in the main a Hymn to Zeus, preceded by a short cosmogony. The Muses, 'uttering their immortal voice, celebrate with their song (1) first the awful race of the gods from the beginning, the children of Earth and the broad Heaven, and the gods born of these, the givers of good things (Cosmogony). Next (2) in turn, both in the beginning and in the end of their song, they hymn Zeus, Father of gods and men, how he is most excellent of the gods and greatest in power' (43 ff.). They will tell how the gods took possession of Olympus under the supreme kingship of Zeus, who apportioned to them their several provinces and honours (111–13).

For the present we must fix our attention on the brief cosmogony with which the story opens after the prelude. We shall find that it is built upon the same pattern as those we have been considering. It runs as follows:

First of all Chaos came into being, and next broad-bosomed Earth, for all things a seat unshaken for ever, and Eros, fairest among the immortal gods, who looses the limbs and subdues the thought and wise counsel of all gods and of all men.

From Chaos were born Erebus and black Night; and from Night in turn Bright Sky (*Aether*) and Day, whom Night conceived and bore in loving union with Erebus.

And Earth first gave birth to the starry Heaven, equal to herself, that he might cover her all round about, that there might be for the blessed gods a seat unshaken for ever.

And she bore the high Hills, the pleasant haunts of the goddess Nymphs who dwell in the wooded hills.

Also she bore the unharvested deep, with raging flood, the Sea (*Pontos*), without the sweet rites of love.

Here follows the marriage of Heaven and Earth. At this point a change comes over the story: Ouranos and Gaia become supernatural persons, who, with their children, the Titans, the Cyclopes, and the Giants, are involved in a series of biographical adventures. But in the cosmogony itself, which tells how the main divisions of the existing cosmos came into being—earth and the starry sky, the dry land and the sea—the veil of mythological language is so thin as to be quite transparent. Ouranos and Gaia are simply the sky and the earth that we see every day. Apart from the passing mention of the nymphs, the only mythological figure is Eros, and he is evidently no more than a bare

---

8. κλέεα προτέρων ἀνθρώπων, *Theog.* 100.

9. This appears clearly in the oral literature of Polynesia, for example.

personification of the love or attraction uniting in marriage all the parents who figure in the subsequent genealogies. Here, however, until we reach the marriage of Heaven and Earth at the end, the only birth which is (as a birth should be) the result of a marriage is the birth of light out of darkness; and even here the duplication of darkness into Erebus (male) and Night (female) and of light into Aether (male) and Day (female) is transparent allegory. The other births, or becomings—of Chaos, Earth, the Starry Heaven, the Hills, and the Sea—are 'without the sweet rites of love' (ἄτερ φιλότητος ἐφιμέρου, 132). This is a remarkable feature. It means that the cosmogonical process is a separating apart of the great departments of the ordered world, such as we have found in the Orphic and philosophical cosmogonies.

What, then, is the starting-point? 'First of all Chaos came into being.' In the modern mind the word Chaos has come to be associated with a primitive disorder in which, as the Ionian pluralists said, 'all things were together'. This is not the sense of the word in sixth- and fifth-century Greek. 'Chaos' meant the 'yawning gap',[10] between the fiery heaven and the earth, which could be described as 'empty' or as occupied by the air. Hesiod himself uses it in this sense at *Theog.* 700, where, when the ordered world already exists, 'chaos' is filled with a prodigious heat in the battle of Zeus and the Titans (καῦμα θεσπέσιον κάτεχεν χάος). It is so used by Ibycus, Bacchylides, Aristophanes and Euripides,[11] in a way that shows it was familiar to their contemporaries. The later ancients falsely derived *chaos* from χεῖσθαι, but remembered that it meant 'the empty space between heaven and earth.'[12] It is probable that in the sixth and fifth centuries the word *chaos* still carried its true etymological associations with χάσμα 'yawn,' χάσκειν, χασμᾶσθαι 'to gape,' 'yawn.'

Now, if cosmogony begins with the coming into being of a yawning gap between heaven and earth, this surely implies that previously, in accordance with Melanippe's formula, 'Heaven and Earth were once one form,' and the first thing that happened was that they were 'separated apart from one another.'

10. Boisacq, *Dict. Etym.* connects χάος, χάζω, χαίνω, χωρίς, χώρα all with a root meaning 'opening,' 'separation,' 'hollow,' etc.
11. Suidas: Χάος· καὶ ὁ ἀὴρ παρ' Ἀριστοφάνους ἐν Ὄρνισι (192).
    διὰ τῆς πόλεως τῆς ἀλλοτρίας καὶ τοῦ χάους
    τῶν μηρίων τὴν κνίσσαν οὐ διαφρήσετε,
καὶ Ἴβυκος (*frag.* 29) · πωτᾶται δ' ἐν ἀλλοτρίῳ χάει. Bacchyl. v, 27: νωμᾶται δ' ἐν ἀτρώτῳ χάει λεπτότριχα σὺν ζεφύρου πνοαῖσιν ἔθειραν. Aristophanes, *Clouds* 627: μὰ τὴν Ἀναπνοὴν μὰ τὸ Χάος μὰ τὸν Ἀέρα; 424: τὸ χάος τουτί. (Schol. R *ad loc.*: Χάος λέγει τὸν Ἀέρα.) Eurip. *frag.* 448 = Probus in Virg. *Ecl.* 6, 31: accipere debemus aera, quem Euripides in Cadmo χάος appellavit sic . . . τόδ' ἐν μέσῳ τοῦ οὐρανοῦ τε καὶ χθονός, οἱ μὲν ὀνομάζουσι χάος (the reading is uncertain but the meaning is clear). In the sixth century 'air' and 'void' were synonymous. Aristotle (*Phys.* IV, 208 b 30) explains Hesiod's χάος as the empty place (τόπος) or room (χώρα) supposed to be required before anything could exist to occupy it. Cf. [Ar.] *MXG.* 976 b 15: εἶναί τι κενόν, οὐ μέντοι τοῦτό γε τι σῶμα εἶναι, ἀλλ' οἷον καὶ Ἡσίοδος ἐν τῇ γενέσει πρῶτον τὸ χάος φησὶ γενέσθαι, ὡς δέον χώραν πρῶτον ὑπάρχειν τοῖς οὖσι. Most modern discussions of the term are vitiated by the introduction of the later idea of strictly infinite empty space. I do not think chaos is ever called ἄπειρον, and if it were, that would mean no more than 'immense,' as applied to earth and sea.
12. *Etym. Gud.* p. 562: χάος παρὰ τὸ χεῖσθαι, ὁ κενὸς τόπος μεταξὺ γῆς καὶ οὐρανοῦ . . . δηλοῖ δὲ τὸ χάος τὸ μέγα καὶ ἀπέραντον χώρημα.

Hesiod can hardly have meant anything else. He does not say that Earth was born of Chaos, but that Earth came into being 'thereafter' (ἔπειτα). The first distinct body was the earth, 'broad-bosomed,' probably conceived as a broad flat disk. We shall see later[13] why the *starry* Heaven' (filled with the visible heavenly bodies) is said to have arisen afterwards, born from the Earth. Finally the separating process is completed, as in the cosmogonies we reviewed earlier, with the distinction of the dry land, raised up into hills and the sea.

When the gap has come into being, between the sundered opposites appears the figure of Eros, a transparent personification of the mutual attraction which is to reunite them. We have seen how Eros held the same place in the Orphic cosmogony. According to Pherekydes (*frag*. 3) Zeus, when about to fashion the world, was transformed into Eros, because (adds Proclus) he brought into agreement and love the opposites of which he was framing the cosmos.[14]

In the Milesian cosmogony this mythical personality disappears, but only to re-emerge in later systems which again avail themselves of the language of poetry. In the *Symposium* (178 A) Phaedrus argues that Eros is the eldest of the gods, for no writer in poetry or prose has spoken of his having any parents. He quotes Hesiod's lines about Chaos, Earth and Eros, and cites Acusilaus as agreeing that after Chaos, Earth and Eros came into being. 'And Parmenides says of his birth: "First of all the gods she devised Eros."' With this passage in mind, Aristotle remarks that one might suspect that the need for a moving cause was first felt by Hesiod and by 'whoever else posited love or desire as a principle among things, for example Parmenides, on the ground that there must exist some cause which will move things and draw them together' (συνάξει, *Met*. Λ4, 984b 23). The Love of Empedocles has the same function of uniting unlike or opposite elements. Aristotle was not slow to recognize the mythical or poetical antecedents of philosophic concepts. Opening his discussion of friendship, he recalls in the same breath Heracleitus' declaration that the fairest harmony is composed of differing elements, and that all things come into being through strife, and Euripides' lines describing how the parched Earth desires (ἐρᾶν) the rain, and the majestic Heaven, filled with rain, desires to fall upon the Earth.[15] Euripides was imitating Aeschylus (*Danaids, frag*. 44): 'Love moves the pure Heaven to wed the Earth; and Love takes hold on Earth to join in marriage. And the rain, dropping from the husband Heaven, impregnates Earth, and she brings forth for men pasture for flocks and corn, the life of man.'

These fragments, again, are imitated by Lucretius (II, 991 ff.): 'We are all sprung from a heavenly seed; all have that same father by whom mother Earth the giver of increase, when she has taken in from him liquid drops of

---

13. Below, p. 28.

14. Compare the physical account of Eros by Eryximachus in Plato's *Symposium*.

15. Aristotle, *Eth. Nic*. 1155 b 1; Eurip. *frag*. 898: ἐρᾷ μὲν ὄμβρου γαῖ', ὅταν ξηρὸν πέδον | ἄκαρπον αὐχμῷ νοτίδος ἐνδεῶς ἔχῃ· | ἐρᾷ δ' ὁ σεμνὸς οὐρανὸς πληρούμενος | ὄμβρου πεσεῖν ἐς γαῖαν 'Αφροδίτης ὕπο· | ὅταν δὲ συμμιχθῆτον ἐς ταὐτὸν δύο, | φύουσιν ἡμῖν πάντα καὶ τρέφουσ' ἅμα, | δι' ὧν βρότειον ζῇ τε καὶ θάλλει γένος. Cf. Eur. *Chrysippus, frag*. 839.

moisture, conceives and bears goodly crops and joyous trees and the race of man, bears all kinds of brute beasts, in that she supplies food with which all feed their bodies and lead a pleasant life and continue their race; wherefore with good cause she has gotten the name of mother.'

We can now see why the Milesians identified the living stuff of the world with the intermediate element—water, cloud, mist (air)—between the fiery heaven and the solid Earth. They drop the language of poetical personification, substituting for Eros his physical equivalent or medium, the moisture which rises from earth under the sun's heat and falls back in the rain, to fertilize the dry earth and enable it to produce living things. Anaximander dissipates the thin disguise of mythical imagery and keeps the indubitably natural factors—the hot and the cold, the moist and the dry, fire, water and earth. He thinks he has got hold of the real factors and processes which will furnish a prosaic and rational account of what goes on in the sky and of the origin of life. But the earlier history of these factors is revealed by comparison with the mythical cosmogonies, by their behaviour in the philosophic systems, and by the fact that they are used to explain everything in a way that innocent observation of Nature would never suggest. Even in Aristotle's system the two primary pairs of opposites—hot and cold, wet and dry—remain as the basic qualities of which the four simple bodies are composed. All other differences, such as heaviness and lightness, density and rarity, roughness and smoothness, are secondary, 'for it seems clear that these (the four primary qualities) are the causes of life and death, sleeping and waking, maturity and old age, health and disease; while no similar influence belongs to roughness, smoothness and the rest.' A long history, stretching back into the mythical epoch, lies behind the statement that this reason for their primacy 'seems clear.'

To return to Hesiod: a second consequence of the opening of the gap between Heaven and Earth is the birth of light out of darkness. Erebus and Night are the parents of Aether (the bright region of the sky) and Day. In the Orphic system Eros has another name, Phanes, the Bright One; and it is even suggested that the word Eros is primarily to be connected with a root meaning 'light.'[16] In physical terms the lifting up of the sky from the earth lets in the light of day where before there was darkness. We note that this appearance of light precedes the formation of the heavenly bodies. For the next event is that Earth gives birth to the *starry* Heaven.

The order of events in Anaximander's scheme is closely parallel. We first heard of 'a sphere of flame growing round the "air" (dark mist) encompassing the earth.' Thus the earth wrapped in mist was the first solid body, as in Hesiod. Then, when the sphere of flame was torn off into rings the heavenly bodies were formed. The Heaven which was separated from the Earth when the gap came into being was not a heaven of stars. The stars, in

---

16. A. Saric Rebac, *Wiener Studien*, LV, 1937. I take this reference from the summary in *C.Q.* XXXIII, 1939, p. 127.

both accounts, appear to be formed of fire rising from the Earth and afterwards fed by exhalations of the moist element. So Earth gave birth to the starry Heaven 'that he might cover her all round about.' In the last words we recognize that bright Aether on high, which Euripides spoke of as 'holding Earth in the moist embrace of his arms' and identified with Zeus. The Earth mother is to be embraced by her husband Heaven. When the dry land has been separated off from the seas, to complete the world order, this Marriage follows and the eldest gods are born. The whole mythical theogony which begins here has, of course, no place in the Ionian systems, for these ignore the personal gods. In natural philosophy the intercourse of heaven and earth, heat and moisture, results in the birth of plant and animal life.

It is exactly at this point, where the formation of the physical cosmos is complete, that a significant change comes over Hesiod's story. The cosmogony we have reviewed is not a myth, or rather it is no longer a myth. It has advanced so far along the road of rationalization that only a very thin partition divides it from the early Ionian systems. Eros is the only mythological name, having no connexion here with religion or cult. Gaia and Ouranos are simply the earth and the sky, not mythical figures. But no sooner is the cosmos framed, to serve as a stage for the action which follows, than they are transformed into supernatural persons, indulging human passions of jealousy and hate in those 'violent deeds' which caused so much scandal later to religious minds.

In wedlock with Ouranos, Gaia brought forth the Titans, of whom 'the youngest, Kronos of crooked counsel, was the most terrible and hated his lusty father.' At once we are plunged back into the world of myth, and the rest of Hesiod's story moves in this supernatural atmosphere. Before we follow it further, we will pause to note a curiously close parallel, in another literature, to this sudden shift from rationalized cosmogony to myth.

The first three chapters of Genesis contain two alternative accounts of creation. The first account (Gen. i-ii. 3), in its present form, was composed not earlier than the exile; it is considerably later than Hesiod, and may even be later than Anaximander. In this Hebrew cosmogony, moreover, we find nearly the same sequence of events. Let us recall what happened on the six days of creation.

(1) 'The earth was without form and void; and darkness was upon the face of the deep; and the spirit of God moved upon the face of the waters.' Then light appeared, divided from the darkness, as Day from Night.

(So, when Hesiod's gap opened, there was earth and the moving spirit of life, Eros; and then Night gave birth to Day.)

(2) The heaven, as a solid firmament (στερέωμα), is created as a roof to divide the heavenly waters, whence comes the rain, from the waters below.

(This corresponds to Hesiod's Earth generating the starry Heaven as 'an unshaken seat for the blessed gods.' In later Ionian systems the solid crystalline sphere of heaven reappears, in place of Anaximander's sphere of flame which burst into rings. It is the shell of the world-egg.)

(3) The dry land is separated from the seas, and clothed with grass and trees.

(In Hesiod, Earth generates the hills and *Pontos* (the sea). Note that Empedocles made the trees, the first living creatures, spring up from the earth 'like embryos in the mother's womb,' before the sun existed, *Vors.*[5] vol. 1, p. 296.)

(4) The sun, moon, and stars are created to divide day from night and to 'be for signs and for seasons and for days and years.'

(As in the Greek cosmogonies, the heavenly bodies are formed later than the earth. Their function is to serve as 'signs,' both for purposes of divination and to mark the distinctions of time in the calendar. As Plato says, 'the sight of day and night, of months and the revolving years, of equinox and solstice, has caused the invention of number and bestowed on us the notion of time,' *Timaeus* 47A.)

(5) and (6). The waters brought forth the 'moving creatures that have life,' birds and fishes and all the beasts that creep on the earth; and finally man was made, both male and female, to be fruitful and multiply and to have dominion over all living things.

(Thus 'creatures having life,' distinguished from plants and trees by their power of motion, arose from the moist element, when the cosmic frame was complete. So it was in Anaximander, where the action of the sun's heat on moisture reproduces in physical terms Hesiod's marriage of Heaven and Earth.)

The most striking difference from the Greek cosmogonies is that Hebrew monotheism has retained the divine Creator as first cause. Otherwise there are no mythical personifications like Eros or Phanes. And the action of Elohim is limited to the utterance of the creative word. He has become extremely abstract and remote. If we eliminate the divine command, 'Let there be' so-and-so, leaving only the event commanded, 'There was' so-and-so, and then link these events in a chain of natural causation, the whole account becomes a quasi-scientific evolution of the cosmos. The process is the same as in the Greek cosmogonies—separation or differentiation out of a primitive confusion. As measured by the absence of personifications, Genesis i is less mythical than Hesiod, and even closer to the rationalized system of the Milesians.

The foregoing argument will perhaps have made clear the nature of Anaximander's achievement. We can see that his thought was at work on a scheme of cosmogony already provided by Hesiod and other poetical cosmogonies. He took the final step in the process of rationalization, divesting the scheme of the last traces of mythical imagery. It was not for nothing that his book was one of the earliest written in prose, the proper language for literal statements of fact.

In particular the figure of Eros vanishes. This means the elimination of the imagery of sex, and with it goes the representation of cosmogony as a series of births forming a genealogical tree. He speaks in abstract terms of the 'separating out of opposites,' followed by their interaction and recombination. These opposites are the hot and the cold, the moist and the dry. Without the evidence of the mythical cosmogonies we could hardly guess that the primary 'opposites' were male and female, though, as we shall see, the tradition persisted elsewhere. What we claim to have established so far is that the pattern of Ionian cosmogony, for all its appearance of complete rationalism, is not a free construction of the intellect reasoning from direct observation of the existing world. There is nothing in the obvious appearances of Nature to suggest that the sky ever had to be lifted up from the earth, or that the heavenly bodies were formed after the earth, or indeed that the present order of the world has not existed for ever.

THEODOR GOMPERZ:

The Development
of the Pythagorean Doctrine

VOLTAIRE[1] called the later Pythagorean astronomy, connected with the name of Philolaus, a "Gallimathias," and Sir George Cornewall Lewis[2] indicts it as "wild and fanciful." But the great French writer with his frequently over-hasty judgments and the Englishman with his excess of conscientiousness have fallen into the same mistake. It is true that the doctrine in question is a tissue of truth and invention, but its features of truth were its vital and fundamental parts, whereas the fictitious portions were merely a superficial covering which was soon to dissolve like smoke-wreaths. But if we are to understand the motives which inspired the cosmology of Philolaus, we must pause a moment at the commonest phenomena of the heavens.

Each day the sun runs his course from east to west. Simultaneously he climbs higher up the sky to sink at the end of a few months from the height he has reached. The combination of his daily and annual movements has the effect of the windings of a screw or spiral—something like the shell of a snail—and like it, too, the intervals between the circles contract as the zenith is approached. This view was hardly likely to satisfy inquirers who had approached the question of celestial motion in the confident belief that it was "simple, steady, and regular."[3] It may be permissible to blame this belief as a prejudice; but though it was in part a preconceived opinion, yet the closer observation of facts tended generally to confirm it. And even where such

From Theodor Gomperz, *Greek Thinkers*, translated by Laurie Magnus, London, 1901, Volume I, Chapter IV.

1. Voltaire, *Oeuvres Complètes*, edited by Baudouin, Vol. 58, p. 249.

2. Sir George Cornewall Lewis, *An Historical Survey of the Astronomy of the Ancients*, p. 189. The material here employed is for the main part collected in the epoch-making treatise of Schiaparelli (*I Precursori di Copernico Nell' Antichita*). We are considerably indebted, too, to the rich contents of this and of a second masterly work by the same author, *Le sfere omocentriche, etc.*, Milan, 1876. The first to shed light on this confusion was Boeckh, in his *Philolaos des Pythagoreers Lehren*. In another connection we shall have to deal with the personality of this Pythagorean, and with the other doctrines that may with greater certainty be attributed to him.

3. "Simple, steady, and regular:" cf. Geminus, in Simplicius, *Phys.*, 292, 26, 27 D.

confirmation was wanting, the belief was of excellent use as a principle of research, just like the kindred assumption of a teleological purpose in the structure of organisms. It was possible to get rid of the confusing irregularity. For a complex movement may be irregular while the partial movements that compose it are regular. What was needed was an act of mental separation. And the clue was found by separating the daily movement of the sun from its annual movement. At this point our early philosophers had a brilliant flash of inspiration. They conceived the daily movement of the sun and moon, and indeed of all the whole starry heaven, as not real at all, but merely apparent. Their supposition that the earth was moving from west to east enabled them to dispense with the assumption that the sun, moon, planets, and fixed stars were moving in an opposite direction. The question suggests itself here, Did these Pythagoreans recognize and teach the rotation of the earth round its axis? Our answer is: They did not do that, but they did recognize and teach the existence of a movement which operated in a precisely similar manner. It was, so to say, the rotation round its axis of an earth-ball with a considerably enlarged circumference. They represented the earth as circulating in twenty-four hours round a central point, the nature of which will presently occupy us. Here, however, the reader should familiarize himself with a simple feature of this doctrine. A moment's reflection will show him that, for any given point in the earth's surface, and for its shifting relations with the sun, moon, and stars, it makes not the remotest difference whether the ball on which it is situated revolves on its axis in the course of a day, or describes a circular course, while facing the same directions, which brings it back to its starting-point in the same limit of time. We can hardly exaggerate the importance of this discovery. The revelation that there were apparent heavenly motions broke the barrier that obstructed the path to further progress. The central position of the earth and its immobility had both been given up, and the way was open for the Copernican doctrine which followed after an interval the extraordinary brevity of which is hardly sufficiently recognized. Nor need we be at all surprised that an equivalent for the theory of rotation was adopted instead of the theory. For though we never actually see a luminary turning on its axis, yet changes in its position are matters of daily and hourly observation. Nothing, then, could be more natural than that scientific imagination, which had just succeeded by a mighty effort in freeing itself from the delusions of sense, should have been content to replace the apparent immobility of the earth by a movement moulded on familiar models, and not by one unique in its kind and entirely without a parallel.

The centre round which the earth was now admitted to move served equally as the centre of the rest of the luminaries, which had formerly been supposed to revolve round the earth. The moon accomplished its course once a month; the sun once a year; the five planets visible to the naked eye required various periods, which, with the exception of Mercury and Venus, were considerably longer; finally, the firmament of fixed stars, whose daily rotation[4] had been recognized as apparent, was similarly equipped with a circular movement of its own, though of a very much slower order—a conception which may either have been due to the mere desire for conformity, but which is far more probably to be ascribed to that change of position already observed and taken in account which we call the precession of the equinoxes. The daily movement of the sun—or rather, according to this theory, of the earth—took place in a plane which was now recognized to incline towards the plane in which the annual movements of the sun, moon, and planets were situated; in other words, the obliquity, whether of the equator or of the ecliptic, had been recognized, and the new conception was thus completely adequate to explain the changes of the seasons.

We come now to the problem of the central point round which the heavenly bodies were to move in concentric circles. It was no ideal centre, but rather an actual body, consisting of universal or central fire. The enemies of Philolaus call it "a dreary and fantastic fiction," but those who try to throw themselves with temperate judgment into the modes of thought obtaining in the dawn of science will rather call it "the product of analogical inferences, the force of which must have been well-nigh irresistible." The assumption that the heavenly bodies described circles was not merely approximately true, but apart from the circular segments traversed by the sun and moon on the firmament, it appeared that no other conclusion could be derived from the circular courses described before our eyes by the circumpolar fixed stars that never set; and though that movement, like the movement of the whole firmament of fixed stars, had now been recognized as purely apparent, yet the daily motion of the earth that took its place was bound to have the same circular character. Here, accordingly, the type was given, conformably with which all the heavenly bodies had to move. But human experience supplies no example of circular movements without an actual centre. A wheel turns on its axis; a stone, attached to a string for the purpose of slinging, turns

---

4. We are of opinion that Schiaparelli errs in disputing the movement of the firmament of fixed stars in the Philolaic system, *I Precursori di Copernico, etc.* (separate edition), p. 7. For then we should have to credit our authorities, above all Aristotle, who speaks of ten heavenly bodies in motion (*Metaphys.*, i. 5), with a hardly conceivable mistake. It is, further, contrary to the strongly marked sense of symmetry shown by the Pythagoreans, that they should ascribe immobility solely to the firmament of the fixed stars. It is true they could no longer believe in the daily movement of this firmament, since it had been superseded by the movement of the earth. "What then remains," asks Böckh, *op. cit.*, 118, "but to assume that the movement of the firmament of the fixed stars is the precession of the equinoxes?" Later, Böckh renounced this opinion (*Manetho und die Hundssternperiode*, 54); still later he returned to it, though with hesitation (*Das Kosmische System des Platon*, 95). In this we unconditionally agree with him, chiefly on account of the following consideration. The precession of the equinoxes is a phenomenon which, as Martin justly remarks (*Etudes sur le Timée de Platon*, ii. 38), "requires only long and steady observations without any mathematical theory, in order to be recognized." It is in itself hardly credible that a deviation in the position of the luminaries, which in the course of a single year amounts to more than fifty seconds of an arc, could remain unnoticed for long. It becomes quite incredible on the following consideration, to which an expert authority, Dr. Robert Fröbe, of the Vienna Observatory, has directed my attention. The data derived from Philolaus or other early Pythagoreans for the angular velocities of the planetary movements are approximately correct. Only prolonged observations of the stars could have made them so, since there was no other means of eliminating the grossest of the errors then inevitable to observation.

round the hand which holds it and which sets it in motion; and, finally, when divine worship invited Greek men and women to the dance, the altar of the god formed the centre of their solemn and rhythmic paces. It may be asked, however, what need there was of inventing a central fire, when it actually existed and was visible to every man's eye. What was wanted was a centre of motion and a source of vigour and life. But instead of accrediting the universal light of the sun with the rank that belonged to it, a luminous body was invented whose rays no mortal eye had seen, and, considering that the habitable side of the earth was turned away from the central fire, no mortal eye would ever see. It was an hypothesis removed by a perverse ingenuity from every chance of verification, and one wonders why its mistaken authors did not rather jump straight away at the heliocentric doctrine, and rest satisfied therewith.

Three sufficiently valid solutions may be suggested for this problem. Remembering that the delusions of sense are only abandoned by degrees, and that the human mind habitually follows the path of least resistance, we have first to note that the heliocentric theory was bound to be later than that of rotation round an axis. It was obviously impossible to let the earth revolve round the sun in a daily and yearly course simultaneously, and we have already learned to justify the precedence of the Pythagorean equivalent over the rotation theory. A second considerable obstacle to the prompt admission of a heliocentric or Copernican astronomy lay, we conceive, in the exact similarity between the sun and moon. The great luminary of day and his more modest sister of the night were visible to men as two heavenly bodies regularly relieving each other and combining to measure time by their revolutions, and it was plainly impossible that, except by a process of elimination, shutting out every other issue, men would ever be brought to believe that luminaries so closely connected differed in the fundamental point that the moon was condemned to ceaseless wandering while the sun was vowed to eternal rest. But, thirdly and chiefly, universal fire was more satisfactory as the centre of the world than the sun. Our sun is the central point of a system of luminaries by the side of which countless other systems exist without visible design or recognizable order. Human intelligence resists this belief, as it resists every other call to renunciation, till the compulsion of fact leaves it no second alternative. But first it demands a uniform picture of the world instead of a fragmentary view of this kind, and the demand springs from the natural impulse towards lightening and simplifying the intellectual complexus—an impulse assisted in the present instance, indeed, by highly developed æsthetic and religious wants.

It will be readily admitted that this picture of the universe owed no little to the contribution of the emotions and the fancy. The circular course of the divine luminaries which had been raised by the fictitious counter-earth to the sacred number ten was described as a "dance." The rhythm of this starry dance was set to the sounds arising from the motion itself, and making unceasing music, which was recognized and known as the "harmony of the spheres." Next, the universal fire, which was the central point of the celestial procession, was known by many names. It was called the "mother of the gods," the "citadel of Zeus,"[5] and so forth, but two of its titles may be mentioned as especially characteristic. These were the "altar" and the "hearth of the universe." These stars revolved round the sacred source of all life and motion like worshippers round an altar, and the universal hearth was the centre of the world or cosmos as a man's domestic hearth was honoured as the sacred centre of his home, or as the flame that burned and was never extinguished in the civic hearth of the Prytaneum formed the holy rallying-point of every Greek community. Hence streamed the rays of light and heat, hence the sun derived his beams and communicated them again to both earths and to the moon, just as the mother of the bride lighted at a Greek wedding the fire of the new home from the parental hearth, or as a new colony would borrow its fire from the hearth of the mother-city. All the threads of the Greek view of life are combined here. We see the exalted joy in existence, the loving awe for the universe ruled by divine forces, the sublime sense of beauty, symmetry, and harmony, and not least the comfortable affection for civic and domestic peace. Those, then, who held these views, and whose universe was surrounded by the fire-circle of Olympus as by a strong wall, found in it their home, their sanctuary, and the type of their art. Nowhere else do we find a picture of the universe at once so genial and so sublime.

The emotional faculties, then, were satisfied in a truly wonderful degree, though at the cost of the intellect. We have now to estimate the price which reason had to pay, and which will be found to have been by no means exorbitant. Even the "dreams of the Pythagoreans" contained a modicum of truth; or, where that modicum was wanting, there was at least an indication given of the road which would ultimately lead to truth. At first sight, for instance, no doctrine could appear more arbitrary than that of the harmony of the spheres. It obviously sprang in the last resort from an æsthetic demand which was formulated as follows: Our eyes are filled with the grandest sights; how is it, then, that the twin sense of our ears should go empty? But the premise on which the answer rested was not wholly unreasonable. For unless the space in which the stars revolve is completely void, the matter that fills it must undergo vibrations which in themselves are capable of being heard. Even in recent times, no meaner philosopher than Karl Ernst von Baer,[6] the great

---

5. Cf. Stobæus, *Eclogues*, 1. 22 (1. 196 Wachsmuth)= Ætius in *Doxogr.*, 336, 337. It has been conjectured on the best grounds that the torch which the bride's mother waved at the marriage ceremony was "kindled at the parental hearth" (cf. Herman-Blümner, *Griech. Privat-altertümer*, 275, n. 1: "Hence ἀφ' ἑστίας ἄγειν γυναῖκα, Iambl., *Vit. Pythagor.*, c. 18, § 84"). It seems an almost unavoidable assumption that the new hearth was kindled with the same torch, especially in view of the similar custom obtaining at the foundation of colonies. For this last ceremony, cf. Herodotus, i. 146; Scholiast to Aristides, iii. p. 48, 8 Dindorf; *Etymol Magn.*, p. 694, 28 Gaisford.

6. Karl Ernst von Baer: *Reden . . . und Kleinere Aufsätzen*, St. Petersburg, 1864, i. 264. On the harmony of the spheres and the reason why it is inaudible, cf. especially Aristotle, *De Coelo*, ii. 9.

founder of embryology, has asked if there is not "perhaps a murmur in universal space, a harmony of the spheres, audible to quite other ears than ours." Now, it was objected to the Pythagoreans that we do not actually hear such sounds; but they deprecated the astonishment of the cavillers by the following happy analogy. A blacksmith, they said, is deaf to the continuous, regular beat of the hammers in his workshop; and herein they anticipated the teaching of Thomas Hobbes, who argued that the operation of the senses depends on a change in the stimuli; the stimulation must be interrupted, or altered in degree or kind. There was nothing fanciful in the Pythagorean

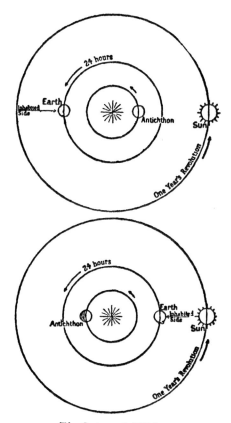

The System of Philolaus

Upper figure: Night on Earth. Only the side turned away from the centre is inhabited; consequently the Central Fire and Antichthon (counter-Earth) are invisible.

Lower figure: Twelve hours later; Day on Earth. Earth has made half a revolution, and her outer side is now lighted by the sun, which has only moved about half a degree forward in its yearly orbit. Antichthon has also made half a revolution, therefore remains invisible.

(From *Dante and the Early Astronomers*, M. A. Orr (Mrs. John Evershed), 1913.)

doctrine except only the belief that the differences of velocity in the movements of the stars were capable of producing a harmonious orchestration and not merely sounds of varying pitch. At this point their artistic imagination had a freer rein, inasmuch as they were completely unable to determine the relative distances of the planets and the absolute velocities that ensued from them, though they could arrive, approximately at least, at the circular segments which the planets described in a given time—in other words, at the angular velocities of their movements.

But here, too, we shall presently find ourselves ready to mitigate our judgment. We have to remember that the premise of law and order, as pervading the universe, could hardly have been applied by the Pythagoreans to any other relations than those of geometry, arithmetic, and music—the last named because of the importance of acoustics in their natural philosophy. Simplicity, symmetry, and harmony were ascribed indifferently to all three. They neither knew nor divined anything of the forces which produce celestial movements, so that, even had they been acquainted with the elliptic orbits of the planets, that knowledge, we may remark, would never have satisfied their demand for order. They would not have recognized the curve as the resultant of two rectilinear forces. Their heaven, says Aristotle,[7] "is all number and harmony," and we may add that a correct intuition of the highest significance was still clothed in unsuitable shape. The seekers were incapable of discovering law where it was really in operation, and it was anyhow better to look for it where it did not exist than not to look for it at all. Further, the assumption that the sun shines with borrowed light may be traced in the main to the parallelism between the sun and the moon which we mentioned just now. Moreover, the homogeneous conception of the universe might conceivably have suffered if a second independent source of light had been assumed so near the centre of the world. But since they could not altogether dispense with such an assumption, they found it in the Olympus alluded to above as the girdle of the universe, containing all the elements in their unsullied beauty. The firmament of fixed stars, and possibly the planets, derived all their light from Olympus, and the sun borrowed a part of his from the same source, to make amends, we presume, for his otherwise too frequent obscurations. The porous and glass-like qualities of the sun, which enabled it to collect the rays of light and to emit them again, should be noted in this connection. Next we come to the second great fiction of the Pythagoreans— that of the counter-earth. We may readily follow Aristotle in believing that the sacredness of the number ten played a part in this conception. But the introduction of a new luminary and its insertion between the earth and the central fire had many important consequences, and there is no reason to doubt that this fiction of the counter-earth was recommended to its inventors as much for the sake of its results as for the reason alleged by the Stagirite. The lacunæ in the information at our disposal do not permit us to pass definite

---

7. Aristotle, *Metaphys.*, i. 5.

judgment; but Boeckh's opinion that the counter-earth was to act as a screen of the central fire, so as to explain its invisibility, is certainly defective. For the supposition that the uninhabited western hemisphere of the earth was turned towards the fire was a quite sufficient explanation. It is more probable that the counter-earth was invented partly as an ostensible cause for the eclipses of the moon which occurred so frequently[8] as to seem to require the shadow of the counter-earth in addition to the shadow of the earth.

The facts of history, however, are more eloquent than all the arguments. Historically considered, the theory of central fire promoted and did not retard the progress of scientific research. In less than a century and a half it engendered the heliocentric doctrine. The fantastic excrescences of the Philolaic system fell away piece by piece. The counter-earth was the first to go: the death-blow was struck at this fiction by the extension of the geographical horizon.[9] The foundations of the hypothetical structure built by the Pythagoreans began to give way in the fourth century at latest. At that time exacter news reached Greece of discoveries in the west and in the east. Hanno, the Carthaginian, had made his great voyage of discovery, and had passed the barrier of the Pillars of Hercules, where the Straits of Gibraltar now are, which had ranked till then as the furthest limit of the Western world; and shortly afterwards the outline of the East was more clearly defined by Alexander's march in India. A coign of observation had been reached from which the counter-earth should have been visible, and since neither the counter-earth nor the central fire, thus robbed of its last protection, came in view at that point, this portion of the Pythagorean cosmology was spontaneously shattered. Nor was this all: the daily circular movement of the earth disappeared with the fictitious centre that conditioned it, and the doctrine of rotation took the place of the theory we have described as its equivalent. Ecphantus, one of the youngest of the Pythagoreans, taught that the earth turned on its own axis. The second step on the road to the heliocentric doctrine followed swiftly on the first. The marked increase in luminosity which the planets occasionally display was first noticed in Mercury and Venus, and the true cause of the phenomenon suggested itself inevitably as the occasional closer propinquity of these wandering stars to the earth. Thus it was clearly impossible that they could revolve concentrically round the earth. These two nearest neighbours of the sun had plainly confessed their dependence on that luminary by the revolution they respectively accomplished in the course of a solar year. Accordingly they were the first of the planets whose movements were combined with the sun's. This was the masterly discovery of Heraclides

---

8. "Eclipses of the moon, which occurred so frequently." As a matter of fact, eclipses of the sun are more frequent; thus in the period of time comprised in Oppolzer's *Canon der Finsternisse* there are 8000 eclipses of the sun against 5200 of the moon. At every single point of the earth, however, very many more of the latter than of the former are visible.

9. "Extension of the geographical horizon:" On Hanno's *Periplus* and the influence of that voyage of discovery on the transformation of the doctrine of a central fire, cf. Schiaparelli, *I Precursori, etc.* (separate edition), p. 25, and H. Berger, *Wissenschaftliche Erdkunde*, ii, 387.

of Heraclia[10] on the Black Sea, a man whose powerful genius, contained in a misshapen body, was familiar with the most diverse regions of science and literature, who had visited the schools of Plato and Aristotle, and had kept up a lively intercourse with the latest Pythagoreans. But here, again, there was no finality. Mars likewise displayed a conspicuous change in his degree of brightness even to the incomplete observation which obtained in that age, and thus a link was forged to unite the two inner planets with one at least of the outer ones. Philosophy was approaching the point of view reached in later times by Tycho de Brahe, who represented all the planets with the exception of the earth as revolving round the sun, while the sun with his train of planets revolved round the earth. The last and final step was taken by Aristarchus of Samos, the Copernicus of antiquity, about 280 b.c., who completed what the astronomer from the Pontus, to whom allusion has just been made, had less definitely begun. Eudoxus had given the clue to this great intellectual achievement by his discovery that the size of the sun is considerably greater than the earth's. Aristarchus computed their relative proportion at seven to one; and inadequate as this estimate was in comparison with the actual fact, it was sufficient to expose the absurdity of setting the great ball of fire to revolve like a satellite round the small world that we inhabit. The earth had to lay down the sceptre which had recently been restored to it; heliocentricity superseded geocentricity, and the goal was reached for which Pythagoras and his disciples had smoothed and pointed the way. As things turned out, however, it was soon to be abandoned again, and its place to be taken for another long series of centuries by the immemorial delusions fostered in the name of religion.

---

10. Heraclides: cf. chiefly Laert. Diog., v. ch. 6. The view taken in the text of Heraclides as the immediate precursor of Aristarchus, is based on the account by Geminus, in Simplicius, *Phys.*, 292, 20 *ff.* D.—a passage not without its difficulties. After the most ample consideration, I find myself compelled to dissent from Diels' view of the passage (*Uber das physik. System des Straton*, in the *Berliner Sitzungs-Berichte*, 1893, p. 18, n. 1). Either the passage must be emended, precisely or similarly as Bergk proposed (*Fünf Abhandlungen zur Gesch. der griech. Philos. u. Astronomie*, 149), or the words Ἡρακλείδης ὁ Ποντικός must be taken as inserted by a (well-informed) reader. The evidence for the progress of astronomy described in the text, and likewise the explanation of that progress, are given by Schiaparelli, *op. cit.* The doctrine of Aristarchus was mentioned by Copernicus, in a passage which he afterwards suppressed: "Credibile est hisce similibusque causis Philolaum mobilitatem terræ sensisse, quod etiam nonnulli Aristarchum Samium ferunt in eadem fuisse sententia," etc. (*De Revolut. Coelest.*, ed. Thorun., 1873, p. 34 n.).

LUCRETIUS:

The Nature of the Universe

MOTHER OF AENEAS and his race, delight of men
and gods, life-giving Venus, it is your doing that under the wheeling constellations of the sky all nature teems with life, both the sea that buoys up our
ships and the earth that yields our food. Through you all living creatures are
conceived and come forth to look upon the sunlight. Before you the winds
flee, and at your coming the clouds forsake the sky. For you the inventive
earth flings up sweet flowers. For you the ocean levels laugh, the sky is
calmed and glows with diffused radiance. When first the day puts on the
aspect of spring, when in all its force the fertilizing breath of Zephyr is
unleashed, then, great goddess, the birds of air give the first intimation of your
entry; for yours is the power that has pierced them to the heart. Next the
cattle run wild, frisk through the lush pastures and swim the swift-flowing
streams. Spell-bound by your charm, they follow your lead with fierce desire.
So throughout seas and uplands, rushing torrents, verdurous meadows and
the leafy shelters of the birds, into the breasts of one and all you instil alluring
love, so that with passionate longing they reproduce their several breeds.

Since you alone are the guiding power of the universe and without you
nothing emerges into the shining sunlit world to grow in joy and loveliness,
yours is the partnership I seek in striving to compose these lines *On the Nature
of the Universe* for my noble Memmius. For him, great goddess, you have
willed outstanding excellence in every field and everlasting fame. For his
sake, therefore, endow my verse with everlasting charm.

Meanwhile, grant that this brutal business of war by sea and land may
everywhere be lulled to rest. For you alone have power to bestow on mortals
the blessing of quiet peace. In your bosom Mars himself, supreme commander
in this brutal business, flings himself down at times, laid low by the irre-
mediable wound of love. Gazing upward, his neck a prostrate column, he
fixes hungry eyes on you, great goddess, and gluts them with love. As he lies
outstretched, his breath hangs upon your lips. Stoop, then, goddess most
glorious, and enfold him at rest in your hallowed bosom and whisper with

From *The Nature of the Universe* by Lucretius, translated by R. E. Latham, Penguin Books,
1951, pp. 27-45, 55-59, 90-95. Reprinted by kind permission of the publishers, Penguin Books, Ltd.

those lips sweet words of prayer, beseeching for the people of Rome untroubled peace. In this evil hour of my country's history, I cannot pursue my task with a mind at ease, as an illustrious scion of the house of Memmius cannot at such a crisis withhold his service from the common weal.

For what is to follow, my Memmius, lay aside your cares and lend undistracted ears and an attentive mind to true reason. Do not scornfully reject, before you have understood them, the gifts I have marshalled for you with zealous devotion. I will set out to discourse to you on the ultimate realities of heaven and the gods. I will reveal those *atoms* from which nature creates all things and increases and feeds them and into which, when they perish, nature again resolves them. To these in my discourse I commonly give such names as the 'raw material,' or 'generative bodies' or 'seeds' of things. Or I may call them 'primary particles,' because they come first and everything else is composed of them.

When human life lay grovelling in all men's sight, crushed to the earth under the dead weight of superstition whose grim features loured menacingly upon mortals from the four quarters of the sky, a man of Greece was first to raise mortal eyes in defiance, first to stand erect and brave the challenge. Fables of the gods did not crush him, nor the lightning flash and the growling menace of the sky. Rather, they quickened his manhood, so that he, first of all men, longed to smash the constraining locks of nature's doors. The vital vigour of his mind prevailed. He ventured far out beyond the flaming ramparts of the world and voyaged in mind throughout infinity. Returning victorious, he proclaimed to us what can be and what cannot: how a limit is fixed to the power of everything and an immovable frontier post. Therefore superstition in its turn lies crushed beneath his feet, and we by his triumph are lifted level with the skies.

One thing that worries me is the fear that you may fancy yourself embarking on an impious course, setting your feet on the path of sin. Far from it. More often it is this very superstition that is the mother of sinful and impious deeds. Remember how at Aulis the altar of the Virgin Goddess was foully stained with the blood of Iphigeneia by the leaders of the Greeks, the patterns of chivalry. The headband was bound about her virgin tresses and hung down evenly over both her cheeks. Suddenly she caught sight of her father standing sadly in front of the altar, the attendants beside him hiding the knife and her people bursting into tears when they saw her. Struck dumb with terror, she sank on her knees to the ground. Poor girl, at such a moment it did not help her that she had been first to give the name of father to a king. Raised by the hands of men, she was led trembling to the altar. Not for her the sacrament of marriage and the loud chant of Hymen. It was her fate in the very hour of marriage to fall a sinless victim to a sinful rite, slaughtered to her greater grief by a father's hand, so that a fleet might sail under happy auspices. Such are the heights of wickedness to which men are driven by superstition.

You yourself, if you surrender your judgment at any time to the blood-curdling declamations of the prophets, will want to desert our ranks. Only think what phantoms they can conjure up to overturn the tenor of your life and wreck your happiness with fear. And not without cause. For, if men saw that a term was set to their troubles, they would find strength in some way to withstand the hocus-pocus and intimidations of the prophets. As it is, they have no power of resistance, because they are haunted by the fear of eternal punishment after death. They know nothing of the nature of the spirit. Is it born, or is it implanted in us at birth? Does it perish with us, dissolved by death, or does it visit the murky depths and dreary sloughs of Hades? Or is it transplanted by divine power into other creatures, as described in the poems of our own Ennius, who first gathered on the delectable slopes of Helicon an evergreen garland destined to win renown among the nations of Italy? Ennius indeed in his immortal verses proclaims that there is also a Hell, which is peopled not by our actual spirits or bodies but only by shadowy images, ghastly pale. It is from this realm that he pictures the ghost of Homer, of unfading memory, as appearing to him, shedding salt tears and revealing the nature of the universe.

I must therefore give an account of celestial phenomena, explaining the movements of sun and moon and also the forces that determine events on earth. Next, and no less important, we must look with keen insight into the makeup of spirit and mind: we must consider those alarming phantasms that strike upon our minds when they are awake but disordered by sickness, or when they are buried in slumber, so that we seem to see and hear before us men whose dead bones lie in the embraces of earth.

I am well aware that it is not easy to elucidate in Latin verse the obscure discoveries of the Greeks. The poverty of our language and the novelty of the theme compel me often to coin new words for the purpose. But your merit and the joy I hope to derive from our delightful friendship encourage me to face any task however hard. This it is that leads me to stay awake through the quiet of the night, studying how by choice of words and the poet's art I can display before your mind a clear light by which you can gaze into the heart of hidden things.

This dread and darkness of the mind cannot be dispelled by the sunbeams, the shining shafts of day, but only by an understanding of the outward form and inner workings of nature. In tackling this theme, our starting-point will be this principle: *Nothing can ever be created by divine power out of nothing.* The reason why all mortals are so gripped by fear is that they see all sorts of things happening on the earth and in the sky with no discernible cause, and these they attribute to the will of a god. Accordingly, when we have seen that nothing can be created out of nothing, we shall then have a clearer picture of the path ahead, the problem of how things are created and occasioned without the aid of the gods.

First then, if things were made out of nothing, any species could spring from any source and nothing would require seed. Men could arise from the sea and scaly fish from the earth, and birds could be hatched out of the sky. Cattle and other domestic animals and every kind of wild beast, multiplying indiscriminately, would occupy cultivated and waste lands alike. The same fruits would not grow constantly on the same trees, but they would keep changing: any tree might bear any fruit. If each species were not composed of its own generative bodies, why should each be born always of the same kind of mother? Actually, since each is formed out of specific seeds, it is born and emerges into the sunlit world only from a place where there exists the right material, the right kind of atoms. This is why everything cannot be born of everything, but a specific power of generation inheres in specific objects.

Again, why do we see roses appear in spring, grain in summer's heat, grapes under the spell of autumn? Surely, because it is only after specific seeds have drifted together at their own proper time that every created thing stands revealed, when the season is favourable and the life-giving earth can safely deliver delicate growths into the sunlit world. If they were made out of nothing, they would spring up suddenly after varying lapses of time and at abnormal seasons, since there would of course be no primary bodies which could be prevented by the harshness of the season from entering into generative unions. Similarly, in order that things might grow, there would be no need of any lapse of time for the accumulation of seed. Tiny tots would turn suddenly into grown men, and trees would shoot up spontaneously out of the earth. But it is obvious that none of these things happens, since everything grows gradually, as is natural, from a specific seed and retains its specific character. It is a fair inference that each is increased and nourished by its own raw material.

Here is a further point. Without seasonable showers the earth cannot send up gladdening growths. Lacking food, animals cannot reproduce their kind or sustain life. This points to the conclusion that many elements are common to many things, as letters are to words, rather than to the theory that anything can come into existence without atoms.

Or again, why has not nature been able to produce men on such a scale that they could ford the ocean on foot or demolish high mountains with their hands or prolong their lives over many generations? Surely, because each thing requires for its birth a particular material which determines what can be produced. It must therefore be admitted that nothing can be made out of nothing, because everything must be generated from a seed before it can emerge into the unresisting air.

Lastly, we see that tilled plots are superior to untilled, and their fruits are improved by cultivation. This is because the earth contains certain atoms which we rouse to productivity by turning the fruitful clods with the plough-share and stirring up the soil. But for these, you would see great improvements arising spontaneously without any aid from our labours.

The second great principle is this: *nature resolves everything into its component atoms and never reduces anything to nothing.* If anything were perishable in all its parts, anything might perish all of a sudden and vanish from sight. There would be no need of any force to separate its parts and loosen their links. In actual fact, since everything is composed of indestructible seeds, nature obviously does not allow anything to perish till it has encountered a force that shatters it with a blow or creeps into chinks and unknits it.

If the things that are banished from the scene by age are annihilated through the exhaustion of their material, from what source does Venus bring back the several races of animals into the light of life? And, when they are brought back, where does the inventive earth find for each the special food required for its sustenance and growth? From what fount is the sea replenished by its native springs and the streams that flow into it from afar? Whence does the ether draw nutriment for the stars? For everything consisting of a mortal body must have been exhausted by the long day of time, the illimitable past. If throughout this bygone eternity there have persisted bodies from which the universe has been perpetually renewed, they must certainly be possessed of immortality. Therefore things cannot be reduced to nothing.

Again, all objects would regularly be destroyed by the same force and the same cause, were it not that they are sustained by imperishable matter more or less tightly fastened together. Why, a mere touch would be enough to bring about destruction supposing there were no imperishable bodies whose union could be dissolved only by the appropriate force. Actually, because the fastenings of the atoms are of various kinds while their matter is imperishable, compound objects remain intact until one of them encounters a force that proves strong enough to break up its particular constitution. Therefore nothing returns to nothing, but everything is resolved into its constituent bodies.

Lastly, showers perish when father ether has flung them down into the lap of mother earth. But the crops spring up fresh and gay; the branches on the trees burst into leaf; the trees themselves grow and are weighed down with fruit. Hence in turn man and brute draw nourishment. Hence we see flourishing cities blest with children and every leafy thicket loud with new broods of songsters. Hence in lush pastures cattle wearied by their bulk fling down their bodies, and the white milky juice oozes from their swollen udders. Hence a new generation frolic friskily on wobbly legs through the fresh grass, their young minds tipsy with undiluted milk. Visible objects therefore do not perish utterly, since nature repairs one thing from another and allows nothing to be born without the aid of another's death.

Well, Memmius, I have taught you that things cannot be created out of nothing nor, once born, be summoned back to nothing. Perhaps, however, you are becoming mistrustful of my words, because these atoms of mine are not visible to the eye. Consider, therefore, this further evidence of *bodies whose existence you must acknowledge though they cannot be seen*. First, wind, when its force is roused, whips up waves, founders tall ships and scatters cloud-rack. Sometimes scouring plains with hurricane force it strews them with huge trees and batters mountain peaks with blasts that hew down forests. Such is wind in its fury, when it whoops aloud with a mad menace in its shouting. Without question, therefore, there must be invisible particles of wind which sweep sea and land and the clouds in the sky, swooping upon them and whirling them along in a headlong hurricane. In the way they flow and the havoc they spread they are no different from a torrential flood of water when it rushes down in a sudden spate from the mountain heights, swollen by heavy rains, and heaps together wreckage from the forest and entire trees. Soft though it is by nature, the sudden shock of oncoming water is more than even stout bridges can withstand, so furious is the force with which the turbid, storm-flushed torrent surges against their piers. With a mighty roar it lays them low, rolling huge rocks under its waves and brushing aside every obstacle from its course. Such, therefore, must be the movement of blasts of wind also. When they have come surging along some course like a rushing river, they push obstacles before them and buffet them with repeated blows; and sometimes, eddying round and round, they snatch them up and carry them along in a swiftly circling vortex. Here then is proof upon proof that winds have invisible bodies, since in their actions and behaviour they are found to rival great rivers, whose bodies are plain to see.

Then again, we smell the various scents of things though we never see them approaching our nostrils. Similarly, heat and cold cannot be detected by our eyes, and we do not see sounds. Yet all these must be composed of bodies, since they are able to impinge upon our senses. For nothing can touch or be touched except body.

Again, clothes hung out on a surf-beaten shore grow moist. Spread in the sun they grow dry. But we do not see how the moisture has soaked into them, nor again how it has been dispelled by the heat. It follows that the moisture is split up into minute parts which the eye cannot possibly see.

Again, in the course of many annual revolutions of the sun a ring is worn thin next to the finger with continual rubbing. Dripping water hollows a stone. A curved ploughshare, iron though it is, dwindles imperceptibly in the furrow. We see the cobble-stones of the highway worn by the feet of many wayfarers. The bronze statues by the city gates show their right hands worn thin by the touch of travellers who have greeted them in passing. We see that all these are being diminished, since they are worn away. But to perceive what particles drop off at any particular time is a power grudged to us by our ungenerous sense of sight.

To sum up, whatever is added to things gradually by nature and the passage of days, causing a cumulative increase, eludes the most attentive scrutiny of our eyes. Conversely, you cannot see what objects lose by the wastage of age—sheer sea-cliffs, for instance, exposed to prolonged erosion by the mordant brine—or at what time the loss occurs. It follows that nature works through the agency of invisible bodies.

On the other hand, things are not hemmed in by the pressure of solid bodies in a tight mass. This is because *there is vacuity in things.* A grasp of this fact will be helpful to you in many respects and will save you from much bewildered doubting and questioning about the universe and from mistrust of my teaching. Well then, by vacuity I mean intangible and empty space. If it did not exist, things could not move at all. For the distinctive action of matter, which is counteraction and obstruction, would be in force always and everywhere. Nothing could proceed, because nothing would give it a starting-point by receding. As it is, we see with our own eyes at sea and on land and high up in the sky that all sorts of things in all sorts of ways are on the move. If there were no empty space, these things would be denied the power of restless movement—or rather, they could not possibly have come into existence, embedded as they would have been in motionless matter.

Besides, there are clear indications that things that pass for solid are in fact porous. Even in rocks a trickle of water seeps through into caves, and copious drops ooze from every surface. Food percolates to every part of an animal's body. Trees grow and bring forth their fruit in season, because their food is distributed throughout their length from the tips of the roots through the trunk and along every branch. Noises pass through walls and fly into closed buildings. Freezing cold penetrates to the bones. If there were no vacancies through which the various bodies could make their way, none of these phenomena would be possible.

Again, why do we find some things outweigh others of equal volume? If there is as much matter in a ball of wool as in one of lead, it is natural that it should weigh as heavily, since it is the function of matter to press everything downwards, while it is the function of space on the other hand to remain weightless. Accordingly, when one thing is not less bulky than another but obviously lighter, it plainly declares that there is more vacuum in it, while the heavier object proclaims that there is more matter in it and much less empty space. We have therefore reached the goal of our diligent enquiry: there is in things an admixture of what we call vacuity.

In case you should be misled on this question by the idle imagining of certain theorists, I must anticipate their argument. They maintain that water yields and opens a penetrable path to the scaly bodies of fish that push against it, because they leave spaces behind them into which the yielding water can flow together. In the same way, they suppose, other things can move by mutually changing places, although every place remains filled. This theory has

been adopted utterly without warrant. For how can the fish advance till the
water has given way? And how can the water retire when the fish cannot
move? There are thus only two alternatives: either all bodies are devoid of
movement, or you must admit that things contain an admixture of vacuity
whereby each is enabled to make the first move.

Lastly, if two bodies suddenly spring apart from contact on a broad surface, all the intervening space must be void until it is occupied by air. However
quickly the air rushes in all round, the entire space cannot be filled instanta-
neously. The air must occupy one spot after another until it has taken possession of the whole space. If anyone supposes that this consequence of such
springing apart is made possible by the condensation of air, he is mistaken.
For condensation implies that something that was full becomes empty, or
vice versa. And I contend that air could not condense so as to produce this
effect; or at any rate, if there were no vacuum, it could not thus shrink into
itself and draw its parts together.

However many pleas you may advance to prolong the argument, you must
end by admitting that there is vacuity in things. There are many other proofs
I could add to the pile in order to strengthen conviction; but for an acute
intelligence these small clues should suffice to enable you to discover the rest
for yourself. As hounds that range the hills often smell out the lairs of wild
beasts screened in thickets, when once they have got on to the right trail, so
in such questions one thing will lead on to another, till you can succeed by
yourself in tracking down the truth to its lurking-places and dragging it
forth. If you grow weary and relax from the chase, there is one thing,
Memmius, that I can safely promise you: my honeyed tongue will pour from
the treasury of my breast such generous draughts, drawn from inexhaustible
springs, that I am afraid slow-plodding age may creep through my limbs and
unbolt the bars of my life before the full flood of my arguments on any single
point has flowed in verse through your ears.

To pick up the thread of my discourse, all nature as it is in itself consists
of two things—bodies and the vacant space in which the bodies are situated
and through which they move in different directions. The existence of bodies
is vouched for by the agreement of the senses. If a belief resting directly on
this foundation is not valid, there will be no standard to which we can refer
any doubt on obscure questions for rational confirmation. If there were no
place and space, which we call vacuity, these bodies could not be situated
anywhere or move in any direction whatever. This I have just demonstrated.
It remains to show that *nothing exists that is distinct both from body and
from vacuity* and could be ranked with the others as a third substance. For
whatever *is* must also be something. If it offers resistance to touch, however
light and slight, it will increase the mass of body by such amount, great or
small, as it may amount to, and will rank with it. If, on the other hand, it is
intangible, so that it offers no resistance whatever to anything passing through

it, then it will be that empty space which we call vacuity. Besides, whatever it may be in itself, either it will act in some way, or react to other things acting upon it, or else it will be such that things can be and happen in it. But without body nothing can act or react; and nothing can afford a place except emptiness and vacancy. Therefore, besides matter and vacuity, we cannot include in the number of things any third substance that can either affect our senses at any time or be grasped by the reasoning of our minds.

You will find that anything that can be named is either a property or an accident of these two. A *property* is something that cannot be detached or separated from a thing without destroying it, as weight is a property of rocks, heat of fire, fluidity of water, tangibility of all bodies, intangibility of vacuum. On the other hand, servitude and liberty, poverty and riches, war and peace, and all other things whose advent or departure leaves the essence of a thing intact, all these it is our practice to call by their appropriate name, *accidents*.

Similarly, time by itself does not exist; but from things themselves there results a sense of what has already taken place, what is now going on and what is to ensue. It must not be claimed that anyone can sense time by itself apart from the movement of things or their restful immobility.

Again, when men say it *is* a fact that Helen was ravished or the Trojans were conquered, do not let anyone drive you to the admission that any such event *is* independently of any object, on the ground that the generations of men of whom these events were accidents have been swept away by the irrevocable lapse of time. For we could put it that whatever has taken place is an accident of a particular tract of earth or of the space it occupied. If there had been no matter and no space or place in which things could happen, no spark of love kindled by the beauty of Tyndareus' daughter would ever have stolen into the breast of Phrygian Paris to light that dazzling blaze of pitiless war; no Wooden Horse, unmarked by the sons of Troy, would have set the towers of Ilium aflame through the midnight issue of Greeks from its womb. So you may see that events cannot be said to *be* by themselves like matter or in the same sense as space. Rather, you should describe them as accidents of matter, or of the place in which things happen.

Material objects are of two kinds, atoms and compounds of atoms. The atoms themselves cannot be swamped by any force, for they are preserved indefinitely by their absolute solidity. Admittedly, it is hard to believe that anything can exist that is absolutely solid. The lightning stroke from the sky penetrates closed buildings, as do shouts and other noises. Iron glows molten in the fire, and hot rocks are cracked by untempered scorching. Hard gold is softened and melted by heat; and bronze, ice-like, is liquefied by flame. Both heat and piercing cold seep through silver, since we feel both alike when a cooling shower of water is poured into a goblet that we hold ceremonially in our hands. All these facts point to the conclusion that nothing is really solid. But sound reasoning and nature itself drive us to the opposite conclusion.

Pay attention, therefore, while I demonstrate in a few lines that there exist certain bodies that are absolutely solid and indestructible, namely those atoms which according to our teaching are the seeds or prime units of things from which the whole universe is built up.

In the first place, we have found that nature is twofold, consisting of two totally different things, matter and the space in which things happen. Hence each of these must exist by itself without admixture of the other. For, where there is empty space (what we call vacuity), there matter is not; where matter exists, there cannot be a vacuum. Therefore the prime units of matter are solid and free from vacuity.

Again, since composite things contain some vacuum, the surrounding matter must be solid. For you cannot reasonably maintain that anything can hide vacuity and hold it within its body unless you allow that the container itself is solid. And what contains the vacuum in things can only be an accumulation of matter. Hence matter, which possesses absolute solidity, can be everlasting when other things are decomposed.

Again, if there were no empty space, everything would be one solid mass; if there were no material objects with the property of filling the space they occupy, all existing space would be utterly void. It is clear, then, that there is an alternation of matter and vacuity, mutually distinct, since the whole is neither completely full nor completely empty. There are therefore solid bodies, causing the distinction between empty space and full. And these, as I have just shown, can be neither decomposed by blows from without nor invaded and unknit from within nor destroyed by any other form of assault. For it seems that a thing without vacuum can be neither knocked to bits nor snapped nor chopped in two by cutting; nor can it let in moisture or seeping cold or piercing fire, the universal agents of destruction. The more vacuum a thing contains within it, the more readily it yields to these assailants. Hence, if the units of matter are solid and without vacuity, as I have shown, they must be everlasting.

Yet again, if the matter in things had not been everlasting, everything by now would have gone back to nothing, and the things we see would be the product of rebirth out of nothing. But, since I have already shown that nothing can be created out of nothing nor any existing thing be summoned back to nothing, the atoms must be made of imperishable stuff into which everything can be resolved in the end, so that there may be a stock of matter for building the world anew. The atoms, therefore, are absolutely solid and unalloyed. In no other way could they have survived throughout infinite time to keep the world in being.

Furthermore, if nature had set no limit to the breaking of things, the particles of matter in the course of ages would have been ground so small that nothing could be generated from them so as to attain in the fullness of time to the summit of its growth. For we see that anything can be more speedily disintegrated than put together again. Hence, what the long day of time, the bygone eternity, has already shaken and loosened to fragments could never in the residue of time be reconstructed. As it is, there is evidently a limit set to breaking, since we see that everything is renewed and each according to its kind has a fixed period in which to grow to its prime.

Here is a further argument. Granted that the particles of matter are absolutely solid, we can still explain the composition and behaviour of soft things—air, water, earth, fire—by their intermixture with empty space. On the other hand, supposing the atoms to be soft, we cannot account for the origin of hard flint and iron. For there would be no foundation for nature to build on. Therefore there must be bodies strong in their unalloyed solidity by whose closer clustering things can be knit together and display unyielding toughness.

If we suppose that there is no limit set to the breaking of matter, we must still admit that material objects consist of particles which throughout eternity have resisted the forces of destruction. To say that these are breakable does not square with the fact that they have survived throughout eternity under a perpetual bombardment of innumerable blows.

Again, there is laid down for each thing a specific limit to its growth and its tenure of life, and the laws of nature ordain what each can do and what it cannot. No species is ever changed, but each remains so much itself that every kind of bird displays on its body its own specific markings. This is a further proof that their bodies are composed of changeless matter. For, if the atoms could yield in any way to change, there would be no certainty as to what could arise and what could not, at what point the power of everything was limited by an immovable frontier-post; nor could successive generations so regularly repeat the nature, behaviour, habits and movements of their parents.

To proceed with our argument, there is an ultimate point in visible objects which represents the smallest thing that can be seen. So also there must be an ultimate point in objects that lie below the limit of perception by our senses. This point is without parts and is the smallest thing that can exist. It never has been and never will be able to exist by itself, but only as one primary part of something else. It is with a mass of such parts, solidly jammed together in order, that matter is filled up. Since they cannot exist by themselves, they must needs stick together in a mass from which they cannot by any means be pried loose. The atoms therefore are absolutely solid and unalloyed, consisting of a mass of least parts tightly packed together. They are not compounds formed by the coalescence of their parts, but bodies of absolute and everlasting solidity. To these nature allows no loss or diminution, but guards them as seeds for things. If there are no such least parts, even the smallest bodies will consist of an infinite number of parts, since they can always be halved and their halves halved again without limit. On this showing, what difference will there be between the whole universe and the very least of things? None at all. For, however endlessly infinite the universe may be, yet the smallest things will equally consist of an infinite number of parts. Since true reason cries out against this and denies that the mind can believe it, you must needs give in and admit that there are least parts which themselves are partless. Granted that these parts exist, you must needs admit that the atoms they compose are also solid and everlasting. But, if all things were compelled by all-creating nature to be broken up into these least parts, nature would lack the power to rebuild anything out of them. For partless objects cannot have the essential properties of generative matter—those varieties of attachment, weight, impetus, impact and movement on which everything depends. . . .

Well then, since I have shown that there are completely solid indestructible particles of matter flying about through all eternity, let us elucidate whether or not there is any limit to their number. Similarly, as we have found that there is a vacuum, the place or space in which things happen, let us see whether its whole extent is limited or whether it stretches far and wide into immeasurable depths.

Learn, therefore, that *the universe is not bounded in any direction*. If it were, it would necessarily have a limit somewhere. But clearly a thing cannot have a limit unless there is something outside to limit it, so that the eye can follow it up to a certain point but not beyond. Since you must admit that there is nothing outside the universe, it can have no limit and is accordingly without end or measure. It makes no odds in which part of it you may take your stand: whatever spot anyone may occupy, the universe stretches away from him just the same in all directions without limit. Suppose for a moment that the whole of space were bounded and that someone made his way to its uttermost boundary and threw a flying dart. Do you choose to suppose that the missile, hurled with might and main, would speed along the course on which it was aimed? Or do you think something would block the way and stop it? You must assume one alternative or the other. But neither of them leaves you a loophole. Both force you to admit that the universe continues without end. Whether there is some obstacle lying on the boundary line that prevents the dart from going farther on its course or whether it flies on beyond, it cannot in fact have started from the boundary. With this argument I will pursue you. Wherever you may place the ultimate limit of things, I will ask you: 'Well then, what does happen to the dart?' The upshot is that the boundary cannot stand firm anywhere, and final escape from this conclusion is precluded by the limitless possibility of running away from it.

It is a matter of observation that one thing is limited by another. The hills are demarcated by air, and air by the hills. Land sets bounds to sea, and sea to every land. But the universe has nothing outside to limit it.

Further, if all the space in the universe were shut in and confined on every side by definite boundaries, the supply of matter would already have accumulated by its own weight at the bottom, and nothing could happen under the dome of the sky—indeed, there would be no sky and no sunlight, since all

THE NATURE OF THE UNIVERSE

the available matter would have settled down and would be lying in a heap throughout eternity. As it is, no rest is given to the atoms, because there is no bottom where they can accumulate and take up their abode. Things go on happening all the time through ceaseless movement in every direction; and atoms of matter bouncing up from below are supplied out of the infinite. There is therefore a limitless abyss of space, such that even the dazzling flashes of the lightning cannot traverse it in their course, racing through an interminable tract of time, nor can they even shorten the distance still to be covered. So vast is the scope that lies open to things far and wide without limit in any dimension.

The universe is restrained from setting any limit to itself by nature, which compels body to be bounded by vacuum and vacuum by body. Thus nature either makes them both infinite in alternation, or else one of them, if it is not bounded by the other, must extend in a pure state without limit. Space, however, being infinite, so must matter be. Otherwise neither sea nor land nor the bright zones of the sky nor mortal beings nor the holy bodies of the gods could endure for one brief hour of time. The supply of matter would be shaken loose from combination and swept through the vastness of the void in isolated particles; or rather, it would never have coalesced to form anything, since its scattered particles could never have been driven into union.

Certainly the atoms did not post themselves purposefully in due order by an act of intelligence, nor did they stipulate what movements each should perform. As they have been rushing everlastingly throughout all space in their myriads, undergoing myriad changes under the disturbing impact of collisions, they have experienced every variety of movement and conjunction till they have fallen into the particular pattern by which this world of ours is constituted. This world has persisted many a long year, having once been set going in the appropriate motions. From these everything else follows. The rivers replenish the thirsty sea with profuse streams of water. Incubated by the sun's heat, the earth renews its fruits, and the brood of animals that springs from it grows lustily. The gliding fires of ether sustain their life. None of these results would be possible if there were not an ample supply of matter to bounce up out of infinite space in replacement of all that is lost. Just as animals deprived of food waste away through loss of body, so everything must decay as soon as its supply of matter goes astray and is cut off.

Whatever world the atoms have combined to form, impacts from without cannot preserve it at every point. By continual battering they can hold back part of it till others come along to make good the deficiency. But they are compelled now and then to bounce back and in so doing to leave space and time for the atoms to break loose from combination. It is thus essential that there should be great numbers of atoms coming up. Indeed, the impacts themselves could not be maintained without an unlimited supply of matter from all quarters.

There is one belief, Memmius, that you must beware of entertaining—
*the theory that everything tends towards what they call 'the centre of the
world.'* On this theory, the world stands fast without any impacts from
without, and top and bottom cannot be parted in any direction, because
everything has been tending towards the centre—if you can believe that anything rests upon itself. Whatever heavy bodies there may be under the earth
must then tend upwards and rest against the surface upside down, like the
images of things which we now see reflected in water. In the same way they
would have it that animals walk about topsy-turvy and cannot fall off the
earth into the nether quarters of the sky any more than our bodies can soar
up spontaneously into the heavenly regions. When they are looking at the
sun, we see the stars of night; so they share the hours with us alternately and
experience nights corresponding to our days. But this is an idle fancy of fools
who have got hold of the wrong end of the stick. There can be no centre in
infinity. And, even if there were, nothing could stand fast there rather than
flee from it. For all place or space, at the centre no less than elsewhere, must
give way to heavy bodies, no matter in what direction they are moving. There
is no place to which bodies can come where they lose the property of weight
and stand still in the void. And vacuum cannot stand in the way of anything
so as not to allow it free passage, as its own nature demands. Therefore things
cannot be held in combination by this means through surrender to a craving
for the centre.

Besides, they do not claim that all bodies have this tendency towards the
centre, but only those of moisture and earth—the waters of the deep and the
floods that pour down from the hills and in general whatever is composed of
a more or less earthy body. But according to their teaching the light breaths
of air and hot fires are simultaneously wafted outwards away from the centre.
The reason why the encircling ether twinkles with stars and the sun feeds its
flames in the blue pastures of the sky is supposed to be that fire all congregates
there in its flight from the centre. Similarly, the topmost branches of trees
could not break into leaf unless their food had this same upward urge. But,
if you allow matter to escape from the world in this way, you are leaving the
ramparts of the world at liberty to crumble of a sudden and take flight with
the speed of flame into the boundless void. The rest will follow. The thunder-
breeding quarters of the sky will rush down from aloft. The ground will fall
away from our feet, its particles dissolved amid the mingled wreckage of
heaven and earth. The whole world will vanish into the abyss, and in the
twinkling of an eye no remnant will be left but empty space and invisible
atoms. At whatever point you first allow matter to fall short, this will be the
gateway to perdition. Through this gate the whole concourse of matter will
come streaming out. . . .

Give your mind now to the true reasoning I have to unfold. A new fact
is battling strenuously for access to your ears. A new aspect of the universe

is striving to reveal itself. But no fact is so simple that it is not harder to believe than to doubt at the first presentation. Equally, there is nothing so mighty or so marvellous that the wonder it evokes does not tend to diminish in time. Take first the pure and undimmed lustre of the sky and all that it enshrines: the stars that roam across its surface, the moon and the surpassing splendour of the sunlight. If all these sights were now displayed to mortal view for the first time by a swift unforeseen revelation, what miracle could be recounted greater than this? What would men before the revelation have been less prone to conceive as possible? Nothing, surely. So marvellous would have been that sight—a sight which no one now, you will admit, thinks worthy of an upward glance into the luminous regions of the sky. So has satiety blunted the appetite of our eyes. Desist, therefore, from thrusting out reasoning from your mind because of its disconcerting novelty. Weigh it, rather, with discerning judgment. Then, if it seems to you true, give in. If it is false, gird yourself to oppose it. For the mind wants to discover by reasoning what exists in the infinity of space that lies out there, beyond the ramparts of this world— that region into which the intellect longs to peer and into which the free projection of the mind does actually extend its flight.

Here, then, is my first point. In all dimensions alike, on this side or that, upward or downward through the universe, there is no end. This I have shown, and indeed the fact proclaims itself aloud and the nature of space makes it crystal clear. Granted, then, that empty space extends without limit in every direction and that seeds innumerable in number are rushing on countless courses through an unfathomable universe under the impulse of perpetual motion, *it is in the highest degree unlikely that this earth and sky is the only one to have been created* and that all those particles of matter outside are accomplishing nothing. This follows from the fact that our world has been made by nature through the spontaneous and casual collision and the multifarious, accidental, random and purposeless congregation and coalescence of atoms whose suddenly formed combinations could serve on each occasion as the starting-point of substantial fabrics—earth and sea and sky and the races of living creatures. On every ground, therefore, you must admit that there exist elsewhere other congeries of matter similar to this one which the ether clasps in ardent embrace.

When there is plenty of matter in readiness, when space is available and no cause or circumstance impedes, then surely things must be wrought and effected. You have a store of atoms that could not be reckoned in full by the whole succession of living creatures. You have the same natural force to congregate them in any place precisely as they have been congregated here. You are bound therefore to acknowledge that in other regions there are other earths and various tribes of men and breeds of beasts.

Add to this the fact that nothing in the universe is the only one of its kind, unique and solitary in its birth and growth; everything is a member of a species comprising many individuals. Turn your mind first to the animals.

You will find the rule apply to the brutes that prowl the mountains, to the children of men, the voiceless scaly fish and all the forms of flying things. So you must admit that sky, earth, sun, moon, sea and the rest are not solitary, but rather numberless. For a firmly established limit is set to their lives also and their bodies also are a product of birth, no less than that of any creature that flourishes here according to its kind.

Bear this well in mind, and you will immediately perceive that *nature is free and uncontrolled by proud masters* and runs the universe by herself without the aid of gods. For who—by the sacred hearts of the gods who pass their unruffled lives, their placid aeon, in calm and peace!—who can rule the sum total of the measureless? Who can hold in coercive hand the strong reins of the unfathomable? Who can spin all the firmaments alike and foment with the fires of ether all the fruitful earths? Who can be in all places at all times, ready to darken the clear sky with clouds and rock it with a thunderclap —to launch bolts that may often wreck his own temples, or retire and spend his fury letting fly at deserts with that missile which often passes by the guilty and slays the innocent and blameless?

After the natal season of the world, the birthday of sea and lands and the uprising of the sun, many atoms have been added from without, many seeds contributed on every side by bombardment from the universe at large. From these the sea and land could gather increase; the dome of heaven could gain more room and lift its rafters high above the earth, and the air could climb upwards. For to each are allotted its own atoms from every quarter under the impact of blows. They all rejoin their own kind: water goes to water, earth swells with earthy matter; fire is forged by fires, ether by ether. At length everything is brought to its utmost limit of growth by nature, the creatress and perfectress. This is reached when what is poured into its vital veins is no more than what flows and drains away. Here the growing-time of everything must halt. Here nature checks the increase of her own strength. The things you see growing merrily in stature and climbing step by step the stairs of maturity—these are gaining more atoms than they lose. The food is easily introduced into all their veins; and they themselves are not so widely expanded as to shed much matter and squander more than their age absorbs as nourishment. It must, of course, be conceded that many particles ebb and drain away from things. But more particles must accrue, until they have touched the topmost peak of growth. Thereafter the strength and vigour of maturity is gradually broken, and age slides down the path of decay. Obviously the bulkier anything is and the more expanded when it begins to wane, the more particles it sheds and gives off from every surface. The food is not easily distributed through all its veins, or supplied in sufficient quantities to make good the copious effluences it exudes. For everything must be restored and renewed by food, and by food buttressed and sustained. And the process is doomed to failure, because the veins do not admit enough and nature does not supply all that is needed. It is natural, therefore, that everything should perish when it is thinned out by the ebbing of matter and succumbs to blows from without. The food supply is no longer adequate for its aged frame, and the deadly bombardment of particles from without never pauses in the work of dissolution and subdual.

In this way the ramparts of the great world also will be breached and collapse in crumbling ruin about us. Already it is far past its prime. The earth, which generated every living species and once brought forth from its womb the bodies of huge beasts, has now scarcely strength to generate animalcules. For I assume that the races of mortal creatures were not let down into the fields from heaven by a golden cord, nor generated from the sea or the rock-beating surf, but born of the same earth that now provides their nurture. The same earth in her prime spontaneously generated for mortals smiling crops and lusty vines, sweet fruits and gladsome pastures, which now can scarcely be made to grow by our toil. We wear down the oxen and wear out the strength of husbandmen, and the ploughshare is scarcely a match for fields that grudge their fruits and multiply our toil. Already the ploughman of ripe years shakes his head with many a sigh that his heavy labours have gone for nothing; and, when he compares the present with the past, he often cries up his father's luck and grumbles that past generations, when men were old-fashioned and god-fearing, supported life easily enough on their small farms, though one man's holding was then far less than now. In the same despondent vein, the cultivator of old and wilted vines decries the trend of the times and rails at heaven. He does not realize that everything is gradually decaying and nearing its end, worn out by old age.

The Classic View of a
Geocentric Finite Universe

Introduction

THE CONCEPTION of the universe that prevailed in the minds of men throughout classical antiquity and the Middle Ages was one according to which the earth occupied the fixed center of the entire universe. The universe itself, bounded at its outermost spatial limits by the spherical shell of fixed stars, was regarded as performing its diurnal revolution about the immovable earth. Many variations came to be made within this schema of thought throughout its long career, variations sometimes suggested by astronomical data, mathematical convenience, philosophical or theological considerations. Three of the chief architects of this classic cosmology, supplying many of its most important and characteristic details, were Plato, Aristotle and Ptolemy.

Plato's great dialogue devoted to cosmology, *Timaeus,* is cast in the form of a myth. In terms of it Plato presents a view which reacts sharply to the materialist views of the Atomists. Plato makes serious use of the analogy of creation or craftsmanship to express his conviction that the universe is not simply the outcome of blind necessity or the chance collocations of atoms. The universe according to him reveals, rather, the presence of purposive order and rational choice. Such order and rationality, as in any work of art, are limited by the recalcitrant and irrational materials in which they are embodied. The resultant product is one which/ evinces a mixture of Reason (purpose) and Necessity (chance). Plato also makes use of another analogical pattern of thought in describing the universe as an all-inclusive Living Creature, one whose body is perfectly spherical and whose soul animates the whole world. In addition to this World-Soul, the various individual heavenly bodies are regarded by Plato as divine beings. It is the perfection of their minds which maintains their observed orderly and regular motions. The principle motion of the World-Soul is what Plato expresses as the rotation of the "circle of the Same." This is to be identified with the celestial equator with respect to which all the stars perform their daily orbits. Contrasted with this motion are the motions along the "circle of the Different" in which are comprised the several motions of the sun, moon, and the five planets. While the daily revolution of the heavens carries along with it these seven bodies in addition to all the other stars, these seven bodies possess their own separate motions along the inclined common plane of the "Different." At the center of the entire scheme of circles is the earth which (despite the suggestions of an obscure and controversial

61

passage in the *Timaeus*) is regarded by Plato to be at rest and wholly immovable with respect to "absolute space."

Plato's rationalism was not only of a teleological kind, but gave ample
room to a deep appreciation of the role of mathematics in lending intelligibility
to the facts of sense experience. Here the Pythagorean strain in his thought was
responsible for the requirement that astronomy make use of circles in describing the paths of the heavenly bodies. For Plato, according to a reliable
tradition, proposed as the essential task of astronomy that of showing how the
apparently irregular motions of the planets can be accounted for in terms of
combinations of uniform circular motions. It was this insistence on the
primacy of the circular pattern, as alone befitting the unvarying and divinely
regulated motions of the heavenly bodies, which was not only to support
Aristotle's sharp division between celestial and terrestrial physics, but was to
shackle astronomy for a long time to come in its search for suitable kinematic
patterns for the planets. First the technique of homocentric (concentric)
spheres devised by Plato's pupil, Eudoxus of Cnidus, and later the techniques
of the eccentric circle and epicycle, as employed by Hipparchus and Ptolemy,
were based on this Platonic axiom on how to "save the appearances." Indeed,
even Copernicus's system, for all its revolutionary significance, continued to
make use of the same idea. It was left for Kepler to show that, at least in the
case of the planetary orbits, one obtains a better "fit" by the use of elliptical
paths.

The basic technique introduced by Eudoxus for making intelligible the
motions of the heavenly bodies may be briefly described as follows. The visible
body whose motion is to be accounted for, for example a planet, is conceived
as attached to the surface of an imaginary invisible sphere. Thus the planet
has no motion of its own, but participates in the uniform circular motion of
the sphere to which it is attached. (See Fig. 1.) Eudoxus posits for each planet
a combination or nest of spheres arranged in such a fashion that the planet
is pictured as attached to the equator of the innermost sphere. The axis of
this sphere is attached at its poles to points in the sphere immediately surrounding it. Each sphere is taken as having its own rate of uniform motion
about its axis and a distinctive angle of inclination for that axis. By positing
for each planet a particular combination of rates of motion and inclination of
axes for the set of interlocking spheres, it was possible to calculate and predict
the observed positions of the planets. The entire system of sets of spheres for
each planet, the sun, moon, and the enveloping sphere of fixed stars were all
centered on a common center (hence the name "homocentric") namely, the
earth. In Eudoxus's scheme twenty-seven spheres in all were considered sufficient, three each for the sun and moon, four for each of the five planets, and
one for the fixed stars. Calippus, a pupil of Eudoxus, improved upon the
accuracy of the entire scheme, by adding to the number of spheres, arriving at
a total of thirty-three in addition to the sphere of the fixed stars.

Aristotle was enormously indebted to Plato for some of the major presuppositions in his own thought. For him, the biological model of the cosmos, that is the conception of the cosmos as a living organism proved, however, to be of greater interest and philosophical affinity than the creational one. He thinks of the universe as a "particular material thing," a single substance that is unique, finite, and spherically shaped. The echoes of Plato's conception of a World-Soul are present in Aristotle's attempt to explain the ultimate source of motion in the revolving heavens. For his conception of God as an unmoved mover is that of a being who functions as an object of love and desire for the soul that presumably animates the body of the outermost heaven. When moved in this way, the outermost heaven communicates its motion to the other bodies in the successive layers of inner spheres that comprise the world of celestial phenomena. In his astronomy, Aristotle builds upon the ideas of Eudoxus and Calippus in their use of the method of homocentric spheres. What were for these astronomers however, merely mathematical devices of representation, useful for purposes of calculation and prediction, and not to be invested with any further reality, are regarded by Aristotle as physical existences. In accordance with the requirements of his general physical theory, which he brings to bear on the domain of cosmological problems, the heavenly bodies must be thought of as composed of a distinctive ethereal substance, different from the objects found below the sphere of the moon. The latter alone are made up of varying combinations of earth, air, fire and water. Unlike these terrestrial objects and phenomena, which are subject to corruption and change, the heavenly bodies are subject neither to generation, corruption, violent motion, or alteration of any kind. Composed of pure ethereal substance, both the invisible spheres to which the various heavenly bodies are rigidly attached, and the visible heavenly bodies themselves, are engaged in the unceasing and unvarying circular motions that are alone befitting their supreme status in the qualitative hierarchy of natural substances. Because of his materialization of the homocentric spheres, Aristotle finds it necessary to go further than Calippus in his development of the theory. In the *Metaphysics* he introduces the complication of a series of "unrolling" spheres which bring the total number of spheres in his own system to fifty-five.

A crucial aspect of Aristotle's cosmology is the claim that the universe is finite. He does not seem to have been acquainted with the view of Heracleides that the earth performs a daily rotation on its axis. For had he taken seriously the import of such a proposal, the claim that it is the heavenly sphere instead which performs its daily revolution would, of course, be unnecessary. At the same time, this would destroy what is, for Aristotle, a main argument for the finitude of the universe, since he argues that the infinite cannot be traversed in the finite time of twenty-four hours. In the same vein, Aristotle sets about rejecting one of the chief items in the Atomist cosmology, the belief in an infinite universe that contains innumerable worlds. Like his master Plato, Aristotle insists there is but one world, that is, a central body like the earth surrounded by a finite number of planets and stars. This one world of ours which makes up the entire universe contains all existent matter. It is a world beyond whose spherical boundary there is, he argues, neither place, nor void, nor time, since there are no material bodies outside the heavens. And where there are no bodies, space and time have no existence of their own. Likewise, in opposition to the suggestion of the Atomists that a world be thought of as having a finite duration, with a beginning and an end in time, Aristotle argues that the one world or universe we know is eternal, without beginning and without end. Finally, in opposition to the Atomist view that celestial as well as terrestrial phenomena and objects are the products of the combination of similar, underlying atomic particles, Aristotle upholds the belief that the two domains of the terrestrial and the celestial are wholly distinct both with respect to the matter and the scheme of motions that belong to each. Below the sphere of the moon all objects are composed of the elements earth, air, fire and water. Such elements have natural rectilinear paths, moving either downward toward the center or upward away from the center. The heavenly bodies being composed of ethereal substance have neither gravity nor levity. They move neither downward nor upward, but maintain their unchanging motions around the fixed center of their orbs. It was for this reason too that such phenomena as comets and meteors, being evanescent and changeful, were taken in the Aristotelian cosmology as belonging to the domain of sublunary phenomena.

Ptolemy's work *The Almagest,* as it has come to be known, the greatest and most comprehensive astronomical treatise of antiquity, refines upon the work of that other great ancient astronomer, Hipparchus (*fl.* 140–129 B.C.). It brings to a systematic culmination all the efforts of Greek astronomy. In it Ptolemy abandons the use of the technique of homocentric spheres and shows how a more efficient representation of planetary motions can be achieved through the use of the devices of eccentric circles and epicycles. The chief defect in the system of homocentric spheres, for all its ingenuity, was its ultimate failure to account for the observed variations in the brightness of the planets as they traversed their several orbits. Such a variation could not be understood if one adhered, as in the Eudoxian scheme, to the belief that each planet remains at a fixed distance from the earth. It was this objection which the use of eccentric circles and epicycles was designed to overcome. Very briefly, the use of the device of the eccentric circle consists in having the body whose motion is to be accounted for, whether sun, moon or planet, represented as moving around a center other than the earth. In the case of the epicyclic technique, the moving body is represented as traveling around a small circle whose center is itself on a larger one, known as the deferent, with the latter having its center in the earth. (See Fig. 2.) A variety of complications and modifications in both schemes, whose equivalence for certain purposes was also mathematically established, were introduced, into which it is unnecessary to go here. Suffice it to say that when so employed, the Ptolemaic scheme made it possible to account for the various observed facts of observational astronomy

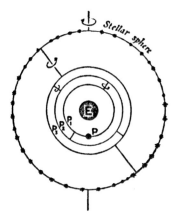

FIG. 1. The system of concentric planetary spheres adopted by Aristotle, showing the system of spheres (p_1, p_2, p_3) for one planet P with the axes all placed in the plane of the paper. If P were Saturn the other planets' spheres would come inside this. The stellar sphere rotates about an axis passing through the centre of the stationary earth E.

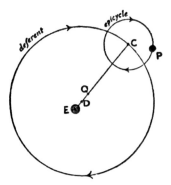

FIG. 2. The device of the epicycle in Ptolemy's system for the motion of a planet P. The centre of the deferent is D which does not coincide with E, the centre of the earth and of the universe. The centre of the epicycle, C, does not rotate uniformly about D but moves so that the line CQ, connecting C with the equant Q, moves through equal angles in equal times. If necessary for the better 'saving of appearances' the system could be complicated by adding further epicycles, as, for example, a second whose centre C_1 revolves round C, a third whose centre C_2 revolves round C_1, and so on, the planet itself always being on the outermost epicycle.

(From A. C. Crombie, *Augustine to Galileo*.)

known to the ancient world with a remarkable degree of accuracy. It was for this reason that it remained superior to any other available scheme until it too was eventually abandoned in modern times.

The cessation of all significant scientific work following upon the decline of the Alexandrian epoch left the cosmology of a geocentric and finite universe as worked out in classic Greek and Hellenistic thought to be incorporated within the framework of Hebraic and Christian schemes of thought. Such modifications as came to be made were principally of a type dictated by theological considerations. Among these the principal modification was the claim that the universe was created *ex nihilo* and came into existence at a finite epoch in the past. Dreyer's account below of medieval cosmology gives a brilliant synopsis of the fortunes of our topic from the earliest and most corrupt backslidings of the Dark Ages up to the "final" synthesis worked out at the height of the Middle Ages in the thirteenth century.

PLATO:

Timaeus

The Nature and Scope of Physics

[TIMAEUS' "prelude," marked off from what follows by Socrates' expression of approval, lays down the principles of the whole discourse and defines the limitations of any treatment of physics. It is constructed with great care. After the opening invocation of the gods, the second paragraph states three general premises concerning anything that is not eternal, but comes to be. These premises are then applied successively to the visible universe: (1) The eternal is the intelligible; what comes to be is the sensible. Since the world is sensible, it must be a thing that comes to be. (2) Whatever comes to be must have a cause. Therefore the world has a cause—a maker and father; but he is hard to find. (3) The work of any maker will be good only if he fashions it after an eternal model. The world is good; so its model must have been eternal. Finally, the conclusion is drawn: any account that can be given of the physical world can be no better than a "likely story," because the world itself is only a "likeness" of unchanging reality.]

TIMAEUS. . . . We who are now to discourse about the universe—how it came into being, or perhaps had no beginning of existence—must, if our senses be not altogether gone astray, invoke gods and goddesses with a prayer that our discourse throughout may be above all pleasing to them and in consequence satisfactory to us. Let this suffice, then, for our invocation of the gods; but we must also call upon our own powers, so that you may follow most readily and I may give the clearest expression to my thought on the theme proposed.

We must, then, in my judgment, first make this distinction: what is that which is always real and has no becoming, and what is that which is always becoming and is never real? That which is apprehensible by thought with a rational account is the thing that is always unchangeably real; whereas that

From F. M. Cornford, *Plato's Cosmology: The Timaeus of Plato*, translated with a running commentary, London, 1937, lines 27C-40D, 47E-53B. Reprinted with the kind permission of the publishers, The Humanities Press, Inc., New York.

which is the object of belief together with unreasoning sensation is the thing that becomes and passes away, but never has real being. Again, all that becomes must needs become by the agency of some cause; for without a cause nothing can come to be. Now whenever the maker of anything looks to that which is always unchanging and uses a model of that description in fashioning the form and quality of his work, all that he thus accomplishes must be good.[1] If he looks to something that has come to be and uses a generated model, it will not be good.

So concerning the whole Heaven or World—let us call it by whatsoever name may be most acceptable to it[2]—we must ask the question which, it is agreed, must be asked at the outset of inquiry concerning anything: Has it always been, without any source of becoming; or has it come to be, starting from some beginning? It has come to be; for it can be seen and touched and it has body, and all such things are sensible; and, as we saw, sensible things, that are to be apprehended by belief together with sensation, are things that become and can be generated. But again, that which becomes, we say, must necessarily become by the agency of some cause. The maker and father of this universe it is a hard task to find, and having found him it would be impossible to declare him to all mankind. Be that as it may, we must go back to this question about the world: After which of the two models did its builder frame it—after that which is always in the same unchanging state, or after that which has come to be? Now if this world is good and its maker is good, clearly he looked to the eternal; on the contrary supposition (which cannot be spoken without blasphemy), to that which has come to be. Everyone, then, must see that he looked to the eternal; for the world is the best of things that have become, and he is the best of causes. Having come to be, then, in this way, the world has been fashioned on the model of that which is comprehensible by rational discourse and understanding and is always in the same state.

Again, these things being so, our world must necessarily be a likeness of something. Now in every matter it is of great moment to start at the right point in accordance with the nature of the subject. Concerning a likeness, then, and its model we must make this distinction: an account is of the same order as the things which it sets forth—an account of that which is abiding and stable and discoverable by the aid of reason will itself be abiding and unchangeable (so far as it is possible and it lies in the nature of an account to be incontrovertible and irrefutable, there must be no falling short of that); while an account of what is made in the image of that other, but is only a likeness, will itself be but likely, standing to accounts of the former kind in a proportion: as reality is to becoming, so is truth to belief. If then, Socrates, in many

---

1. καλόν, 'good,' 'satisfactory,' as at Gen. i. 8, 'God saw that it was good' (εἶδεν ὁ θεὸς ὅτι καλόν, LXX). The Greek word means also 'desirable,' 'beautiful,' and will be sometimes so translated.

2. 'Heaven' (οὐρανός) is used throughout the dialogue as a synonym of *cosmos*, the entire world, not the sky.

respects concerning many things—the gods and the generation of the universe —we prove unable to render an account at all points entirely consistent with itself and exact, you must not be surprised. If we can furnish accounts no less likely than any other, we must be content, remembering that I who speak and you my judges are only human, and consequently it is fitting that we should, in these matters, accept the likely story and look for nothing further.

SOCRATES. Excellent, Timaeus; we must certainly accept it as you say. Your prelude we have found exceedingly acceptable; so now go on to develop your main theme.

THE MOTIVE OF CREATION

[Foreshadowing the contrast between rational purpose and the blind operation of Necessity, Plato opens with the creator's motive, the true reason (αἰτία) for the existence of an *ordered* world in the realm of Becoming.]

TIMAEUS. Let us, then, state for what reason becoming and this universe were framed by him who framed them. He was good; and in the good no jealousy in any matter can ever arise. So, being without jealousy, he desired that all things should come as near as possible to being like himself. That this is the supremely valid principle of becoming and of the order of the world, we shall most surely be right to accept from men of understanding. Desiring, then, that all things should be good and, so far as might be, nothing imperfect, the god took over all that is visible—not at rest, but in discordant and un-ordered motion—and brought it from disorder into order, since he judged that order was in every way the better.

Now it was not, nor can it ever be, permitted that the work of the supremely good should be anything but that which is best. Taking thought, therefore, he found that, among things that are by nature visible, no work that is without intelligence will ever be better than one that has intelligence, when each is taken as a whole, and moreover that intelligence cannot be present in anything apart from soul. In virtue of this reasoning, when he framed the universe, he fashioned reason within soul and soul within body, to the end that the work he accomplished might be by nature as excellent and perfect as possible. This, then, is how we must say, according to the likely account, that this world came to be, by the god's providence, in very truth a living creature with soul and reason.

THE CREATOR'S MODEL

[The visible world has been declared to be a living creature made after the likeness of an eternal original. This model is now further described. It can only be the ideal Living Creature in the world of Forms, not to be identified with any species of animate being, but embracing the ideal types of all such species, 'all the intelligible living creatures.']

This being premised, we have now to state what follows next: What was the living creature in whose likeness he framed the world? We must not suppose that it was any creature that ranks only as a species; for no copy of that which is incomplete can ever be good. Let us rather say that the world is like, above all things, to that Living Creature of which all other living creatures, severally and in their families, are parts. For that embraces and contains within itself all the intelligible living creatures, just as this world contains ourselves and all other creatures that have been formed as things visible. For the god, wishing to make this world most nearly like that intelligible thing which is best and in every way complete, fashioned it as a single visible living creature, containing within itself all living things whose nature is of the same order.

One World, Not Many

[The concluding words of the last paragraph spoke of the world as a *single* living creature. This suggests the possibility that there should be more than one copy of the model—a plurality of visible worlds.]

Have we, then, been right to call it one Heaven, or would it have been true rather to speak of many and indeed of an indefinite number? One we must call it, if we are to hold that it was made according to its pattern. For that which embraces all the intelligible living creatures that there are, cannot be one of a pair; for then there would have to be yet another Living Creature embracing those two, and they would be parts of it; and thus our world would be more truly described as a likeness, not of them, but of that other which would embrace them. Accordingly, to the end that this world may be like the complete Living Creature in respect of its uniqueness, for that reason its maker did not make two worlds nor yet an indefinite number; but this Heaven has come to be and is and shall be hereafter one and unique.

The Body of the World

Why This Consists of Four Primary Bodies

[The next section is concerned with the body of the Universe. Although soul is later declared to be prior to body, the making of the body is taken first for convenience. The present paragraph explains why not less than four primary bodies—fire, air, water, earth—were required, in order to give it the highest measure of unity. This attribute of internal unity follows naturally after the unity, in the sense of uniqueness, asserted in the previous paragraph. The primary bodies are here imagined as materials ready to be "put together" (συνιστάναι) by the builder's hand. The formation of them by the imposition of regular geometrical shape upon their unordered motions and powers belongs

to the second part of the dialogue. There is no reference here to those geometrical shapes, of which nothing has yet been heard. All that the Demiurge
does now is to fix their quantities in a certain definite proportion. This is an
element of rational design in the structure of the world's body, and it belongs
here among the works of Reason.]

Now that which comes to be must be bodily, and so visible and tangible;
and nothing can be visible without fire, or tangible without something solid,
and nothing is solid without earth. Hence the god, when he began to put
together the body of the universe, set about making it of fire and earth. But
two things alone cannot be satisfactorily united without a third; for there
must be some bond between them drawing them together. And of all bonds
the best is that which makes itself and the terms it connects a unity in the
fullest sense; and it is of the nature of a continued geometrical proportion to
effect this most perfectly. For whenever, of three numbers, the middle one
between any two that are either solids (cubes?) or squares is such that, as the
first is to it, so is it to the last, and conversely as the last is to the middle, so is
the middle to the first, then since the middle becomes first and last, and again
the last and first become middle, in that way all will necessarily come to play
the same part towards one another, and by so doing they will all make a unity.

Now if it had been required that the body of the universe should be a
plane surface with no depth, a single mean would have been enough to connect its companions and itself; but in fact the world was to be solid in form,
and solids are always conjoined, not by one mean, but by two. Accordingly
the god set water and air between fire and earth, and made them, so far as
was possible, proportional to one another, so that as fire is to air, so is air to
water, and as air is to water, so is water to earth, and thus he bound together
the frame of a world visible and tangible.

For these reasons and from such constituents, four in number, the body
of the universe was brought into being, coming into concord by means of
proportion, and from these it acquired Amity,[3] so that coming into unity
with itself it became indissoluble by any other save him who bound it together.

The World's Body Contains the Whole of All the Four Primary Bodies

[The next paragraph explicitly rejects the old Ionian conception of an
indefinite circumambient mass of body, surrounding the cosmos and providing
a reservoir of materials from which a series of successive worlds could be
formed; and also the Atomists' conception of an unlimited quantity of matter
scattered throughout an infinite void. In this respect the body of the world is
once more all-inclusive, like its model. It must be (1) a whole and complete,

---

3. A reference to the *Philia* of Empedocles' system. But there is no contrary principle of
Neikos in Plato's scheme, and hence no periodic destruction of the world. Cf. *Gorg.* 508A: the wise
say that heaven and earth, gods and men, are held together by φιλία and κοσμιότης—a truth which
has escaped Callicles because he has neglected geometry and not perceived the significance of
geometrical proportion (ἡ ἰσότης ἡ γεωμετρική).

consisting of parts each of which is whole and complete; (2) single or unique (not one of many coexistent worlds); (3) everlasting (not destroyed and superseded by another world), which it could hardly be, if it were exposed to assaults from outside.]

Now the frame of the world took up the whole of each of these four; he who put it together made it consist of all the fire and water and air and earth, leaving no part or power of any one of them outside. This was his intent: first, that it might be in the fullest measure a living being whole and complete, of complete parts; next, that it might be single, nothing being left over, out of which such another might come into being; and moreover that it might be free from age and sickness. For he perceived that, if a body be composite, when hot things and cold and all things that have strong powers beset that body and attack it from without, they bring it to untimely dissolution and cause it to waste away by bringing upon it sickness and age. For this reason and so considering, he fashioned it as a single whole consisting of all these wholes, complete and free from age and sickness.

It Is a Sphere, Without Organs or Limbs, Rotating on Its Axis

[In the second part of the dialogue we shall be told how Necessity co-operates with Reason by the working of mechanical causes which keep the world's body in spherical shape. Here we are concerned only with the rational desire of the Demiurge to give it the most perfect of forms and motions. The sphere is the most uniform of all solid figures, and the only one which, by rotating on its axis, can move within its own limits without change of place. This axial rotation symbolises the movement of Reason and is superior to all rectilinear motions.]

And for shape he gave it that which is fitting and akin to its nature. For the living creature that was to embrace all living creatures within itself, the fitting shape would be the figure that comprehends in itself all the figures there are; accordingly, he turned its shape rounded and spherical, equidistant every way from centre to extremity—a figure the most perfect and uniform of all; for he judged uniformity to be immeasurably better than its opposite.

And all round on the outside he made it perfectly smooth, for several reasons. It had no need of eyes, for nothing visible was left outside; nor of hearing, for there was nothing outside to be heard. There was no surrounding air to require breathing, nor yet was it in need of any organ by which to receive food into itself or to discharge it again when drained of its juices. For nothing went out or came into it from anywhere, since there was nothing: it was designed to feed itself on its own waste and to act and be acted upon entirely by itself and within itself; because its framer thought that it would be better self-sufficient, rather than dependent upon anything else.

It had no need of hands to grasp with or to defend itself, nor yet of feet or anything that would serve to stand upon; so he saw no need to attach to it these limbs to no purpose. For he assigned to it the motion proper to its bodily form, namely that one of the seven which above all belongs to reason and intelligence; accordingly, he caused it to turn about uniformly in the same place and within its own limits and made it revolve round and round; he took from it all the other six motions and gave it no part in their wanderings. And since for this revolution it needed no feet, he made it without feet or legs.

The World-Soul

[The next section, on the World-Soul, opens with a short summary enumerating the perfections which the world's body owes to divine forethought, and adding that its circular motion, already mentioned, is due to its soul, extending from centre to circumference. The soul is coeval with the body; both exist everlastingly. The composition of the soul is next described: it consists of certain intermediate kinds of Existence, Sameness, and Difference. When these constituents have been compounded, the mixture is divided in the proportions of a musical *harmonia*. Out of the stuff so compounded and divided the Demiurge then constructs a system of circles, representing the principal motions of the stars and planets. The addition of these motions of soul to the bodily frame previously described starts the world upon its unceasing course of intelligent life. Finally, it is explained that, on the principle that like knows like, the composition of the World-Soul out of three elements, Existence, Sameness, and Difference, enables it both to know unchangeably real objects and to have true beliefs about changing things of the lower order of existence.]

SUMMARY. TRANSITION TO THE WORLD-SOUL

All this, then, was the plan of the god who is for ever for the god who was sometime to be. According to this plan he made it smooth and uniform, everywhere equidistant from its centre, a body whole and complete, with complete bodies for its parts. And in the centre he set a soul and caused it to extend throughout the whole and further wrapped its body round with soul on the outside; and so he established one world alone, round and revolving in a circle, solitary but able by reason of its excellence to bear itself company, needing no other acquaintance or friend but sufficient to itself. On all these accounts the world which he brought into being was a blessed god.

SOUL IS PRIOR TO BODY

Now this soul, though it comes later in the account we are now attempting, was not made by the god younger than the body; for when he joined them together, he would not have suffered the elder to be ruled by the younger.

There is in us too much of the casual and random,[4] which shows itself in our speech; but the god made soul prior to body and more venerable in birth and excellence, to be the body's mistress and governor.

COMPOSITION OF THE WORLD-SOUL

[We now come to the composition and structure of the World-Soul. The next sentence states that it is compounded of three ingredients, which are described. The sentence (which, for convenience, I have divided into three numbered parts) is one of the most obscure in the whole dialogue, but not so obscure as it has been made by critics, who have altered the text and thereby dislocated the grammar and the sense. Proclus construed it in the only possible way, and his interpretation, once disengaged from the irrelevant intricacies of his own theology, is obviously correct.]

The things of which he composed soul and the manner of its composition were as follows: (1) Between the indivisible Existence that is ever in the same state and the divisible Existence that becomes in bodies, he compounded a third form of Existence composed of both. (2) Again, in the case of Sameness and in that of Difference, he also on the same principle made a compound intermediate between that kind of them which is indivisible and the kind that is divisible in bodies. (3) Then, taking the three, he blended them all into a unity, forcing the nature of Difference, hard as it was to mingle, into union with Sameness, and mixing them together with Existence.

DIVISION OF THE WORLD-SOUL INTO HARMONIC INTERVALS

[In the figurative language of the myth the compound of three ingredients is spoken of as if it were a piece of malleable stuff—say, an amalgam of three soft metals—forming a long strip, which will presently be slit along its whole length and bent round into circles. But first the strip is marked off into divisions, corresponding to the intervals of a musical scale (*harmonia*). The intention is the same as in the previous paragraph. The soul must partake of harmony as well as of reason. Like knows like; and just as the soul can recognise existence, sameness, and difference because these are elements in its own composition, so the World-Soul must contain the harmonious order which individual souls ought to learn and reproduce in themselves.

The Demiurge begins by dividing the entire length into 'portions' measured by the numbers forming two geometrical proportions of four terms each: 1, 2, 4, 8 and 1, 3, 9, 27.]

And having made a unity of the three, again he divided this whole into as many parts as was fitting, each part being a blend of Sameness, Difference, and Existence.

---

4. Because we are not wholly rational, but partly subject to those wandering causes which, 'being devoid of intelligence, produce their effects casually and without order.'

And he began the division in this way. First he took one portion (1) from
the whole, and next a portion (2) double of this; the third (3) half as much
again as the second, and three times the first; the fourth (4) double of the
second; the fifth (9) three times the third; the sixth (8) eight times the first;
and the seventh (27) twenty-seven times the first.

Next, he went on to fill up both the double and the triple intervals, cutting
off yet more parts from the original mixture and placing them between the
terms, so that within each interval there were two means, the one (harmonic)
exceeding the one extreme and being exceeded by the other by the same
fraction of the extremes, the other (arithmetic) exceeding the one extreme by
the same number whereby it was exceeded by the other.[5]

These links gave rise to intervals of $\frac{3}{2}$ and $\frac{4}{3}$ and $\frac{9}{8}$ within the original
intervals.

And he went on to fill up all the intervals of $\frac{4}{3}$ (i.e. fourths) with the
interval $\frac{9}{8}$ (the tone), leaving over in each a fraction.

This remaining interval of the fraction had its terms in the numerical
proportion of 256 to 243 (semitone).

By this time the mixture from which he was cutting off these portions was
all used up.

CONSTRUCTION OF THE CIRCLES OF THE SAME AND THE DIFFERENT AND THE PLANETARY CIRCLES

[Timaeus now speaks as if the Demiurge had made a long band of soul-
stuff, marked off by the intervals of his scale. This he proceeds to slit lengthwise into two strips, which he puts together by their middles and bends round
into two circles or rings, corresponding to the sidereal equator and the Zodiac.]

This whole fabric, then, he split lengthwise into two halves; and making
the two cross one another at their centres in the form of the letter X, he bent
each round into a circle and joined it up, making each meet itself and the
other at a point opposite to that where they had been brought into contact.

He then comprehended them in the motion that is carried round uniformly
in the same place, and made the one the outer, the other the inner circle. The
outer movement he named the movement of the Same; the inner, the movement of the Different. The movement of the Same he caused to revolve to
the right by way of the side; the movement of the Different to the left by
way of the diagonal.

And he gave the supremacy to the revolution of the Same and uniform;
for he left that single and undivided; but the inner revolution he split in six
places into seven unequal circles, severally corresponding with the double and
triple intervals, of each of which there were three. And he appointed that the

---

5. If we take for illustration the extremes 6 and 12, the harmonic mean is 8, exceeding the
one extreme (6) by one-third of 6 and exceeded by the other extreme (12) by one-third of
12. The arithmetic mean is 9, exceeding 6 and falling short of 12 by the same number, 3.

circles should move in opposite senses to one another; while in speed three should be similar, but the other four should differ in speed from one another and from the three, though moving according to ratio.

THE WORLD'S BODY FITTED TO ITS SOUL

[The structure of the World's Soul is now complete. Plato has described its composition out of the three intermediate kinds of Existence, Sameness, and Difference; its division according to the intervals of the cosmic harmony; and its rational motions, represented by the two main circles. Nothing has yet been said about the bodies which display these motions and the additional motions of the seven circles. The intention is to emphasize the superior dignity of soul and the truth that the self-moving soul is the source of all physical motions. The next step is to fit the World's body, previously described, into the frame of the soul. This means imparting to the body the motions symbolised by the soul circles.]

When the whole fabric of the soul had been finished to its maker's mind, he next began to fashion within the soul all that is bodily, and brought the two together, fitting them centre to centre. And the soul, being everywhere inwoven from the centre to the outermost heaven and enveloping the heaven all round on the outside,[6] revolving within its own limit, made a divine beginning of ceaseless and intelligent life for all time.

DISCOURSE IN THE WORLD-SOUL

[The cognitive activity of the soul's ceaseless and intelligent life is based on the principle that like knows like. As Proclus says, 'Since the soul consists of three parts, Existence, Sameness, and Difference, in a form intermediate between the indivisible things and the divisible, by means of these she knows both orders of things; . . . for all knowing is accomplished by means of likeness between the knower and the known.']

Now the body of the heaven has been created visible; but she is invisible, and, as a soul having part in reason and harmony, is the best of things brought into being by the most excellent of things intelligible and eternal. Seeing, then, that soul had been blended of Sameness, Difference, and Existence, these three portions, and had been in due proportion divided and bound together, and moreover revolves upon herself, whenever she is in contact with anything that has dispersed existence or with anything whose existence is indivisible, she is set in motion all through herself and tells in what respect precisely, and how, and in what sense, and when, it comes about that something is qualified as

---

6. Adam compares our passage to *Rep.* 616c, where the light passes through the centre of the universe and round the outer surface of the heavenly sphere, acting as a bond that holds together all the revolving firmament, like the undergirders of a man-of-war.

either the same or different with respect to any given thing, whatever it may be, with which it is the same or from which it differs, either in the sphere of things that become or with regard to things that are always changeless.

Now whenever discourse that is alike true, whether it takes place concerning that which is different or that which is the same, being carried on without speech or sound within the thing that is self-moved,[7] is about that which is sensible, and the circle of the Different, moving aright, carries its message throughout all its soul—then there arise judgments and beliefs that are sure and true. But whenever discourse is concerned with the rational, and the circle of the Same, running smoothly, declares it, the result must be rational understanding and knowledge. And if anyone calls that in which this pair come to exist by any name but 'soul,' his words will be anything rather than the truth.

TIME, THE MOVING LIKENESS OF ETERNITY

[We turn now from the spiritual motions of the World-Soul—its thoughts and judgments—to the physical motions of perceptible bodies in the Heaven. Planets, stars, and Earth have yet to be created and set in the revolutions symbolised earlier by the eight circles of the celestial mechanism. This work is prefaced by a description of Time, which cannot exist apart from the heavenly clock whose movements are the measure of Time.]

When the father who had begotten it saw it set in motion and alive, a shrine brought into being for the everlasting gods, he rejoiced and being well pleased he took thought to make it yet more like its pattern. So as that pattern is the Living Being that is for ever existent, he sought to make this universe also like it, so far as might be, in that respect. Now the nature of that Living Being was eternal, and this character it was impossible to confer in full completeness on the generated thing. But he took thought to make, as it were, a moving likeness of eternity; and, at the same time that he ordered the Heaven, he made, of eternity that abides in unity, an everlasting likeness moving according to number—that to which we have given the name Time.

For there were no days and nights, months and years, before the Heaven came into being; but he planned that they should now come to be at the

---

7. The self-moved thing is the Heaven as a whole, which, as a living creature, is self-moved by its own self-moving soul. That an animal (soul and body) is self-moved is a commonplace. Ar., *Phys.* 265b, 34, 'Witness to this truth (that locomotion is prior to other motions) is borne by those who make soul the cause of motion, for they say that what moves itself is the source of motion and the *animal or anything that has a soul does move itself locally*'. This explains αὐτοῦ τὴν ψυχὴν below; and the world (κινηθὲν καὶ ζῶν) is again referred to as αὐτό. The passive (κινούμενον ὑφ' αὐτοῦ) is more appropriate to the animal which *is moved* by its soul than to the soul which *moves itself* (τὸ ἑαυτὸ κινοῦν). Commenting on the statement that the Demiurge gave the world 'the motion proper to its body,' Pr. (ii, 92³¹) says that it refers to the peculiar constitution of the cosmos, in virtue of which it is so moved by itself (ὑφ' ἑαυτοῦ), ἔχει γάρ τι καὶ αὐτὸς καὶ κατὰ τὴν ζωὴν αὐτοκίνητον καὶ κατὰ τὸ σῶμα σφαιροειδὲς ὂν πρὸς τὴν κύκλῳ κίνησιν οἰκεῖον(where αὐτοκίνητον and οἰκεῖον are both epithets of τι, and the insertion of τὴν after ζωήν is unnecessary).

same time that the Heaven was framed. All these are parts of Time, and 'was' and 'shall be' are forms of time that have come to be; we are wrong to transfer them unthinkingly to eternal being. We say that it was and is and shall be; but 'is' alone really belongs to it and describes it truly; 'was' and 'shall be' are properly used of becoming which proceeds in time, for they are motions. But that which is for ever in the same state immovably cannot be becoming older or younger by lapse of time, nor can it ever become so; neither can it now have been, nor will it be in the future; and in general nothing belongs to it of all that Becoming attaches to the moving things of sense; but these have come into being as forms of time, which images eternity and revolves according to number. And besides we make statements like these: that what is past *is* past, what happens now *is* happening now, and again that what will happen *is* what will happen, and that the non-existent *is* non-existent: no one of these expressions is exact. But this, perhaps, may not be the right moment for a precise discussion of these matters.[8]

Be that as it may, Time came into being together with the Heaven, in order that, as they were brought into being together, so they may be dissolved together, if ever their dissolution should come to pass; and it is made after the pattern of the ever-enduring nature, in order that it may be as like that pattern as possible; for the pattern is a thing that has being for all eternity, whereas the Heaven has been and is and shall be perpetually throughout all time.

The Planets as Instruments of Time

[Before proceeding to the creation of all the everlasting heavenly gods who are to be enshrined in the system of revolutions already prepared, Plato takes first those among their number, namely the Planets, whose special utility to mankind lies in their marking off the periods of time and so teaching men to count and calculate. He remarks later that the observation of these regular periods led to the discovery of number, to all inquiry into nature, and to philosophy itself.]

In virtue, then, of this plan and intent of the god for the birth of Time, in order that Time might be brought into being, Sun and Moon and five other stars—'wanderers,' as they are called—were made to define and preserve the numbers of Time. Having made a body for each of them, the god set them in the circuits in which the revolution of the Different was moving—in seven circuits seven bodies: the Moon in the circle nearest the Earth; the Sun in the second above the Earth; the Morning Star (Venus) and the one called sacred

---

8. The objection is to using the word *'is'* in statements about things that become or happen in time or are non-existent. 'Being', in contrast here with Becoming, ought strictly to be reserved for the real unchanging Being of eternal things. Its application to Becoming is at least ambiguous, not 'exact'. The last sentence hints that a discussion of the ambiguity of *'is'* will be found in the *Sophist*. 'The non-existent' means (as in ordinary speech) the absolutely non-existent, of which, as the *Sophist* shows, nothing whatever can be truly asserted.

to Hermes (Mercury) in circles revolving so as, in point of speed, to run their race with the Sun, but possessing the power contrary to his; whereby the Sun and the star of Hermes and the Morning Star alike overtake and are overtaken by one another. As for the remainder,[9] where he enshrined them and for what reasons—if one should explain all these, the account, though only by the way, would be a heavier task than that for the sake of which it was given. Perhaps these things may be duly set forth later at our leisure.

To resume: when each one of the beings that were to join in producing Time had come into the motion suitable to it, and, as bodies bound together with living bonds, they had become living creatures and learnt their appointed task,[10] then they began to revolve by way of the motion of the Different, which was aslant, crossing the movement of the Same and subject to it: some moving in greater circles, some in lesser; those in the lesser circles moving faster, those in the greater more slowly.

So, by reason of the movement of the Same, those which revolve most quickly appeared to be overtaken by the slower, though really overtaking them. For the movement of the Same, which gives all their circles a spiral twist because they have two distinct forward motions in opposite senses, made the body which departs most slowly from itself—the swiftest of all movements —appear as keeping pace with it most closely.

And in order that there might be a conspicuous measure for the relative speed and slowness with which they moved in their eight revolutions, the god kindled a light in the second orbit from the Earth—what we now call the Sun— in order that he might fill the whole heaven with his shining and that all living things for whom it was meet might possess number, learning it from the revolution of the Same and uniform. Thus and for these reasons day and night came into being, the period of the single and most intelligent revolution.[11]

The month comes to be when the Moon completes her own circle and overtakes the Sun; the year, when the Sun has gone round his own circle. The periods of the rest have not been observed by men, save for a few; and men have no names for them, nor do they measure one against another by numerical reckoning. They barely know that the wanderings of these others are time at all, bewildering as they are in number and of surprisingly intricate pattern. None the less it is possible to grasp that the perfect number of time fulfils the perfect year at the moment when the relative speeds of all the eight revolutions have accomplished their courses together and reached their con- summation, as measured by the circle of the Same and uniformly moving.

In this way, then, and for these ends were brought into being all those

9. The three outer planets, Mars, Jupiter, Saturn. 'Enshrined' rather overtranslates ἱδρύσατο, but the planets are gods and ἱδρύεσθαι θεόν means 'setting up (a statue of) a god' for cult purposes.
10. Here, as at *Laws* 898, it is clearly stated that every planet, like the other heavenly gods, is a living creature with a body and an intelligent soul. So Pr. iii, 70¹; Chalc., p. 179²⁸.
11. The single (undivided) revolution of the Same, which is the only motion of translation possessed by the fixed stars.

stars that have turnings[12] on their journey through the Heaven; in order that this world may be as like as possible to the perfect and intelligible Living Creature, in respect of imitating its ever-enduring nature.

THE FOUR KINDS OF LIVING CREATURE. THE HEAVENLY GODS

[So far, the planets are the only living creatures, within the universal frame, whose creation has been described. Among the everlasting gods who were to take up their positions in that frame, the planets were singled out because they are, in a special way, the 'instruments of Time'; and Plato wished first to define Time in order to contrast the temporal existence of even the ever-lasting gods with the unchanging duration of the eternal model. Time cannot exist without the clock. Plato, accordingly, had to anticipate the creation of the heavenly gods by mentioning the planets. He now repeats the statement that the Demiurge designed to make his image as like as possible to the model. This is to be done by making all the four chief families of living creature, corresponding to the four regions of fire, air, water, and earth.]

Now so far, up to the birth of Time, the world had been made in other respects in the likeness of its pattern; but it was still unlike in that it did not yet contain all living creatures brought into being within it. So he set about accomplishing this remainder of his work, making the copy after the nature of the model. He thought that this world must possess all the different forms that intelligence discerns contained in the Living Creature that truly is. And there are four: one, the heavenly race of gods; second, winged things whose path is in the air; third, all that dwells in the water; and fourth, all that goes on foot on the dry land.

The form of the divine kind he made for the most part of fire, that it might be most bright and fair to see; and after the likeness of the universe he gave them well-rounded shape, and set them in the intelligence of the supreme to keep company with it, distributing them all round the heaven, to be in very truth an adornment (*cosmos*) for it, embroidered over the whole. And he assigned to each two motions: one uniform in the same place, as each always thinks the same thoughts about the same things; the other a forward motion, as each is subjected to the revolution of the Same and uniform. But in respect of the other five motions he made each motionless and still, in order that each might be as perfect as possible.

For this reason came into being all the unwandering stars, living beings divine and everlasting, which abide for ever revolving uniformly upon themselves; while those stars that having turnings and in that sense[13] 'wander' came to be in the manner already described.

---

12. τροπαί. The Sun, for instance, 'turns back' at the top of its spiral when it touches the *tropic* of Cancer at midsummer.

13. τοιαύτην. But only in that sense. They are not really 'wanderers,' but keep to their regular paths, though they 'turn' back at the limits of their spiral tracks.

Rotation of the Earth

[The Earth is now included, with the stars and planets, as 'the most venerable of all the gods within the heaven.' She, too, is a 'living being, divine and everlasting'; as such, she must possess a soul as well as a body, and Soul being defined as 'the self-moving thing,' she may be expected to possess a proper movement of axial rotation, in the same right as the stars and planets. But is this consistent with the rest of Plato's astronomical scheme in the *Timaeus?* The question has been debated by ancient and modern critics without reaching any agreement. It turns on the interpretation of the word ἰλλομένην ('winds') in the following sentence:]

And Earth he designed to be at once our nurse and, as she winds round the axis that stretches right through, the guardian and maker of night and day, first and most venerable of all the gods that are within the heaven.

The Further Movements of the Heavenly Bodies Are Too
Complicated for Description Here

[With the creation of Earth the list of the heavenly gods is complete. The astronomical chapter is now closed with the remark that, without a visible model, all the complicated movements cannot be described.]

To describe the evolutions in the dance of these same gods, their juxtapositions, the counter-revolutions of their circles relatively to one another, and their advances; to tell which of the gods come into line with one another at their conjunctions, and which in opposition, and in what order they pass in front of or behind one another, and at what periods of time they are severally hidden from our sight and again reappearing send to men who cannot calculate panic fears and signs of things to come—to describe all this without visible models of these same would be labour spent in vain. So this much shall suffice on this head, and here let our account of the nature of the visible and generated gods come to an end....

Necessity. The Errant Cause

[The opening paragraph is of fundamental importance for the understanding of the whole discourse. It describes the relations between Reason and Necessity, and how they co-operate to produce the visible world.]

Now our foregoing discourse, save for a few matters, has set forth the works wrought by the craftsmanship of Reason; but we must now set beside them the things that come about of Necessity. For the generation of this universe was a mixed result of the combination of Necessity and Reason.

Reason overruled Necessity by persuading her to guide the greatest part of the things that become towards what is best; in that way and on that principle this universe was fashioned in the beginning by the victory of reasonable persuasion over Necessity. If, then, we are really to tell how it came into being on this principle, we must bring in also the Errant Cause—in what manner its nature is to cause motion. So we must return upon our steps thus, and taking, in its turn, a second principle concerned in the origin of these same things, start once more upon our present theme from the beginning, as we did upon the theme of our earlier discourse.

We must, in fact, consider in itself the nature of fire and water, air and earth, before the generation of the Heaven, and their condition before the Heaven was. For to this day no one has explained their generation, but we speak as if men knew what fire and each of the others is, positing them as original principles, elements (as it were, letters) of the universe; whereas one who has ever so little intelligence should not rank them in this analogy even so low as syllables. On this occasion, however, our contribution is to be limited as follows. We are not now to speak of the 'first principle' or 'principles'—or whatever name men choose to employ—of all things, if only on account of the difficulty of explaining what we think by our present method of exposition. You, then, must not demand the explanation of me; nor could I persuade myself that I should be right in taking upon myself so great a task; but holding fast to what I said at the outset—the worth of a probable account—I will try to give an explanation of all these matters in detail, no less probable than another, but more so, starting from the beginning in the same manner as before. So now once again at the outset of our discourse let us call upon a protecting deity to grant us safe passage through a strange and unfamiliar exposition to the conclusion that probability dictates; and so let us begin once more.

The Receptacle of Becoming

[For his fresh starting-point, Timaeus goes back here to the very beginning of his discourse: the distinction between the two orders of existence, the intelligible and unchanging model and the changing and visible copy. We now learn that the copy is not self-subsistent; it needs the support of a medium, just as a reflection requires a mirror to hold it. Accordingly, a third factor has now to be added—a factor which had no place in the first part among the creations of Reason.]

Our new starting-point in describing the universe must, however, be a fuller classification than we made before. We then distinguished two things; but now a third must be pointed out. For our earlier discourse the two were sufficient: one postulated as model, intelligible and always unchangingly real; second, a copy of this model, which becomes and is visible. A third we did not then distinguish, thinking that the two would suffice; but now, it seems, the argument compels us to attempt to bring to light and describe a form difficult and obscure. What nature must we, then, conceive it to possess and what part does it play? This, more than anything else: that it is the Receptacle—as it were, the nurse—of all Becoming.

FIRE, AIR, ETC., ARE NAMES OF QUALITIES, NOT OF SUBSTANCES

[This question is first approached by a consideration of fire, air, etc., as the contents of the Receptacle. The point is that these are not permanent irreducible elements, not 'things' with a constant nature. Plato rejects the old Milesian doctrine of a single fundamental form of matter, which was to serve both as the original state of things (ἀρχή) and as the permanent ground (φύσις) underlying change. He also rejects the belief of the pluralists who, in reply to Parmenides, had reduced all change to the rearrangement in space of the four elements (Empedocles) or of 'seeds' (Anaxagoras) or of atoms (Leucippus and Democritus). Plato's position was nearer to that of Heraclitus, who alone had rejected the notion of substance underlying change and had taught the complete transformation of every form of body into every other. We are now to think of qualities which are not also 'things' or substances, but transient appearances in the Receptacle. The Receptacle itself alone has some sort of permanent being.]

True, however, as this statement is, it needs to be put in clearer language; and that is hard, in particular because to that end it is necessary to raise a previous difficulty about fire and the things that rank with fire. It is hard to say, with respect to any one of these, which we ought to call really water rather than fire, or indeed which we should call by any given name rather than by all the names together or by each severally, so as to use language in a sound and trustworthy way. How, then, and in what terms are we to speak of this matter, and what is the previous difficulty that may be reasonably stated?

In the first place, take the thing we now call water. This, when it is compacted, we see (as we imagine) becoming earth and stones, and this same thing, when it is dissolved and dispersed, becoming wind and air; air becoming fire by being inflamed; and, by a reverse process, fire, when condensed and extinguished, returning once more to the form of air, and air coming together again and condensing as mist and cloud; and from these, as they are yet more closely compacted, flowing water; and from water once more earth and stones: and thus, as it appears, they transmit in a cycle the process of passing into one another. Since, then, in this way no one of these things ever makes its appearance as the *same* thing, which of them can we steadfastly affirm to be *this*—whatever it may be—and not something else, without blushing for ourselves? It cannot be done; but by far the safest course is to speak of them in the following terms. Whenever we observe a thing perpetually changing—fire, for example—in every case we should speak of fire, not as 'this,' but as 'what is of such and such a quality,' nor of water as 'this,' but always as 'what is of such and such a quality'; nor must we speak of anything else as having some permanence, among all the things we indicate by the expressions 'this' or 'that,' imagining we are pointing out some definite thing. For they slip away and do not wait to be described as 'that' or 'this' or by any phrase that exhibits them as having permanent being. We should not use these expressions of any of them, but 'that which is of a certain quality and has the same sort of quality as it perpetually recurs in the cycle'—that is the description we should use in the case of each and all of them. In fact, we must give the name 'fire' to that which is at all times of such and such a quality; and so with anything else that is in process of becoming. Only in speaking of that *in* which all of them are always coming to be, making their appearance and again vanishing out of it, may we use the words 'this' or 'that'; we must not apply any of these words to that which is of some quality—hot or cold or any of the opposites—or to any combination of these opposites.

The Receptacle Compared to a Mass of Plastic Material

[Turning now from the contents to the Receptacle, Plato begins to illustrate its nature by an image which, as he admits, is in some respects misleading. It is compared to a mass of plastic material, moulded and remoulded into various shapes. The nature of the material (gold) is permanent; the shapes are formed only to be obliterated and give place to others.]

But I must do my best to explain this thing once more in still clearer terms.

Suppose a man had moulded figures of all sorts out of gold, and were unceasingly to remould each into all the rest: then, if you should point to one of them and ask what it was, much the safest answer in respect of truth would be to say 'gold,' and never to speak of a triangle or any of the other figures that were coming to be in it as things that have being, since they are changing even while one is asserting their existence. Rather one should be content if they so much as consent to accept the description 'what is of such and such a quality' with any certainty. Now the same thing must be said of that nature which receives all bodies. It must be called always the same; for it never departs at all from its own character; since it is always receiving all things, and never in any way whatsoever takes on any character that is like any of the things that enter it: by nature it is there as a matrix for everything, changed and diversified by the things that enter it, and on their account it *appears* to have different qualities at different times; while the things that pass in and out are to be called copies of the eternal things, impressions taken from them in a strange manner that is hard to express: we will follow it up on another occasion.

The Receptacle Has No Qualities of Its Own

[The illustration of the man moulding all sorts of figures out of gold was sufficient for its purpose, to illustrate the contrast between the permanent nature of the Receptacle and the shifting qualities. Its defect is that gold is a stuff that has sensible qualities of its own, persisting through all the variations of shape. Aristotle's objections to the illustration turn partly on this point. But Plato himself proceeds to correct the defect. He has already said that the Receptacle does not in itself possess any of the characters that pass in and out, any more than gold as such possesses any of the shapes. It is now added that the Receptacle has no characters of its own 'before' the qualities enter it, unlike the gold which has its own sensible properties.

Before making this point, Plato introduces the image of the father, the mother, and the child, to illustrate the relations of the eternal Form, the Receptacle, and Becoming.]

Be that as it may, for the present we must conceive three things: that which becomes; that in which it becomes; and the model in whose likeness that which becomes is born.[14] Indeed we may fittingly compare the Recipient to a mother, the model to a father, and the nature that arises between them to their offspring. Further we must observe that, if there is to be an impress presenting all diversities of aspect, the thing itself in which the impress comes to be situated, cannot have been duly prepared unless it is free from all those characters which it is to receive from elsewhere. For if it were like any one of the things that come in upon it, then, when things of contrary or entirely different nature came, in receiving them it would reproduce them badly, intruding its own features alongside. Hence that which is to receive in itself all kinds must be free from all characters; just like the base which the makers of scented ointments skilfully contrive to start with: they make the liquids that are to receive the scents as odourless as possible. Or again, anyone who sets about taking impressions of shapes in some soft substance, allows no shape to show itself there beforehand, but begins by making the surface as smooth and level as he can. In the same way, that which is duly to receive over its whole extent and many times over all the likenesses of the intelligible and eternal things ought in its own nature to be free of all the characters. For this reason, then, the mother and Receptacle of what has come to be visible and otherwise sensible must not be called earth or air or fire or water, nor any of their compounds or components; but we shall not be deceived if we call it a nature invisible and characterless, all-receiving, partaking in some very puzzling way of the intelligible and very hard to apprehend. So far as its nature can be arrived at from what has already been said, the most correct account of it would be this: that part of it which has been made fiery appears at any time as fire; the part that is liquefied as water; and as earth or air such parts as receive likenesses of these.

IDEAL MODELS OF FIRE, AIR, WATER, EARTH

[Plato has just spoken of 'copies' (μιμήματα) of Fire, Air, Water, and Earth being 'received' by the Receptacle. This leads to the next question: Are there models to serve as originals for these copies?]

---

[14] φύεται, 'born.' The next sentence takes up this metaphor as furnishing an appropriate image, which replaces that of the craftsman.

But in pressing our inquiry about them, there is a question that must rather be determined by argument. Is there such a thing as 'Fire just in itself' or any of the other things which we are always describing in such terms, as things that 'are just in themselves'? Or are the things we see or otherwise perceive by the bodily senses the only things that have such reality, and has nothing else, over and above these, any sort of being at all? Are we talking idly whenever we say that there is such a thing as an intelligible Form of anything? Is this nothing more than a word?

Now it does not become us either to dismiss the present question without trial or verdict, simply asseverating that it is so, nor yet to insert a lengthy digression into a discourse that is already long. If we could see our way to draw a distinction of great importance in few words, that would best suit the occasion. My own verdict, then, is this. If intelligence and true belief are two different kinds, then these things—Forms that we cannot perceive but only think of—certainly exist in themselves; but if, as some hold, true belief in no way differs from intelligence, then all the things we perceive through the bodily senses must be taken as the most certain reality. Now we must affirm that they are two different things, for they are distinct in origin and unlike in nature. The one is produced in us by instruction, the other by persuasion; the one can always give a true account of itself, the other can give none; the one cannot be shaken by persuasion, whereas the other can be won over; and true belief, we must allow, is shared by all mankind, intelligence only by the gods and a small number of men.

SUMMARY DESCRIPTION OF THE THREE FACTORS: FORM, COPY, AND SPACE AS THE RECEPTACLE

[In the foregoing sections we started with the notion of a Receptacle of Becoming; then passed to its contents, the sensible qualities and their combinations, and finally to the ideal models. Next follows a summary description of these three factors, in the reverse order.]

This being so, we must agree that there is, first, the unchanging Form, ungenerated and indestructible, which neither receives anything else into itself from elsewhere nor itself enters into anything else anywhere, invisible and otherwise imperceptible; that, in fact, which thinking has for its object.

Second is that which bears the same name and is like that Form; is sensible; is brought into existence; is perpetually in motion, coming to be in a certain place and again vanishing out of it; and is to be apprehended by belief involving perception.

Third is Space, which is everlasting, not admitting destruction; providing a situation for all things that come into being, but itself apprehended without the senses by a sort of bastard reasoning, and hardly an object of belief.

This, indeed, is that which we look upon as in a dream and say that anything that is must needs be in some place and occupy some room, and

that what is not somewhere in earth or heaven is nothing. Because of this
dreaming state, we prove unable to rouse ourselves and to draw all these
distinctions and others akin to them, even in the case of the waking and truly
existing nature, and so to state the truth: namely that, whereas for an image,
since not even the very principle on which it has come into being belongs to
the image itself, but it is the ever moving semblance of something else, it is
proper that it should come to be *in* something else, clinging in some sort to
existence on pain of being nothing at all, on the other hand that which has
real being has the support of the exactly true account, which declares that, so
long as the two things are different, neither can ever come to be in the other
in such a way that the two should become at once one and the same thing
and two.

DESCRIPTION OF CHAOS

[So far we have been almost wholly concerned with the Receptacle of
Becoming and the shifting qualities that appear in it and disappear, considered,
so far as is possible, in abstraction from the element of rational design contributed by Reason. The Forms of the four primary bodies were only introduced towards the end, because a copy must have an original; but it has been
emphasised that the Forms remain apart and cannot themselves enter the
region of Becoming. Plato now sums up the three factors required for the
production of a visible world, to which, as we have just seen, we must add the
'Demiurge' to produce it. He then passes to a description of the Receptacle
and its contents, imagined as existing 'before' the ordered world came into
being. We are now to hear what the Demiurge does when he 'takes over' this
chaos.]

Let this, then, be given as the tale summed according to my judgment:
that there are Being, Space, Becoming—three distinct things—even before the
Heaven came into being.

Now the nurse of Becoming, being made watery and fiery and receiving
the characters of earth and air, and qualified by all the other affections that
go with these, had every sort of diverse appearance to the sight; but because
it was filled with powers that were neither alike nor evenly balanced, there
was no equipoise in any region of it; but it was everywhere swayed unevenly
and shaken by these things, and by its motion shook them in turn. And they,
being thus moved, were perpetually being separated and carried in different
directions; just as when things are shaken and winnowed by means of
winnowing-baskets and other instruments for cleaning corn, the dense and
heavy things go one way, while the rare and light are carried to another place
and settle there. In the same way at that time the four kinds were shaken by
the Recipient, which itself was in motion like an instrument for shaking, and
it separated the most unlike kinds farthest apart from one another, and thrust
the most alike closest together; whereby the different kinds came to have

different regions, even before the ordered whole consisting of them came to be. Before that, all these kinds were without proportion or measure. Fire, water, earth, and air possessed indeed some vestiges of their own nature, but were altogether in such a condition as we should expect for anything when deity is absent from it. Such being their nature at the time when the ordering of the universe was taken in hand, the god then began by giving them a distinct configuration by means of shapes and numbers. That the god framed them with the greatest possible perfection, which they had not before, must be taken, above all, as a principle we constantly assert. . . .

ARISTOTLE:

On the Heavens

THE QUESTION of the nature of this Whole, whether it is of infinite magnitude or its total bulk is limited, must be left until later.[1] We have now to speak of its formally distinct parts,[2] and we may start from this, that all natural bodies and magnitudes are capable of moving of themselves in space; for nature we have defined as the principle of motion in them.[3] Now all motion in space (locomotion) is either straight or circular or a compound of the two, for these are the only simple motions, the reason being that the straight and circular lines are the only simple magnitudes. By "circular motion" I mean motion around the centre, by "straight," motion up and down. "Up" means away from the centre, "down" towards the centre. It follows that all simple locomotion is either away from the centre or towards the centre or around the centre. This appears to follow consistently on what was said at the beginning: body was completed by the number three, and so now is its motion.

Of bodies some are simple, and some are compounds of the simple. By "simple" I mean all bodies which contain a principle of natural motion, like fire and earth and their kinds,[4] and the other bodies of the same order. Hence motions also must be similarly divisible, some simple and others compound in one way or another; simple bodies will have simple motions and composite

---

From Aristotle, *On the Heavens,* translated by W. K. C. Guthrie, Loeb Classical Library, Harvard University Press, 1939, lines 268b 11–269a 32, 270a 15–270b 25, 271b 1-10, 271b 28–272a 20, 276a 18–276b 22, 278b 5–279a 18, 286b 10–287a 22, 296a 24–298b 20. Reprinted by kind permission of the publishers, Harvard University Press. From Aristotle, *Metaphysics,* translated by John Warrington, Everyman Library, Book Α, Chapter 8. Reprinted by kind permission of the publishers, E. P. Dutton and Co., Inc.

1. It is taken up on pp. 92-94.
2. *i.e.* the elements, which are the "immediate" parts of the whole, the rest being parts of parts (Simpl.). The elements are also the *summa genera* or ultimate distinctions of kind among bodies (Stocks).
3. *Phys.* ii. 1. 192 b 20.
4. This mention of the "kinds" of an element is probably a reference to the *Timaeus* (58C ff.), where there are said to be different varieties of each element, all pure but owing their differences to the different sizes of their elementary pyramids. (I owe this suggestion to Professor Cornford). There seems no point in Stocks's demand for "a variety of movement corresponding to variety of kind."

bodies composite motions, though the movement may be according to the prevailing element in the compound.[5]

If we take these premises, (a) that there is such a thing as simple motion, (b) that circular motion is simple, (c) that simple motion is the motion of a simple body (for if a composite body moves with a simple motion, it is only by virtue of a simple body prevailing and imparting its direction to the whole), then it follows that there exists a simple body naturally so constituted as to move in a circle in virtue of its own nature. By force it can be brought to move with the motion of another, different body, but not naturally, if it is true that each of the simple bodies has one natural motion only. Moreover, granted that (a) unnatural motion is the contrary of natural, (b) a thing can have only one contrary, then circular motion, seeing it is one of the simple motions,[6] must, if it is not the motion natural to the moved body, be contrary to its nature. Suppose now that the body which is moving in a circle be fire or some other of the four elements, then its natural motion must be contrary to the circular. But a thing can have only one contrary, and the contrary of upward is downward, and *vice versa*. Suppose on the other hand that this body which is moving in a circle contrary to its own nature is something other than the elements, there must be some other motion which is natural to it. But that is impossible: for if the motion were upward, the body would be fire or air, if downward, water or earth.

Furthermore, circular motion must be primary.[7] That which is complete is prior in nature to the incomplete, and the circle is a complete figure, whereas no straight line can be so. An infinite straight line cannot, for to be complete it would have to have an end or completion, nor yet a finite, for all finite lines have something beyond them: any one of them is capable of being extended. Now if (a) a motion which is prior to another is the motion of a body prior in nature, (b) circular motion is prior to rectilinear, (c) rectilinear motion is the motion of the simple bodies (as *e.g.* fire moves in a straight line upwards and earthy bodies move downwards towards the centre), then circular motion also must of necessity be the motion of some simple body. (We have already

---

5. In saying "bodies which contain a principle of natural motion," A. is not thinking of natural beings in the wide sense defined in *Phys.* ii, where the term includes plants and animals, but in the more restricted sense of the elements only. Strictly speaking (ἀκριβέστερον, Simpl.), only these can be said to have a principle of natural motion (the motion of a simple natural substance left to itself), since the motions of plants and animals are determined by the life-principle in them, which again is dependent on their possessing a certain complicated structure. Hence Simpl. is probably right in saying that even the phrase "composite bodies" in this sentence refers to the popular elements, earth, water, etc., as they appear to the senses. We never see them in a perfectly pure form, but they each conform to the natural motions of the pure element, because that prevails sufficiently in the compound to govern the direction of the whole. The argument however does not require that it be limited to these. It could at least include inanimate compounds of the elements (*e.g.* metals), and where the clause occurs again (269 a 28), Simpl. himself illustrates it by the example of a man falling off a roof.

6. This caution is necessary because, if circular motion were composite, then the axiom "one thing one contrary" could not be applied. A composite motion would be neither natural to a simple body nor directly contrary to its nature, but only, as Simpl. says, "not according to its nature."

7. This is demonstrated also, from similar premises but more fully, in *Phys.* viii, 9.

made the reservation that the motion of composite bodies is determined by whatever simple body predominates in the mixture.) From all these premises therefore it clearly follows that there exists some physical substance besides the four in our sublunary world, and moreover that it is more divine than, and prior to, all these. . . .

With equal reason we may regard it as ungenerated and indestructible, and susceptible neither to growth nor alteration. (a) Everything that is generated comes into being out of an opposite and a substrate, and is destroyed only if it has a substrate, and through the agency of an opposite, and passes into its opposite, as has been explained in our first discussions.[8] (b) Opposites have opposite motions. (c) There cannot be an opposite to the body under discussion, because there cannot be an opposite motion to the circular. It looks then as if nature had providently abstracted from the class of opposites that which was to be ungenerated and indestructible, because generation and destruction take place among opposites. Moreover anything which is subject to growth [or diminution] grows [or diminishes] in consequence of substance of the same kind being added to it and dissolving into its matter;[9] but this body has no such matter. And if it is subject neither to growth nor to destruction, the same train of thought leads us to suppose that it is not subject to alteration either. Alteration is movement in respect of quality, and the temporary or permanent states of quality, health and disease for example, do not come into being without changes of affection. But all physical bodies which possess changing affections may be seen to be subject also to growth and diminution. Such are, for example, the bodies of animals and plants and their parts, and also those of the elements. If then the body whose natural motion is circular cannot be subject to growth or diminution, it is a reasonable supposition that it is not subject to alteration either.

From what has been said it is clear why, if our hypotheses are to be trusted, the primary body of all is eternal, suffers neither growth nor diminution, but is ageless, unalterable and impassive. I think too that the argument bears out experience and is borne out by it. All men have a conception of gods, and all assign the highest place to the divine, both barbarians and Hellenes, as many as believe in gods, supposing, obviously, that immortal is closely linked with immortal. It could not, they think, be otherwise. If then—and it is true—there is something divine, what we have said about the primary bodily substance is well said. The truth of it is also clear from the evidence of the senses, enough at least to warrant the assent of human faith; for throughout all past time, according to the records handed down from generation to generation,[10] we find no trace of change either in the whole of the outermost heaven or in

---

8. A.'s reference is to the *Physics* (i, 7-9), but the point is perhaps put most concisely in *Met.* A 1069b 2-9.

9. *i.e.* growth and diminution are really only particular examples of generation and destruction (Simpl.).

10. According to Simplicius, it was believed that the astronomical records of the Egyptians went back for 630,000 years, and those of the Babylonians for 1,440,000.

any one of its proper parts. It seems too that the name of this first body has been passed down to the present time by the ancients, who thought of it in the same way as we do, for we cannot help believing that the same ideas recur to men not once nor twice but over and over again. Thus they, believing that the primary body was something different from earth and fire and air and water, gave the name *aither* to the uppermost region, choosing its title from the fact that it "runs always" (ἀεὶ θεῖν) and eternally. . . .

This, then, is now clear, and we must turn to consider the rest of our problems, of which the first is whether there exists any infinite body, as most of the early philosophers believed, or whether that is an impossibility. This is a point whose settlement one way or the other makes no small difference, in fact all the difference, to our investigation of the truth. It is this, one might say, which has been, and may be expected to be, the origin of all the contradictions between those who make pronouncements in natural science, since a small initial deviation from the truth multiplies itself ten-thousandfold as the argument proceeds. . . .

The following arguments make it plain that every body which revolves in a circle must be finite. If the revolving body be infinite, the straight lines radiating from the centre[11] will be infinite. But if they are infinite, the intervening space must be infinite. "Intervening space" I am defining as space beyond which there can be no magnitude in contact with the lines. This must be infinite. In the case of finite lines it is always finite, and moreover it is always possible to take more than any given quantity of it, so that this space is infinite in the sense in which we say that number is infinite, because there exists no greatest number. If then it is impossible to traverse an infinite space, and in an infinite body the space between the radii is infinite, the body cannot move in a circle. But we ourselves see the heaven revolving in a circle, and also we established by argument that circular motion is the motion of a real body.

Again, if a finite time be subtracted from a finite time, the remainder must also be finite and have a beginning. But if the time of the journey has a beginning, so also must the movement, and hence the distance which is traversed. This applies equally to everything else. Let ACE be a straight line infinite in the direction of E, and BB another straight line infinite in both directions. If the line ACE describes a circle about C[12] as centre, it will be expected to cut BB in its revolution for a certain finite time (for the whole time taken by the heaven to complete its revolution is finite, therefore the subtracted time, during which the line in its movement cuts the other, is also finite). There will therefore be a point of time at which the line ACE first

---

11. That is, the centre of the circular path of the supposed revolving body.

12. So Bekker, with all MSS. but F. Allan reads "A," following F and Simpl. This is more satisfactory, for it adds a needless confusion to the problem to suppose that the line ACE revolves not about its starting-point but about some other point farther along it.

cuts the line BB. But this is impossible. Therefore it is impossible for an infinite body to revolve in a circle. Neither then could the heaven, if it were infinite. . . .

We must now explain why there cannot even be more than one world. This was a question which we noted for consideration, to meet the objection that no general proof has been given that no body whatsoever can exist beyond this world, since the foregoing discussion applied only to those with no definite situation.[13]

All bodies both rest and move naturally and by constraint. A body moves naturally to that place where it rests without constraint, and rests without constraint in that place to which it naturally moves. It moves by constraint to that place in which it rests by constraint, and rests by constraint in that place to which it moves by constraint. Further, if a certain movement is enforced, then its opposite is natural. Thus if it is by constraint that earth moves to the centre here from wherever else it is, its movement thither from here will be natural: and if, having come from there, it remains here without constraint, its movement hither will be natural also. And the natural movement of each is one. Further, all the worlds must be composed of the same bodies, being similar in nature. But at the same time each of these bodies must have the same potentialities, fire, that is to say, and earth, and the bodies intermediate between them; for if the bodies of another world resemble our own in name only, and not in virtue of having the same form, then it would only be in name that the whole which they compose could be pronounced a world. It clearly follows that one of them will be of a nature to move away from the centre, and another towards the centre, seeing that all fire must have the same form as other fire, just as the different portions of fire in this world have the same form; and the same may be said about each of the other simple bodies. The necessity for this emerges clearly from our assumptions about the motions of simple bodies, namely that they are limited in number and that each of the elements has a particular motion assigned to it. Consequently if the motions are the same, the elements also must be the same wherever they are. It must be natural therefore for the particles of earth in another world to move towards the centre of this one also, and for the fire in that world to move towards the circumference of this. This is impossible, for if it were to happen the earth would have to move upwards in its own world and the fire to the centre; and similarly earth from our own world would have to move naturally away from the centre, as it made its way to the centre of the other, owing to the assumed situation of the worlds relatively to each other. Either, in fact, we must deny that the simple bodies of the several worlds have the same natures, or if we

---

13. κεῖσθαι, as Professor Cornford pointed out to me, is more likely to refer to situation than extent, and this sense fits the context well. In infinite space, bodies (e.g. the atoms of Democritus) can have no proper region and hence no natural motion (such as A. is discussing in this and the previous chapters). The view now to be considered is that, even though the whole be not infinite, it may yet contain several cosmoi, in each of which the elements will have their definite places as in ours.

admit it we must, as I have said, make the centre and the circumference one for all; and this means that there cannot be more worlds than one. . . .

Now the world must be counted among particulars and things made from matter; but if it is composed, not of a portion of matter, but of all matter whatsoever, then we may admit that its essential nature as "world" and as "this world" are distinct, but nevertheless there will not be another world, nor could there be more than one, for the reason that all the matter is contained in this one.

This therefore remains to be demonstrated, that our own world is composed of the whole sum of natural perceptible body. Let us first establish what we mean by ouranos,[14] and in how many senses the word is used, in order that we may more clearly understand the object of our questions. (1) In one sense we apply the word ouranos to the substance of the outermost circumference of the world, or to the natural body which is at the outermost circumference of the world; for it is customary to give the name of ouranos especially to the outermost and uppermost region, in which also we believe all divinity to have its seat. (2) Secondly we apply it to that body which occupies the next place to the outermost circumference of the world, in which are the moon and the sun and certain of the stars;[15] for these, we say, are in the ouranos. (3) We apply the word in yet another sense to the body which is enclosed by the outermost circumference; for it is customary to give the name of ouranos to the world as a whole.

The word, then, is used in these three senses, and the whole which is enclosed by the outermost circumference must of necessity be composed of the whole sum of natural perceptible body, for the reason that there is not, nor ever could be, any body outside the heaven. For if there is a natural body beyond the outermost circumference, it must be either simple or composite, and its position there must be either natural or unnatural. It cannot be one of the simple bodies, for (a) with regard to the body which revolves it has been shown that it cannot change its place; (b) but no more can it be either the body which moves away from the centre or that which settles towards it. They could not be there naturally (for their proper places are elsewhere), but if they are there unnaturally, then this outside region will belong naturally to some other body; for the place which is unnatural to one must be natural to another. But we have seen that there is no other body besides these three. Therefore it is impossible that any of the simple bodies should lie outside the heaven. And if this is true of the simple bodies, it is true also of composite, for where the composite body is the simple bodies must be also. It is equally impossible that a body should ever come to be there, for its coming to be there will be

---

14. This is the word which so far in this chapter has been translated "world." Elsewhere in the treatise it is rendered as "world," "heavens" or "sky" according to that one of the three senses here enumerated which A. is employing at any particular moment. In this passage a repetition of the Greek is unavoidable, since no one English word covers all the three senses which ouranos is here stated to possess.

15. i.e. the planets. The fixed stars are in ouranos no. 1.

either natural or unnatural, and it will be either simple or composite, in fact the same argument will recur: it makes no difference whether we ask "Is it there?" or "Can it come to be there?"

It is plain, then, from what has been said, that there is not, nor do the facts allow there to be, any bodily mass beyond the heaven. The world in its entirety is made up of the whole sum of available matter (for the matter appropriate to it is, as we saw, natural perceptible body), and we may conclude that there is not now a plurality of worlds, nor has there been, nor could there be. This world is one, solitary and complete. It is clear in addition that there is neither place nor void nor time beyond the heaven; for (a) in all place there is a possibility of the presence of body, (b) void is defined as that which, although at present not containing body, can contain it, (c) time is the number of motion, and without natural body there cannot be motion. It is obvious then that there is neither place nor void nor time outside the heaven, since it has been demonstrated that there neither is nor can be body there. . . .

The shape of the heaven must be spherical. That is most suitable to its substance, and is the primary shape in nature. But let us discuss the question of what is the primary shape, both in plane surfaces and in solids. Every plane figure is bounded either by straight lines or by a circumference; the rectilinear is bounded by several lines, the circular by one only. Thus since in every genus the one is by nature prior to the many, and the simple to the composite, the circle must be the primary plane figure. Also, if the term "perfect" is applied, according to our previous definition, to that outside which no part of itself can be found,[16] and addition to a straight line is always possible, to a circle never, the circumference of the circle must be a perfect line: granted therefore that the perfect is prior to the imperfect, this argument too demonstrates the priority of the circle to other figures. By the same reasoning the sphere is the primary solid, for it alone is bounded by a single surface, rectilinear solids by several. The place of the sphere among solids is the same as that of the circle among plane figures. Even those who divide bodies up into surfaces and generate them out of surfaces[17] seem to agree with this, for the sphere is the one solid which they do not divide, holding that it has only one surface, not a plurality; for their division into surfaces does not mean division in the manner of one cutting a whole into its parts, but division into elements specifically different.[18]

It is clear, then, that the sphere is the first solid figure, and it would also

---

16. This is not quite the same as the definition of "perfect" at *Phys.* iii. 207 a 8, to which Stocks refers (οὗ μηδὲν ἔξω, τοῦτο τέλειον. *Cf.* also *Met.* 1055 a 12). In order to bring the two into agreement, Allan would omit the words τῶν αὐτοῦ, but the evidence of Simpl. is strongly in favour of retaining them.

17. The theory of Plato in the *Timaeus*.

18. *i.e.* it is not that a sphere is indivisible, but any division of it can only be into parts belonging to the same kind (*sc.* body) as the whole, whereas the "division" which these thinkers are seeking means theoretical analysis into elements of a simpler kind—solid bodies into surfaces and surfaces into lines. Thus a cube can be "divided" into six rectangles, a rectangle into four lines, but no similar analysis can be made of a sphere, if it has only one surface bounding it.

be most natural to give it that place if one ranked figures according to number, the circle corresponding to one and the triangle to two, on account of its two right angles—for if one gives unity to the triangle, the circle will cease to be a figure. But the primary figure belongs to the primary body, and the primary body is that which is at the farthest circumference, hence it, the body which revolves in a circle, must be spherical in shape.

The same must be true of the body which is contiguous[19] to it, for what is contiguous to the spherical is spherical, and also of those bodies which lie nearer the centre, for bodies which are surrounded by the spherical and touch it at all points must themselves be spherical, and the lower bodies are in contact with the sphere above. It is, then, spherical through and through, seeing that everything in it is in continuous contact with the spheres.

Again, since it is an observed fact, and assumed in these arguments, that the whole revolves in a circle, and it has been shown that beyond the outermost circumference there is neither void nor place, this provides another reason why the heaven must be spherical. For if it is bounded by straight lines, that will involve the existence of place, body, and void. A rectilinear body revolving in a circle will never occupy the same space, but owing to the change in position[20] of the corners there will at one time be no body where there was body before, and there will be body again where now there is none. It would be the same if it were of some other shape whose radii were unequal, that of a lentil or an egg for example. All will involve the existence of place and void outside the revolution, because the whole does not occupy the same space throughout. . . .

For ourselves, let us first state whether the earth is in motion or at rest. Some, as we have said, make it one of the stars, whereas others put it at the centre but describe it as winding and moving about the pole as its axis. But the impossibility of these explanations is clear if we start from this, that if the earth moves, whether at the centre or at a distance from it, its movement must be enforced: it is not the motion of the earth itself, for otherwise each of its parts would have the same motion, but as it is their motion is invariably in a straight line towards the centre. The motion therefore, being enforced and unnatural, could not be eternal; but the order of the world is eternal.

Secondly, all the bodies which move with the circular movement are observed to lag behind and to move with more than one motion, with the exception of the primary sphere: the earth therefore must have a similar double motion, whether it move around the centre or as situated at it. But if this were so, there would have to be passings and turnings of the fixed stars. Yet these are not observed to take place: the same stars always rise and set at the same places on the earth.[21]

19. I do not translate συνεχής consistently as = "continuous," because Aristotle does not seem to be consistent in his use of it. It seems better to keep "continuous" to represent the strict sense of συνεχής.
20. For παράλλαξις cf. Plato, Tim. 22 D and Politicus 269 E.
21. The criticism depends on the analogy with the planets, following which A. assumes that if the earth moved with a motion of its own, as well as being carried round in the motion of

Thirdly, the natural motion of the earth as a whole, like that of its parts, is towards the centre of the Universe: that is the reason why it is now lying at the centre. It might be asked, since the centre of both is the same point, in which capacity the natural motion of heavy bodies, or parts of the earth, is directed towards it; whether as centre of the Universe or of the earth. But it must be towards the centre of the Universe that they move, seeing that light bodies like fire, whose motion is contrary to that of the heavy, move to the extremity of the region which surrounds the centre.[22] It so happens that the earth and the Universe have the same centre, for the heavy bodies do move also towards the centre of the earth, yet only incidentally, because it has its centre at the centre of the Universe. As evidence that they move also towards the centre of the earth, we see that weights moving towards the earth do not move in parallel lines but always at the same angles to it:[23] therefore they are moving towards the same centre, namely that of the earth. It is now clear that the earth must be at the centre and immobile. To our previous reasons we may add that heavy objects, if thrown forcibly upwards in a straight line, come back to their starting-place, even if the force hurls them to an unlimited distance.

From these considerations it is clear that the earth does not move, neither does it lie anywhere but at the centre. In addition the reason for its immobility

the first heaven, its proper motion would be in the plane of the ecliptic and not of the equator. Were this so, the fixed stars would exhibit to our eyes the irregularities which he describes by the words παρόδους καὶ τροπάς; the pole-star would appear to describe a circle in the sky, and the stars would not rise and set as they do. (For the senses of τροπαί, see Heath, *Aristarchus*, p. 33, n. 3. He discusses the present passage *o.c.* p. 241.) The objection is lodged against both the planetary theory and Plato's theory of motion at the centre. If we may accept Prof. Cornford's suggestion (*Plato's Cosmology*, pp. 132 ff.), Plato would reply that he had expressly limited the motion in the ecliptic (= motion of the Different) to the seven planetary circles, and that the motion of the earth, caused by its soul, was independent and around the same axis as the motion of the Same, only in the reverse direction to cancel it.

22. *Sc.* the upward motion of fire is upward in relation to the Universe (*i.e.* towards its extremity), not in relation to itself. Simpl. thinks that "the region which surrounds the centre" is the region occupied by the air, *i.e.* immediately surrounding the earth. Fire moves to the outer extremity of this region. It is true that if Aristotle simply means the circumference of the Universe, the phrase is unusually elaborate.

23. *I.e.* at right angles to a tangent. Stocks explains the Greek as meaning that the angles at each side of the line of fall of any one body are equal. But does it not more naturally mean that the angles made by one falling body with the earth are similar to those made by another? See Fig.

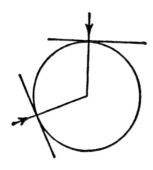

is clear from our discussions. If it is inherent in the nature of earth to move from all sides to the centre (as observation shows), and of fire to move away from the centre towards the extremity, it is impossible for any portion of earth to move from the centre except under constraint; for one body has one motion and a simple body a simple motion, not two opposite motions, and motion from the centre is the opposite of motion towards it. If then any particular portion is incapable of moving from the centre, it is clear that the earth itself as a whole is still more incapable, since it is natural for the whole to be in the place towards which the part has a natural motion. If then it cannot move except by the agency of a stronger force, it must remain at the centre. This belief finds further support in the assertions of mathematicians about astronomy: that is, the observed phenomena—the shifting of the figures by which the arrangement of the stars is defined—are consistent with the hypothesis that the earth lies at the centre. This may conclude our account of the situation and the rest or motion of the earth.

Its shape must be spherical. For every one of its parts has weight until it reaches the centre, and thus when a smaller part is pressed on by a larger, it cannot surge round it,[24] but each is packed close to, and combines with, the other until they reach the centre. To grasp what is meant we must imagine the earth as in the process of generation in the manner which some of the natural philosophers describe (except that they make external compulsion responsible for the downward movement: let us rather substitute the true statement that this takes place because it is the nature of whatever has weight to move towards the centre). In these systems, when the mixture existed in a state of potentiality,[25] the particles in process of separation were moving from every side alike towards the centre. Whether or not the portions were evenly distributed at the extremities, from which they converged towards the centre, the same result will be produced. It is plain, first, that if particles are moving from all sides alike towards one point, the centre, the resulting mass must be similar on all sides; for if an equal quantity is added all round, the extremity must be at a constant distance from the centre. Such a shape is a sphere. But it will make no difference to the argument even if the portions of the earth did not travel uniformly from all sides towards the centre. A greater mass must always drive on a smaller mass in front of it, if the inclination of both is to go as far as the centre, and the impulsion of the less heavy by the heavier persists to that point.

A difficulty which might be raised finds its solution in the same considerations. If, the earth being at the centre and spherical in shape, a weight many times its own were added to one hemisphere, the centre of the Universe would no longer coincide with that of the earth. Either, therefore, it would not remain at the centre, or, if it did, it might even as it is be at rest although not

---

24. The verb (cf. *Phys.* iv. 216 b 25) signifies the motion of a wave, and Simpl. notes that the behaviour of the less heavy particles when in contact with the heavier is being contrasted with the behaviour of liquids under similar pressure.
25. Compare the language used of earlier cosmologies in *Met.* Λ, 1069 b 20-23.

occupying the centre, *i.e.* though in a situation where it is natural for it to be in motion. That then is the difficulty. But it is not hard to understand, if we make a little further effort and define the manner in which we suppose any magnitude, possessed of weight, to travel towards the centre. Not, clearly, to the extent of only touching the centre with its edge: the larger portion must prevail until it possesses the centre with its own centre, for its impulse extends to that point. It makes no difference whether we posit this of any chance portion or clod, or of the earth as a whole, for the fact as explained does not depend on smallness or greatness, but applies to everything which has an *impulse* towards the centre. Therefore whether the earth moved as a whole or in parts, it must have continued in motion until it occupied the centre evenly all round, the smaller portions being equalized by the greater under the forward pressure of their common impulse.

If then the earth has come into being, this must have been the manner of its generation, and it must have grown in the form of a sphere: if on the other hand it is ungenerated and everlasting, it must be the same as it would have been had it developed as the result of a process. Besides this argument for the spherical shape of the earth, there is also the point that all heavy bodies fall at similar angles, not parallel to each other; this naturally means that their fall is towards a body whose nature is spherical. Either then it *is* spherical, or at least it is natural for it to be so, and we must describe each thing by that which is its natural goal or its permanent state, not by any enforced or unnatural characteristics.

Further proof is obtained from the evidence of the senses. (i) If the earth were not spherical, eclipses of the moon would not exhibit segments of the shape which they do. As it is, in its monthly phases the moon takes on all varieties of shape—straight-edged, gibbous and concave—but in eclipses the boundary is always convex. Thus if the eclipses are due to the interposition of the earth, the shape must be caused by its circumference, and the earth must be spherical. (ii) Observation of the stars also shows not only that the earth is spherical but that it is of no great size, since a small change of position on our part southward or northward visibly alters the circle of the horizon, so that the stars above our heads change their position considerably, and we do not see the same stars as we move to the North or South. Certain stars are seen in Egypt and the neighbourhood of Cyprus, which are invisible in more northerly lands, and stars which are continuously visible in the northern countries are observed to set in the others. This proves both that the earth is spherical and that its periphery is not large, for otherwise such a small change of position could not have had such an immediate effect. For this reason those who imagine that the region around the Pillars of Heracles joins on to the regions of India, and that in this way the ocean is one, are not, it would seem, suggesting anything utterly incredible. They produce also in support of their contention the fact that elephants are a species found at the extremities of both lands, arguing that this phenomenon at the extremes is due to communication between the two. Mathematicians who try to calculate the circumference put it at 400,000 stades.[26]

From these arguments we must conclude not only that the earth's mass is spherical but also that it is not large in comparison with the size of the other stars.

Metaphysics

The Number of Eternal Moving Principles

WE MUST NOT forget to ask whether it is necessary to recognize one or more than one eternal and unchangeable substance. If there are more than one, what is their number? It is worth remarking that our predecessors have expressed no clear opinion on this subject. The Ideal theory does not discuss it; for those who believe in Ideas identify them with numbers, treating these sometimes as unlimited, sometimes (though with insufficient proof) as limited by the number 10.

We must, however, discuss the subject, beginning with the earlier premises and distinctions. The first principle or primary being is not movable, either in itself or accidentally, but produces the primary eternal and single motion. Now (a) that which is moved is moved by something; (b) the prime mover must be in itself immovable; (c) eternal motion requires an eternal cause; and (d) we see that in addition to the simple spatial movement of the universe[27] (which we say is produced by the primary immovable substance) there are other eternal[28] spatial movements (viz. those of the planets). *Therefore* each of *these* movements must likewise be caused by a substance which is immovable in itself and eternal. For the nature of the stars[29] is eternal, being a kind of substance; the mover is eternal and prior to the moved; and only substance can be prior to substance. Evidently, then, there must be as many such substances as there are motions of the stars; and they must be by nature eternal, immovable in themselves, and without magnitude, for the reason before mentioned.[30]

---

26. *I.e.* 9987 geographical miles. Prantl (p. 319) remarks that this is the oldest recorded calculation of the earth's circumference. He quotes the following estimates for comparison: Archimedes 7495 geogr. miles; Eratosthenes and Hipparchus 6292; Posidonius 5992 or 4494; present day 5400. (The present-day figure in English miles is 24,902.)
   This passage of Aristotle is said to have provided a stimulus to the voyage of Columbus. (Ross, *Aristotle*, p. 96, n. 3.)
27. *I.e.* the apparent daily motion of the whole heavens.
28. They are eternal; for a body which moves in a circle is eternal and unresting, as I have shown in the physical treatises [*Physics*, viii. 8, 9; *De Coelo*, i. 2; ii. 3-8].—(A.)
29. The word here includes the fixed stars and the planets.
30. Chapter vii.

It is therefore clear that the movers are substances, prior and posterior according to the movements of the stars. The number of these movements, however, can be determined only by that one of the mathematical sciences which is most akin to philosophy, i.e. astronomy, which alone deals with substance that is concrete but eternal, while the others (arithmetic and geometry) do not treat of substance at all. That the motions are more numerous than the moved bodies is obvious to anyone who has considered the facts even superficially; for each of the planets has more than one movement. But as to the actual number of those movements, I shall now try to throw a little more light upon the subject by outlining what various mathematicians have to say. For the rest, we must rely partly on our own study and partly on the investigations of others. If someone develops a theory contrary to that which I shall now put forward—well, with all respect to the other party, we shall follow the more accurate.

Eudoxus[31] supposed that the motion of the sun and of the moon involves, in each case, three spheres:

(1) a sphere having the daily rotation of the fixed stars;
(2) a sphere moving in the circle which runs along the middle of the Zodiac;[32]
(3) a sphere in the circle which is inclined across the width of the Zodiac, though the circle in which the moon moves is inclined at a greater angle than that in which the sun moves.

He further assumed that the motion of the planets involves, in each case, four spheres: (1) and (2) above;[33]

(3′) a sphere whose *poles* are in the circle which bisects the Zodiac (the poles being the same for Mercury and Venus, but different for the other planets);

(4′) a sphere moved obliquely to (3).[34]

Callippus[35] made the position of the spheres, i.e. the order of their intervals, the same; but while he assigned the same number as Eudoxus did to Jupiter and Saturn, he considered that, in the light of observation, two more spheres should be added to the sun, two to the moon, and one more to each of the other planets.[36]

But it is necessary, if the combined spheres of Callippus are to explain the phenomena, that for the sun and each of the planets, except the moon, there

31. Eudoxus of Cnidos (408-355 B.C.). For a full study of his system and that of Callippus see Sir T. L. Heath, *Aristarchus of Samos* (1913), pp. 190-224.
32. This second sphere moved in the ecliptic which bisects the Zodiac longitudinally.
33. For the sphere of the fixed stars is that which moves all the other spheres, but that which is placed beneath this and has its movement in the circle which bisects the Zodiac is common to all.—(A.)
34. This gives a total of 27 spheres: three each for the sun and moon, four for each of the five planets, and one for the diurnal rotation.
35. Callippus of Cyzicus (fl. 330 B.C.).
36. Giving a total of 33 spheres.

should be certain spheres which neutralize the action of those already mentioned by rolling back the outer sphere of the planet just nearer to the earth
than the given planet.[37] Only so can all the forces at work produce the
observed motion of the planets. Now the spheres by which the planets themselves are moved are eight for Saturn and Jupiter, and twenty-five for the
other bodies;[38] therefore the spheres which counteract those of Saturn and
Jupiter will be six in number, the spheres which counteract those of Mars,
Venus, Mercury, and the Sun will be sixteen, and the total number of all the
spheres—both those which move the planets and those which counteract these
—will be 55. On the other hand, if we ignore the additional movements
assigned to the Sun and Moon, we get 47.[39] This being the number of the
spheres, the unmoved substances may probably be taken as equal in number.
I say 'probably,' because any definite pronouncement must be left to more
experienced scientists.

All motions in the heavens are such as are required to explain the behaviour
of a star; and every perfect substance—i.e. one which is impassive and in itself
has attained to the *Summum Bonum*—is a final cause. Therefore there can be
no spatial motion other than those mentioned above, and their number must
represent the total of unmoved substances. If there were others they would
cause motion as ends of motion; but we have seen that there cannot be spatial
motions other than those which explain the behaviour of heavenly bodies—a
view which is borne out by study of the moved bodies themselves. For if every
mover exists as such for the sake of what is moved, and every motion belongs
to something moved, it follows that no motion is for its own sake or for that
of another motion, but for the sake of the heavenly bodies. If one movement
were for the sake of another, this latter would have to be for the sake of yet
another; i.e. there would be an infinite regress. But since this is impossible,
the end of every movement must be one of the divine bodies which move
through the heavens.

There is a very ancient tradition in the form of a myth, that the stars are
gods and that the divine embraces the whole of nature.[40] The remaining
features of popular religion were added at a later date in order to frighten
ignorant people, to lend sanction to the laws, and on general utilitarian
grounds:[41] these gods are said to be in the form of men or beasts,[42] and other
stories of that kind are told. But if we strip the original doctrine of its later

---

37. Their purpose is to 'prevent the influence of the forward-moving or deferent spheres of
one planet from affecting the next' (Ross, *ad loc.* II. 391-2).

38. Mars, Venus, Mercury, Sun, Moon.

39. We should expect this number to be 49, and several reasons have been advanced to
account for Aristotle's calculation. The most likely explanation seems to be that in respect of the
sun and moon he proposed returning to the theory of Eudoxus, who assigned three spheres to
each. Accordingly he deducted the extra spheres assigned by Callippus and himself to the sun
and moon, but, being on unfamiliar ground, overlooked the fact that two extra sun-spheres would
still be required to neutralize the action of two of Eudoxus's three positive sun-spheres.

40. Aristotle means that the gods of popular mythology were derived from the underlying
forces of nature, i.e. the divine.

41. Cf. ch. iv.

42. The reference is probably to Egyptian animal-worship.

accretions and consider it alone, we cannot but recognize it as inspired. It teaches that the prime substances are gods, and is a relic of that perfect flowering of the arts and sciences which must have been often achieved and often lost.[43] It is, so to speak, the surviving relic of an ancient treasure, allowing us a fleeting glimpse of what our early ancestors believed.

---

43. Cf. *De Coelo*, 270^{b}19; *Meteor.* 339^{b}27; *Pol.* 1329^{b}25. Cf. also Plato's *Timaeus*, 22 C, 23 A-B; *Crit.* 1090; *Laws*, 676 A-677 D.

CLAUDIUS PTOLEMY:

The Almagest

1. Preface

THOSE who have been true philosophers, Syrus, seem to me to have very wisely separated the theoretical part of philosophy from the practical. For even if it happens the practical turns out to be theoretical prior to its being practical, nevertheless a great difference would be found in them; not only because some of the moral virtues can belong to the everyday ignorant man and it is impossible to come by the theory of whole sciences without learning, but also because in practical matters the greatest advantage is to be had from a continued and repeated operation upon the things themselves, while in theoretical knowledge it is to be had by a progress onward. We accordingly thought it up to us so to train our actions even in the application of the imagination as not to forget in whatever things we happen upon the consideration of their beautiful and well-ordered disposition, and to indulge in meditation mostly for the exposition of many beautiful theorems and especially of those specifically called mathematical.

For indeed Aristotle quite properly divides also the theoretical into three immediate genera: the physical, the mathematical, and the theological. For given that all beings have their existence from matter and form and motion, and that none of these can be seen, but only thought, in its subject separately from the others, if one should seek out in its simplicity the first cause of the first movement of the universe, he would find God invisible and unchanging. And the kind of science which seeks after Him is the theological; for such an act [ἐνέργεια] can only be thought as high above somewhere near the loftiest things of the universe and is absolutely apart from sensible things. But the kind of science which traces through the material and ever moving quality, and has to do with the white, the hot, the sweet, the soft, and such things, would be called physical; and such an essence [οὐσία], since it is only generally what it is, to be found in corruptible things and below the lunar sphere. And

From Ptolemy, *The Almagest*, translated by R. C. Taliaferro, *Great Books of the Western World*, Chicago, 1952, Volume 16, Book I, secs. 1-8. Reprinted with the kind permission of the publishers, Encyclopedia Britannica, Inc.

the kind of science which shows up quality with respect to forms and local motions, seeking figure, number, and magnitude, and also place, time, and similar things, would be defined as mathematical. For such an essence falls, as it were, between the other two, not only because it can be conceived both through the senses and without the senses, but also because it is an accident in absolutely all beings both mortal and immortal, changing with those things that ever change, according to their inseparable form, and preserving unchangeable the changelessness of form in things eternal and of an ethereal nature.

And therefore meditating that the other two genera of the theoretical would be expounded in terms of conjecture rather than in terms of scientific understanding: the theological because it is in no way phenomenal and attainable, but the physical because its matter is unstable and obscure, so that for this reason philosophers could never hope to agree on them; and meditating that only the mathematical, if approached enquiringly, would give its practitioners certain and trustworthy knowledge with demonstration both arithmetic and geometric resulting from indisputable procedures, we were led to cultivate most particularly as far as lay in our power this theoretical discipline [θεωρία]. And especially were we led to cultivate that discipline developed in respect to divine and heavenly things as being the only one concerned with the study of things which are always what they are, and therefore able itself to be always what it is—which is indeed the proper mark of a science—because of its own clear and ordered understanding, and yet to cooperate with the other disciplines no less than they themselves. For that special mathematical theory would most readily prepare the way to the theological, since it alone could take good aim at that unchangeable and separate act, so close to that act are the properties having to do with translations and arrangements of movements, belonging to those heavenly beings which are sensible and both moving and moved, but eternal and impassible. Again as concerns the physical there would not be just chance correspondances. For the general property of the material essence is pretty well evident from the peculiar fashion of its local motion—for example, the corruptible and incorruptible from straight and circular movements, and the heavy and light or passive and active from movement to the center and movement from the center. And indeed this same discipline would more than any other prepare understanding persons with respect to nobleness of actions and character by means of the sameness, good order, due proportion, and simple directness contemplated in divine things, making its followers lovers of that divine beauty, and making habitual in them, and as it were natural, a like condition of the soul.

And so we ourselves try to increase continuously our love of the discipline of things which are always what they are, by learning what has already been discovered in such sciences by those really applying themselves to them, and also by making a small original contribution such as the period of time from them to us could well make possible. And therefore we shall try and set forth as briefly as possible as many theorems as we recognize to have come to light

up to the present, and in such a way that those who have already been initiated somewhat may follow, arranging in proper order for the completeness of
the treatise all matters useful to the theory of heavenly things. And in order
not to make the treatise too long we shall only report what was rigorously
proved by the ancients, perfecting as far as we can what was not fully proved
or not proved as well as possible.

2. On the Order of the Theorems

A view, therefore, of the general relation of the whole earth to the whole
of the heavens will begin this composition of ours. And next, of things in
particular, there will first be an account of the ecliptic's position and of the
places of that part of the earth inhabited by us, and again of the difference, in
order, between each of them according to the inclinations of their horizons.
For the theory of these, once understood, facilitates the examination of the rest.
And, secondly, there will be an account of the solar and lunar movements and
of their incidents. For without a prior understanding of these one could not
profitably consider what concerns the stars. The last part, in view of this
plan, will be an account of the stars. Those things having to do with the sphere
of what are called the fixed stars would reasonably come first, and then those
having to do with what are called the five planets. And we shall try and show
each of these things using as beginnings and foundations for what we wish to
find, the evident and certain appearances from the observations of the ancients
and our own, and applying the consequences of these conceptions by means
of geometrical demonstrations.

And so, in general, we have to state that the heavens are spherical and
move spherically; that the earth, in figure, is sensibly spherical also when
taken as a whole; in position, lies right in the middle of the heavens, like a
geometrical centre; in magnitude and distance, has the ratio of a point with
respect to the sphere of the fixed stars, having itself no local motion at all. And
we shall go through each of these points briefly to bring them to mind.

3. That the Heavens Move Spherically

It is probable the first notions of these things came to the ancients from
some such observation as this. For they kept seeing the sun and moon and
other stars always moving from rising to setting in parallel circles, beginning
to move upward from below as if out of the earth itself, rising little by little
to the top, and then coming around again and going down in the same way
until at last they would disappear as if falling into the earth. And then again
they would see them, after remaining some time invisible, rising and setting
as if from another beginning; and they saw that the times and also the places

of rising and setting generally corresponded in an ordered and regular way.

But most of all the observed circular orbit of those stars which are always visible, and their revolution about one and the same centre, led them to this spherical notion. For necessarily this point became the pole of the heavenly sphere; and the stars nearer to it were those that spun around in smaller circles, and those farther away made greater circles in their revolutions in proportion to the distance, until a sufficient distance brought one to the disappearing stars. And then they saw that those near the always-visible stars disappeared for a short time, and those farther away for a longer time proportionately. And for these reasons alone it was sufficient for them to assume this notion as a principle, and forthwith to think through also the other things consequent upon these same appearances, in accordance with the development of the science. For absolutely all the appearances contradict the other opinions.

If, for example, one should assume the movement of the stars to be in a straight line to infinity, as some have opined, how could it be explained that each star will be observed daily moving from the same starting point? For how could the stars turn back while rushing on to infinity? Or how could they turn back without appearing to do so? Or how is it they do not disappear with their size gradually diminishing, but on the contrary seem larger when they are about to disappear, being covered little by little as if cut off by the earth's surface? But certainly to suppose that they light up from the earth and then again go out in it would appear most absurd. For if anyone should agree that such an order in their magnitudes and number, and again in the distances, places, and times is accomplished in this way at random and by chance, and that one whole part of the earth has an incandescent nature and another a nature capable of extinguishing, or rather that the same part lights the stars up for some people and puts them out for others, and that the same stars happen to appear to some people either lit up or put out and to others not yet so—even if anyone, I say, should accept all such absurdities, what could we say about the always-visible stars which neither rise nor set? Or why don't the stars which light up and go out rise and set for every part of the earth, and why aren't those which are not affected in this way always above the earth for every part of the earth? For in this hypothesis the same stars will not always light up and go out for some people, and never for others. But it is evident to everyone that the same stars rise and set for some parts, and do neither of these things for others.

In a word, whatever figure other than the spherical be assumed for the movement of the heavens, there must be unequal linear distances from the earth to parts of the heavens, wherever or however the earth be situated, so that the magnitudes and angular distances of the stars with respect to each other would appear unequal to the same people within each revolution, now larger now smaller. But this is not observed to happen. For it is not a shorter linear distance which makes them appear larger at the horizon, but the steaming up of the moisture surrounding the earth between them and our eyes, just as things put under water appear larger the farther down they are placed.

The following considerations also lead to the spherical notion: the fact that instruments for measuring time cannot agree with any hypothesis save the spherical one; that, since the movement of the heavenly bodies ought to be the least impeded and most facile, the circle among plane figures offers the easiest path of motion, and the sphere among solids; likewise that, since of different figures having equal perimeters those having the more angles are the greater, the circle is the greatest of plane figures and the sphere of solid figures, and the heavens are greater than any other body.

Moreover, certain physical considerations lead to such a conjecture. For example, the fact that of all bodies the ether has the finest and most homogeneous parts [ὁμοιομερέστερος]; but the surfaces of homogeneous parts must have homogeneous parts, and only the circle is such among plane figures and the sphere among solids. And since the ether is not plane but solid, it can only be spherical. Likewise the fact that nature has built all earthly and corruptible bodies wholly out of rounded figures but with heterogeneous parts, and all divine bodies in the ether out of spherical figures with homogeneous parts, since if they were plane or disc-like they would not appear circular to all those who see them from different parts of the earth at the same time. Therefore it would seem reasonable that the ether surrounding them and of a like nature be also spherical, and that because of the homogeneity of its parts it moves circularly and regularly.

4. That also the Earth, Taken as a Whole, Is Sensibly Spherical

Now, that also the earth taken as a whole is sensibly spherical, we could most likely think out in this way. For again it is possible to see that the sun and moon and the other stars do not rise and set at the same time for every observer on the earth, but always earlier for those living towards the orient and later for those living towards the occident. For we find that the phenomena of eclipses taking place at the same time, especially those of the moon, are not recorded at the same hours for everyone—that is, relatively to equal intervals of time from noon; but we always find later hours recorded for observers towards the orient than for those towards the occident. And since the differences in the hours is found to be proportional to the distances between the places, one would reasonably suppose the surface of the earth spherical, with the result that the general uniformity of curvature would assure every part's covering those following it proportionately. But this would not happen if the figure were any other, as can be seen from the following considerations.

For, if it were concave, the rising stars would appear first to people towards the occident; and if it were flat, the stars would rise and set for all people together and at the same time; and if it were a pyramid, a cube, or any other polygonal figure, they would again appear at the same time for all observers on the same straight line. But none of these things appears to happen. It is further clear that it could not be cylindrical with the curved surface turned to the risings and settings and the plane bases to the poles of the universe, which some think more plausible. For then never would any of the stars be always visible to any of the inhabitants of the curved surface, but either all the stars would both rise and set for observers or the same stars for an equal distance from either of the poles would always be invisible to all observers. Yet the more we advance towards the north pole, the more the southern stars are hidden and the northern stars appear. So it is clear that here the curvature of the earth covering parts uniformly in oblique directions proves its spherical form on every side. Again, whenever we sail towards mountains or any high places from whatever angle and in whatever direction, we see their bulk little by little increasing as if they were arising from the sea, whereas before they seemed submerged because of the curvature of the water's surface.

5. That the Earth Is in the Middle of the Heavens

Now with this done, if one should next take up the question of the earth's position, the observed appearances with respect to it could only be understood if we put it in the middle of the heavens as the centre of the sphere. If this were not so, then the earth would either have to be off the axis but equidistant from the poles, or on the axis but farther advanced towards one of the poles, or neither on the axis nor equidistant from the poles.

The following considerations are opposed to the first of these three positions—namely, that if the earth were conceived as placed off the axis either above or below in respect to certain parts of the earth, those parts, in the right sphere, would never have any equinox since the section above the earth and the section below the earth would always be cut unequally by the horizon. Again, if the sphere were inclined with respect to these parts, either they would have no equinox or else the equinox would not take place midway between the summer and winter solstices. The distances would be unequal because the equator which is the greatest of those parallel circles described about the poles would not be cut in half by the horizon; but one of the circles parallel to it, either to the north or to the south, would be so cut in half. It is absolutely agreed by all, however, that these distances are everywhere equal because the increase from the equinox to the longest day at the summer tropic are equal to the decreases to the least days at the winter tropic. And if the deviation for certain parts of the earth were supposed either towards the orient or the occident, it would result that for these parts neither the sizes and angular distances of the stars would appear equal and the same at the eastern and western horizons, nor would the time from rising to the meridian be equal to the time from the meridian to setting. But these things evidently are altogether contrary to the appearances.

As to the second position where the earth would be on the axis but farther advanced towards one of the poles, one could again object that, if this were so, the plane of the horizon in each latitude would always cut into uneven parts the sections of the heavens below the earth and above, different with respect to each other and to themselves for each different deviation. And the horizon could cut into two even parts only in the right sphere. But in the case of the inclined sphere with the nearer pole ever visible, the horizon would always make the part above the earth less and the part below the earth greater with the result that also the great circle through the centre of the signs of the zodiac [ecliptic] would be cut unequally by the plane of the horizon. But this has never been seen, for six of the twelve parts are always and everywhere visible above the earth, and the other six invisible; and again when all these last six are all at once visible, the others are at the same time invisible. And so—from the fact that the same semicircles are cut off entirely, now above the earth, now below—it is evident that the sections of the zodiac are cut in half by the horizon.

And, in general, if the earth did not have its position under the equator but lay either to the north or south nearer one of the poles, the result would be that, during the equinoxes, the shadows of the gnomons at sunrise would never perceptibly be on a straight line with those at sunset in planes parallel to the horizon. But the contrary is everywhere seen to occur. And it is immediately clear that it is not possible to advance the third position since each of the obstacles to the first two would be present here also.

In brief, all the observed order of the increases and decreases of day and night would be thrown into utter confusion if the earth were not in the middle. And there would be added the fact that the eclipses of the moon could not take place for all parts of the heavens by a diametrical opposition to the sun, for the earth would often not be interposed between them in their diametrical oppositions, but at distances less than a semicircle.

6. *That the Earth Has the Ratio of a Point to the Heavens*

Now, that the earth has sensibly the ratio of a point to its distance from the sphere of the so-called fixed stars gets great support from the fact that in all parts of the earth the sizes and angular distances of the stars at the same times appear everywhere equal and alike, for the observations of the same stars in the different latitudes are not found to differ in the least.

Moreover, this must be added: that sundials placed in any part of the earth and the centres of armillary spheres can play the role of the earth's true centre for the sightings and the rotations of the shadows, as much in conformity with the hypotheses of the appearances as if they were at the true midpoint of the earth.

And the earth is clearly a point also from this fact: that everywhere the planes drawn through the eye, which we call horizons, always exactly cut in half the whole sphere of the heavens. And this would not happen if the magnitude of the earth with respect to its distance from the heavens were perceptible; but only the plane drawn through the point at the earth's centre would exactly cut the sphere in half, and those drawn through any other part of the earth's surface would make the sections below the earth greater than those above.

7. That the Earth Does Not in any Way Move Locally

By the same arguments as the preceding it can be shown that the earth can neither move in any one of the aforesaid oblique directions, nor ever change at all from its place at the centre. For the same things would result as if it had another position than at the centre. And so it also seems to me superfluous to look for the causes of the motion to the centre when it is once for all clear from the very appearances that the earth is in the middle of the world and all weights move towards it. And the easiest and only way to understand this is to see that, once the earth has been proved spherical considered as a whole and in the middle of the universe as we have said, then the tendencies and movements of heavy bodies (I mean their proper movements)[1] are everywhere and always at right angles to the tangent plane drawn through the falling body's point of contact with the earth's surface. For because of this it is clear that, if they were not stopped by the earth's surface, they too would go all the way to the centre itself, since the straight line drawn to the centre of a sphere is always perpendicular to the plane tangent to the sphere's surface at the intersection of that line.

All those who think it paradoxical that so great a weight as the earth should not waver or move anywhere seem to me to go astray by making their judgment with an eye to their own affects and not to the property of the whole. For it would not still appear so extraordinary to them, I believe, if they stopped to think that the earth's magnitude compared to the whole body surrounding it is in the ratio of a point to it. For thus it seems possible for that which is relatively least to be supported and pressed against from all sides equally and at the same angle by that which is absolutely greatest and homogeneous. For there is no "above" and "below" in the universe with respect to the earth, just as none could be conceived of in a sphere. And of the compound bodies in

---

1. All local motions or movements according to place are divided by Aristotle into natural and violent local motions. In the case of compound bodies (that is, those bodies subject to generation and corruption and consisting of all those, and only those, bodies lying below the lunar sphere), the natural local motions are those of unimpeded and unpropelled fall; the violent local motions are any propelled or interrupted motions. In the case of simple bodies (that is, the heavenly bodies within and above the lunar sphere), there are only natural local motions: the regular or uniform circular motions. Ptolemy here calls the natural local motions of compound bodies their proper motions. This distinction between natural and violent motions is preserved by Galileo. For in his *Two New Sciences*, natural motion is treated in the "Third Day" and violent motion in the "Fourth Day." In the Newtonian system, the distinction is dissolved in a general mathematical treatment, a treatment more in line with the Platonic myth of the *Timaeus*, and so it loses all meaning.

the universe, to the extent of their proper and natural motion, the light and subtle ones are scattered in flames to the outside and to the circumference, and they seem to rush in the upward direction relative to each one because we too call "up" from above our heads to the enveloping surface of the universe; but the heavy and coarse bodies move to the middle and centre and they seem to fall downwards because again we all call "down" the direction from our feet to the earth's centre. And they properly subside about the middle under the everywhere-equal and like resistance and impact against each other. Therefore the solid body of the earth is reasonably considered as being the largest relative to those moving against it and as remaining unmoved in any direction by the force of the very small weights, and as it were absorbing their fall. And if it had some one common movement, the same as that of the other weights, it would clearly leave them all behind because of its much greater magnitude. And the animals and other weights would be left hanging in the air, and the earth would very quickly fall out of the heavens. Merely to conceive such things makes them appear ridiculous.

Now some people, although they have nothing to oppose to these arguments, agree on something, as they think, more plausible. And it seems to them there is nothing against their supposing, for instance, the heavens immobile and the earth as turning on the same axis from west to east very nearly one revolution a day; or that they both should move to some extent, but only on the same axis as we said, and conformably to the overtaking of the one by the other.

But it has escaped their notice that, indeed, as far as the appearances of the stars are concerned, nothing would perhaps keep things from being in accordance with this simpler conjecture, but that in the light of what happens around us in the air such a notion would seem altogether absurd. For in order for us to grant them what is unnatural in itself, that the lightest and subtlest bodies either do not move at all or no differently from those of contrary nature, while those less light and less subtle bodies in the air are clearly more rapid than all the more terrestrial ones; and to grant that the heaviest and most compact bodies have their proper swift and regular motion, while again these terrestrial bodies are certainly at times not easily moved by anything else—for us to grant these things, they would have to admit that the earth's turning is the swiftest of absolutely all the movements about it because of its making so great a revolution in a short time, so that all those things that were not at rest on the earth would seem to have a movement contrary to it, and never would a cloud be seen to move toward the east nor anything else that flew or was thrown into the air. For the earth would always outstrip them in its eastward motion, so that all other bodies would seem to be left behind and to move towards the west.

For if they should say that the air is also carried around with the earth in the same direction and at the same speed, none the less the bodies contained in it would always seem to be outstripped by the movement of both. Or if they should be carried around as if one with the air, neither the one nor the other would appear as outstripping, or being outstripped by, the other. But these bodies would always remain in the same relative position and there would be no movement or change either in the case of flying bodies or projectiles. And yet we shall clearly see all such things taking place as if their slowness or swiftness did not follow at all from the earth's movement.

8. That There Are Two Different Prime Movements in the Heavens

It will be sufficient for these hypotheses, which have to be assumed for the detailed expositions following them, to have been outlined here in such a summary way since they will finally be established and confirmed by the agreement of the consequent proofs with the appearances. In addition to those already mentioned, this general assumption would also be rightly made that there are two different prime movements in the heavens. One is that by which everything moves from east to west, always in the same way and at the same speed with revolutions in circles parallel to each other and clearly described about the poles of the regularly revolving sphere. Of these circles the greatest is called the equator, because it alone is always cut exactly in half by the horizon which is a great circle of the sphere, and because everywhere the sun's revolution about it is sensibly equinoctial. The other movement is that according to which the spheres of the stars make certain local motions in the direction opposite to that of the movement just described and around other poles than those of that first revolution. And we assume that it is so because, while, from each day's observation, all the heavenly bodies are seen to move generally in paths sensibly similar and parallel to the equator and to rise, culminate, and set (for such is the property of the first movement), yet from subsequent and more continuous observation, even if all the other stars appear to preserve their angular distances with respect to each other and their properties as regards their places within the first movement, still the sun and moon and planets make certain complex movements unequal to each other, but all contrary to the general movement, towards the east opposite to the movement of the fixed stars which preserve their respective angular distances and are moved as if by one sphere.

If, then, this movement of the planets also took place in circles parallel to the equator—that is, around the same poles as those of the first revolution—it would be sufficient to assume for them all one and the same revolving movement in conformity with the first. For it would then be plausible to suppose that their movement was the result of a lag and not of a contrary movement. But they always seem, at the same time they move towards the east, to deviate towards the north and south poles without any uniform magnitude's being observed in this deviation, so that this seems to befall them through impulsions. But although this deviation is irregular on the hypothesis of one prime movement, it is regular when effected by a circle oblique to the equator. And so such a circle is conceived one and the same for, and proper to, the planets, quite exactly expressed and as it were described by the motion of the sun, but traveled also by the moon and planets which ever turn about it with every deviation from it on the part of any planet either way, a deviation within a prescribed distance and governed by rule. And since this is seen to be a great circle also because of the sun's equal oscillation to the north and south of the equator, and since the eastward movements of all the planets (as we said) take place on one and the same circle, it was necessary to suppose a second movement different from the general one, a movement about the poles of this oblique circle or ecliptic in the direction opposite to that of the first movement.

Then if we think of a great circle described through the poles of both the circles just mentioned, which necessarily cuts each of them—that is, the equator and the circle inclined to it—exactly in half and at right angles, there will be four points on the oblique circle or ecliptic: the two made by the equator diametrically opposite each other and called the equinoxes of which the one guarding the northern approach is called spring, and the opposite one autumn. And the two made by the circle drawn through both sets of poles, also clearly diametrically opposite each other, are called the tropics, of which the one to the south of the equator is called winter, and the one to the north summer.

The one first movement which contains all the others will be thought of then as described and as if defined by the great circle, through both sets of poles, which is carried around and carries with it all the rest from east to west about the poles of the equator. And these poles are as if they were on what is called the meridian, which differs from the circle through both sets of poles in this alone: that it is not always drawn through the poles of the ecliptic, but is conceived as continuously at right angles to the horizon and therefore called the meridian, since such a position cutting in half as it does each of the two hemispheres, that below the earth and that above, provides midday and midnight. But the second movement, consisting of many parts and contained by the first, and embracing itself all the planetary spheres[2] is carried by the first as we said, and revolves about the poles of the ecliptic in the opposite direction. And these poles of the ecliptic being on the circle effecting the first revolution—that is, on the circle drawn through all four poles together—are carried around with it as one would expect; and, moving therefore with a motion opposite to the second prime movement, in this way keep the position of the great circle which is the ecliptic ever the same with respect to the equator.

---

2. These are the two movements of the same and of the other described in Plato's myth of the *Timaeus*.

J. L. E. DREYER:

Medieval Cosmology

THE ROMAN EMPIRE was destroyed in the course of the hundred years following the memorable year (375) when the Huns invaded Europe through the natural gateway between the Caspian Sea and the Ural mountains, and drove Gothic and Germanic races headlong before them over most of the provinces of the Roman Empire. In 476 the last nominal Emperor of the West was deposed by a barbarian chieftain; that part of Europe which had formed the Western Empire had been partitioned among the conquerors; ruin and devastation reigned everywhere. There seemed to be an end of all civilisation, as the conquerors were utterly untouched either by the ancient culture of Asia or by anything they might have learned from their new subjects. To some extent their savage state was doubtless softened by the Christian religion, which they gradually adopted; but most of their teachers were unfortunately devoid of sympathy for anything that emanated from the heathen Greek and Roman world; and it was left to the dying Neo-Platonic school and to pagan commentators like Macrobius and Simplicius and the encyclopædic writer Martianus Capella to keep alive for a while the traditions of the past.

But even before the days when enemies from outside had begun to assail the Roman Empire, a fierce onslaught had commenced on the results of Greek thought. A narrow-minded literal interpretation of every syllable in the Scriptures was insisted on by the leaders of the Church, and anything which could not be reconciled therewith was rejected with horror and scorn. In this way some of the Fathers of the Church lent a hand to the barbarians who wrenched back the hand of time about a thousand years, and centuries were to elapse before their work was to some extent undone and human thought began to free itself from the fetters imposed on it in the days when the ancient world was crumbling to decay. In no branch of knowledge was the desire to sweep away all the results of Greek learning as conspicuous as with regard to the figure of the earth and the motion of the planets. When we turn over the pages of some of these Fathers, we might imagine that we were reading the

From J. L. E. Dreyer, *A History of the Planetary Systems from Thales to Kepler*, Cambridge University Press, 1905, Chapter X. Reprinted by the kind permission of the publishers, Cambridge University Press.

opinions of some Babylonian priest written down some thousands of years before the Christian era; the ideas are exactly the same, the only difference being that the old Babylonian priest had no way of knowing better, and would not have rejected truth when shown to result from astronomical observations.

At first there was no enmity to science exhibited by the followers of the Apostles. Clemens Romanus, in his epistle to the Corinthians,[1] written about A.D. 96, alludes in passing to the Antipodes as dwelling in a part of the earth to which none of our people can approach, and from which no one can cross over to us; and in the beginning of the same chapter he uses an expression often found in classical writings, that "the sun and moon and the dancing stars (ἀστέρων τε χοροί) according to God's appointment circle in harmony within the bounds assigned to them without any swerving aside." In Alexandria, where the leaders of the Christians were familiar with the philosophical speculations of Philo and the Neo-Platonists, it was natural that they should feel no

Pre-Copernican Universe appears in Peter Apian's *Cosmographia* (1539). The earth is in the center. The sun, moon, planets and stars occupy a series of concentric spheres.

1. C. 20. Lightfoot's ed. I. p. 282.

desire to place themselves in opposition to science. Clement of Alexandria (about A.D. 200), who had commenced life as a heathen, is indeed the first to view the Tabernacle and its furniture as representing allegorically the whole world; but he is not thereby led astray into sweeping aside the knowledge gained by the Greeks. The lamp was placed to the south of the table, and by it were represented the motions of the seven planets, as the lamp with three branches on either side signifies the sun set in the midst of the planets. The golden figures, each with six wings, represent either the two Bears, or more likely (he thinks) the two hemispheres, while he believes the ark to signify the eighth region and the world of thought or God.[2] This desire to find allegories in Scripture was carried to excess by Origen (185–254), who was likewise associated with Alexandrian thought, and he managed thereby to get rid of anything which could not be harmonised with pagan learning, such as the separation of the waters above the firmament from those below it, mentioned in Genesis, which he takes to mean that we should separate our spirits from the darkness of the abyss, where the Adversary and his angels dwell.[3]

But this kind of teaching was not to the taste of those who would have nothing to do with anything that came from the pre-Christian world, and to whom even "the virtues of the heathen were but splendid vices." A typical representative of these men was Lactantius, the first and the worst of the adversaries of the rotundity of the earth, whose seven books on *Divine Institutions* seem to have been written between A.D. 302 and 323. In the third book, *On the false wisdom of the philosophers*, the 24th chapter is devoted to heaping ridicule on the doctrine of the spherical figure of the earth and the existence of antipodes.[4] It is unnecessary to enter into particulars as to his remarks about the absurdity of believing that there are people whose feet are above their heads, and places where rain and hail and snow fall upwards, while the wonder of the hanging gardens dwindles into nothing when compared with the fields, seas, towns, and mountains, supposed by philosophers to be hanging without support. He brushes aside the argument of philosophers that heavy bodies seek the centre of the earth as unworthy of serious notice; and he adds that he could easily prove by many arguments that it is impossible for the heavens to be lower than the earth, but he refrains because he has nearly come to the end of his book, and it is sufficient to have counted up some errors, from which the quality of the rest may be imagined.

More moderate in his views was Basil, called the Great, who wrote a lengthy essay on the six days of creation about the year 360.[5] He does not rave against the opinions of philosophers as Lactantius did; he is evidently acquainted with the writings of Aristotle, and generally expresses himself with a certain degree of moderation and caution. Thus he is aware of the fact that there are stars about the south pole of the heaven invisible to us, and he understands

2. *Stromata*, 1. v. cap. 6; *Ante-Nicene Chr. Libr.* xii. pp. 240-242.
3. *In Genesim Homiliæ, Opera*, ed. Delarue, Paris, 1733, t. ii. p. 53.
4. Ed. Dufresnoy, Paris, 1748, t. i. p. 254.
5. *Homiliæ novem in Hexaemeron, Op. omn.*, ed. Garnier, Paris, 1721, t. i.

perfectly well how summer and winter depend on the motion of the sun
through the northern and southern halves of the zodiac.[6] When speaking of
the two "great lights," he says that they are really of an immense size, since
they are seen equally large from all parts of the earth; no one is nearer to the
sun or farther from it, whether it is rising or on the meridian or setting; besides
which the whole earth is illumined by the sun, while all the other stars give
only a feeble light.[7] But though he is aware of the annual motion of the sun,
he does not uphold the spherical form of the heavens or deny that there is
more than one heaven; the words of Genesis about the upper waters are too
distinct for that; and he sets forth the idea, common among patristic writers,
that these waters were placed above the firmament to keep it cool and prevent
the world from being consumed by the celestial fire.[8] As to the figure of the
earth, he says that many have disputed whether the earth is a sphere or a
cylinder or a disc, or whether it is hollow in the middle; but Moses says
nothing about this nor about the circumference of the earth being 180,000
stadia, nor about anything which it is not necessary for us to know.[9] Basil
evidently was too sensible to deny the results of scientific investigation, but
also too timid to advocate them openly, so that he at most merely mentions
them without comment, or endeavours to show that a Christian may accept
them without danger to his faith. But for his acceptance of the upper waters,
he might seem to have been a comparatively unprejudiced thinker.

The ruthlessly literal interpretation of Scripture was especially insisted on
by the leaders of the Syrian Church, who would hear of no cosmogony or
system of the world but that of Genesis. A contemporary of Basil, Cyril of
Jerusalem, lays great stress on the necessity of accepting as real the supercelestial waters,[10] while a younger contemporary of Basil, Severianus, Bishop of
Gabala, speaks out even more strongly and in more detail in his *Six Orations
on the Creation of the World*,[11] in which the cosmical system sketched in the
first chapter of Genesis is explained. On the first day God made the heaven,
not the one we see, but the one above that, the whole forming a house of
two storeys with a roof in the middle and the waters above that. As an angel
is spirit without body, so the upper heaven is fire without matter, while the
lower one is fire with matter, and only by the special arrangement of providence sends its light and heat down to us, instead of upwards as other fires
do.[12] The lower heaven was made on the second day; it is crystalline, congealed water, intended to be able to resist the flame of sun and moon and the
infinite number of stars, to be full of fire and yet not dissolve nor burn, for

6. i. 4, l.c. p. 4 e.
7. vi. 9–10, l.c. p. 58 d–60 a.
8. iii. 3, l.c. p. 23 e–24 e.
9. ix. 1, l.c. p. 80.
10. *Catechesis*, ix., *Opera*, Oxford, 1703, p. 116.
11. Joh. Chrysostomi *Opera*, ed. Montfaucon, t. vii. (Paris, 1724), p. 436 sqq. Compare also
the extracts given by Kosmas, pp. 320–325.
12. i. 4.

which reason there is water on the outside. This water will also come in handy on the last day, when it will be used for putting out the fire of the sun, moon and stars.[13] The heaven is not a sphere, but a tent or tabernacle; "it is He . . . that stretcheth out the heavens as a curtain and spreadeth them out as a tent to dwell in";[14] the Scripture says that it has a top, which a sphere has not, and it is also written: "The sun was risen upon the earth when Lot came unto Zoar."[15] The earth is flat and the sun does not pass under it in the night, but travels through the northern parts "as if hidden by a wall," and he quotes: "The sun goeth down and hasteth to his place where he ariseth."[16] When the sun goes more to the south, the days are shorter and we have winter, as the sun takes all the longer to perform his nightly journey.[17]

The tabernacle shape of the universe was from that time generally accepted by patristic writers; thus by Diodorus, Bishop of Tarsus (died 394), who in his book *Against Fatalism*[18] declaims against those atheists who believe in the geocentric system; and he shows how Scripture tells us that there are two heavens created, one which subsists with the earth, and one above that again, the latter taking the place of a roof, the former being to the earth a roof but to the upper heaven a floor. Heaven is not a sphere but a tent or a vault.[19] This was also the opinion of Theodore, Bishop of Mopsuestia in Cilicia (d. about 428), but his work is lost, and we only know from the sneers of a later writer, Philoponus, that he taught the tabernacle theory and let all the stars be kept in motion by angels. About the same time St. Jerome wrote with great violence against those who followed "the stupid wisdom of the philosophers" and had imagined the Cherubim to represent the two hemispheres, ourselves and the antipodes, and he explained that Jerusalem was the navel of the earth.[20]

Somewhat more sensible opinions seem to have prevailed at that time in the Western Church. Ambrose of Milan (d. 397) says that it is of no use to us to know anything about the quality or position of the earth, or whether heaven is made of the four elements or of a fifth;[21] but still he mentions the heaven repeatedly as a sphere.[22] Driven into a corner by the question how there can be water outside the sphere, he somewhat feebly suggests that a house may be round inside and square outside,[23] or he asks why water should not be

---

13. II. 3-4.
14. Isaiah xl. 22.
15. Gen. xix. 23. The above is from the Revised Version, but Severianus (III. 4) has: "Sol egressus est super terram, et Lot ingressus est in Segor. Quare liquet, Scriptura teste, egressum esse Solem, non ascendisse."
16. Eccles. i. 5.
17. III. 5.
18. This book is lost, but Photius gives a *résumé* of it, *Bibl.*, Codex 223.
19. Compare Chrysostom in his comment. on Hebrews viii. 1: "Where are those who say that the heaven is in motion? Where are those who think it spherical? For both these opinions are here swept away." Quoted by Kosmas p. 328.
20. Comment. on Ezekiel, chs. i. and v.; *Opera*, Benedict. ed., Paris, 1704, II. p. 702 and 726.
21. *Hexaem.* I. 6; Benedict. ed., Paris, 1686, t. I. col. 11, also II. 2, col. 25.
22. Ibid., also I. 3, col. 4.
23. II. 3, col. 26.

suspended in space just as well as the heavy earth, while its use is obviously to keep the upper regions from being burned by the fiery ether.[24] It was natural that Augustine (354–430), who may be considered a disciple of Ambrose, should express himself with similar moderation, as befitted a man who had been a student of Plato as well as of St Paul in his younger days. With regard to antipodes, he says that there is no historical evidence of their existence, but people merely conclude that the opposite side of the earth, which is suspended in the convexity of heaven, cannot be devoid of inhabitants. But even if the earth is a sphere, it does not follow that that part is above water, or, even if this be the case, that it is inhabited; and it is too absurd to imagine that people from our parts could have navigated over the immense ocean to the other side, or that people over there could have sprung from Adam.[25] With regard to the heavens, Augustine was, like his predecessors, bound hand and foot by the unfortunate water above the firmament. He says that those who defend the existence of this water point to Saturn being the coolest planet, though we might expect it to be much hotter than the sun, because it travels every day through a much greater orbit; but it is kept cool by the water above it. The water may be in a state of vapour, but in any case we must not doubt that it is there, for the authority of Scripture is greater than the capacity of the human mind.[26] He devotes a special chapter[27] to the figure of the heaven, but does not commit himself in any way, though he seems to think that the allusions in Scripture to the heaven above us cannot be explained away by those who believe the world to be spherical. But anyhow Augustine did not, like Lactantius, treat Greek science with ignorant contempt; he appears to have had a wish to yield to it whenever Scripture did not pull him the other way, and in times of bigotry and ignorance this is deserving of credit.

We have thus seen that the Fathers of the Church did not all go equally far in their condemnation of Greek astronomy, and that none of them took the trouble to work out in detail a system to take the place of the detested doctrines of the pagan philosophers. This work was undertaken by one who did not hold high office in the Church, but who had travelled a great deal by land and by sea, and might therefore have been expected to be more liberal-minded in his views than a churchman who had not had that advantage. He is known by the name of Kosmas, surnamed Indicopleustes, or the Indian navigator. His book, the *Christian Topography*,[28] contains some passages which throw light on his history and enable us to fix the date at which he wrote. He was prob-

24. *Hexaem.* II. col. 29.
25. *De Civitate Dei*, lib. XVI. cap. 9; Benedict. ed., t. VII. cols. 423–24.
26. *De Genesi ad litteram*, lib. II. cap. 5 (t. III. pp. 134–135). Compare *De Civ. Dei*, XI. 34 (t. VII. col. 299), where he says that some have thought that heavy water could not be above heaven, and have therefore interpreted it as meaning angels; but they should remember that pituita, which the Greeks call φλέγμα, is placed in the head of man! A beautiful comparison.
27. *De Genesi*, II. 9; t. III. cols. 138–39.
28. "The Christian Topography of Kosmas, an Egyptian monk." Translated from the Greek and edited with notes and introduction by J. W. McCrindle, London, printed for the Hakluyt Society, 1897. The pages quoted are those of Montfaucon's edition in Vol. II. of his *Nova Collectio Patrum*.

ably a native of Alexandria, and during the earlier part of his life he was a merchant. He tells us himself (somewhat needlessly) that he was "deficient in the school learning of the pagans,"[29] though on the other hand he alludes to the theory of epicycles, and thereby deprives himself of the excuse for his silly notions which total ignorance of Alexandrian learning might have supplied. He travelled in the Mediterranean, the Red Sea, and the Persian Gulf, and on one occasion he even dared to sail on the dreaded Ocean, which "cannot be navigated on account of the great number of its currents and the dense fogs which it sends up, obscuring the rays of the sun; and because of the vastness of its extent."[30] One of the most interesting parts of his book is that which describes his travels in Abyssinia and adjoining countries. As he must have reached places within ten degrees of the equator, it is very remarkable that he could be blind to the fact that the earth is a sphere. His work on *Christian Topography* consisted originally of five books, to which seven others were subsequently added in order to further elucidate various points; it must have been written between the years 535 and 547, as events which happened in these years are alluded to in the text as occurring while the author was writing.

The first book is *Against those who, while wishing to profess Christianity, think and imagine like the pagans that the heaven is spherical.* The daily revolution of the heaven he thinks he has swept away by saying that the appearance of the Milky Way shows that the heaven must be constituted of more than one element, and that it must therefore either have a motion upwards or downwards, but nothing of the kind has ever been perceived by anybody. He next asks why the planets stand still, and even make retrogressions. "They will, perhaps, in reply assign as the cause those invisible epicycles which they have assumed as vehicles on which, as they will insist, the planets are borne along. But they will be in no better case from this invention, for we shall ask: Why have they need of vehicles? Is it because they are incapable of motion? Then if so, why should you assert them to be animated, and that too even with souls more than usually divine? Or is it that they are capable? The very idea is, methinks, ridiculous. And why have not the moon and the sun their epicycles? Is it that they are not worthy on account of their inferiority? But this could not be said by men in their sober senses. Was it, then, from the scarcity of suitable material the Creator could not construct vehicles for them? On your own head let the blasphemy of such a thought recoil."[31]

The alleged position of the earth in the centre of the universe is also in the eyes of Kosmas utterly absurd, as the earth is so unspeakably heavy that it can only find rest at the bottom of the universe. The usual cheap arguments against the existence of antipodes are next served out, but he does not think these "old wives' fables" worthy of many words. In a note appended to his fourth book[32] he asks, into which of the eight or nine heavens (spheres) of

29. p. 124.
30. p. 132.
31. p. 119.
32. pp. 189–191.

the pagans Christ has ascended, and into which one Christians hope themselves to ascend? "If the sphere which has motion forces the others to revolve along with it from east to west, whence is produced the motion in the contrary direction of the seven planets? Is it the spheres that have the contrary motion, or the stars themselves? If the spheres, how can they at one and the same time move both westward and eastward? And if the stars, how do the stars cut their way through the heavenly bodies?" After which Kosmas quotes various passages of Scripture in order utterly to crush those Christians who wished to listen to the Greek philosophers ("no man can serve two masters"), and winds up by enquiring how a spherical earth situated in the middle of the world could have emerged from the waters on the third day of creation, or how it could have been swamped by the deluge in the days of Noah?

Kosmas' own idea is, that the figure of the universe can only be learned by studying the design of the Tabernacle, which Moses constructed in the wilderness. We have seen how Severianus and others had already assumed that the earth was like a tabernacle, but their suggestion was now worked out in detail by Kosmas, who points out[33] that Moses had pronounced the outer tabernacle to be a pattern of the visible world, while the Epistle to the Hebrews, in explaining the inner tabernacle, or that which was within the veil, declared that it was a pattern of the kingdom of heaven, the veil being the firmament which divides the universe into two parts, an upper and a lower. The table of shew-bread with its wavy border signified the earth surrounded by the ocean, and another border outside the first one represented another earth beyond the ocean, while every other article in the tabernacle similarly had a cosmographical meaning. As the table was placed lengthwise from east to west, we learn that the earth is a rectangular plane, twice as long as it is broad, and its longer dimension extending from east to west. The ocean which encompasses our earth is in its turn surrounded by another earth, which had been the seat of Paradise and the dwelling-place of man until the deluge, when the Ark carried Noah and his family and animals over to this earth, while the old one has since been inaccessible owing to the unnavigable state of the ocean. The walls of heaven are four perpendicular planes joined to the edges of the transoceanic earth, and the roof is shaped like a half cylinder resting on the north and south walls, the whole thing being by Kosmas likened to the vaulted roof of a bathroom, while to the modern mind it looks more like a travelling trunk with a curved lid. The whole structure is divided into two storeys by the firmament, which forms a floor for the upper and a ceiling for the lower storey, the latter being the abode of angels and men until the day of judgment, the upper storey being the future dwelling of the blest.

The earth, the footstool of the Lord, is at the bottom of the structure, while the sun, moon and stars are not attached to its sides or roof, but are carried along in their courses below the firmament by angels, who have to carry on this work until the last day. The rising and setting of the sun required a

33. p. 134.

special explanation. Of course it could not possibly go under the earth, and it was necessary (with Severianus) to assume that it was hidden by the northern part of the earth during the night. Quoting the same passage from the book of Ecclesiastes he states[34] that the earth is much higher in the north and west than in the south and east, and that it is well known that ships sailing to the north and west are called lingerers, because they are climbing up and therefore sail more slowly, while in returning they descend from high places to low, and thus sail fast. The Tigris and Euphrates flowing south have far more rapid currents than the Nile, which is "running, as one may say, up."[35] In the north there is a huge conical mountain, behind which the sun passes in the night, and according as the sun during this passage is more or less close to the mountain, it appears to us as if it passes nearer to the top or nearer to the base of the mountain; in the former case the night is short and we have summer, in the latter it is long and we have winter. All the other heavenly bodies are likewise moved in their orbits by angels and pass behind the northern, elevated part of the earth, while eclipses are produced because "the revolution and the course of the heavenly bodies have some slight obliquity."[36]

When Kosmas had finished his five books, he was asked how the sun could possibly be hidden behind the northern part of the earth, if it is many times larger than the earth? He therefore devotes his sixth book to proving that the sun is in reality quite small, and nowhere does he prove himself as incapable of reasoning on the simplest facts as here. Because at the summer solstice the shadow of a man at Antioch or Rhodes (the beginning of the sixth climate of Ptolemy)[37] was half a foot shorter than it was at Byzantium (a little beyond the beginning of the seventh), he concluded that the sun "has the size of two climates."[38] For at Meroe the man would be shadowless, at Syene (one climate north) the shadow would be half a foot to the north, and in Ethiopia (one climate south) half a foot to the south, therefore the diameter of the sun is two climates![39]

Such was the celebrated system of the world of Kosmas Indicopleustes. He was not a leader in the Church (it is even uncertain whether he was Orthodox or a Nestorian) and his book apparently never rose to be considered a great authority. By the fact that he wrote, so to say, a text-book on the subject, he acquired a certain notoriety; but though it cannot be denied that he displayed a good deal of originality in twisting the innumerable passages from Scripture with which his book bristles into proofs of his assertions, his system had in reality been indicated by the Church Fathers of the preceding two hundred

---

34. p. 133.

35. It was absurd to believe that the rain at the antipodes could fall up, but evidently it was all right to believe that a river was flowing up!

36. p. 156.

37. The first climate begins where the longest day is $12^h\ 0^m$; the second where it is $12^h\ 30^m$, and so on.

38. p. 265.

39. About 1060 miles, or if we take the two climates from Alexandria to Byzantium, about 680 miles.

years. This is acknowledged by Kosmas, who in his tenth book collects a number of quotations from the Fathers, especially from Severianus. Poor Kosmas
has come in for a good deal of ridicule, but in fairness this ought to be
addressed to his predecessors, who had abused the authority of their position
in the Church and their literary ability to propagate ideas which had been
abandoned in Greece eight hundred years before. But what Kosmas does
deserve to be blamed for is his not finding out on his travels that the earth is a
sphere.

The fanatics who desired to clear away as noxious weeds the whole luxuriant
growth of Greek science had, however, not altogether the field to themselves.
Some writers there were, even then, who studied the works of the Greek philosophers, and were not afraid to accept at least some of their doctrines. Among
these was Johannes Philoponus, a grammarian of Alexandria, who seems to
have lived about the end of the sixth century, and who wrote commentaries
to several of the writings of Aristotle as well as a number of treatises showing
a remarkable freedom of thought, which naturally obtained for him the name
of a heretic, and in later times would certainly have caused him to be sent to
the stake. In his book on the creation of the world[40] he argues against the
abuse of Scriptural quotations by Theodore of Mopsuestia to prove that the
heaven is not spherical[41] or that the stars are moved by angels appointed to
this task; and he asks why God should not have endowed the stars with some
motive power. He even goes so far as to compare this power to the tendency
of all bodies, heavy and light, to fall to the earth.[42] Owing to his want of
orthodoxy, the opinions of Philoponus could, however, not influence his contemporaries to any appreciable extent, and it was of much more importance
that a man holding high office in the Western Church, Isidorus Hispalensis,
Bishop of Seville, expressed himself very sensibly on the constitution of the
world. Isidore was born about 570, and became Bishop of Seville already in
601, probably through his high family connections; but he soon became widely
known by his learning and eloquence, and twice presided over Church Coun-
cils. He died in 636. Among his numerous writings is an encyclopædic work:
Etymologiarum libri xx, in the beginning of which he enumerates the seven
free arts, grammar, rhetorics, dialectics (*the trivium*), and arithmetic, music,
geometry, astronomy (*the quadrivium*), which already long before his time
had come to be considered as embracing all human knowledge.[43] The work is
called *Etymologies* (sometimes *Origins*), because Isidore generally explains
the meaning of a word or an expression by means of its supposed derivation.
When dealing with dangerous topics, such as the figure of the world and the

---

40. *Bibliotheca veterum patrum cura Andreæ Gallandii*, Venice, 1776, t. xii. p. 471 sq.
41. iii. 10.
42. i. 12. He probably got the idea from Empedokles, quoted by Aristotle, *De Coelo*, ii. 1,
p. 284 a, 24, and Simplicius, p. 375, 29–34 (Heib.).
43. The names, though not the usual order, of the seven arts may be remembered by the
distich:
"*Gram.* loquitur, *dia.* verba docet, *rhe.* verba ministrat,
Mus. canit, *ar.* numerat, *ge.* ponderat, *as.* colit astra."

earth, he does not lay down the law himself, but quotes "the philosophers" as teaching this or that, though without finding fault with them. Thus he repeatedly mentions that according to them the heaven is a sphere, rotating round an axis and having the earth in the centre.[44] In the same manner he refers to the spherical shape of the earth, saying in the chapter about Africa: "But in addition to the three parts of the orbis (Asia, Europe, Africa) there is a fourth to the south beyond the ocean, which, owing to the heat of the sun, is unknown to us, at the outskirts of which the antipodes are fabulously reported to dwell."[45] This fourth continent has frequently been postulated by geographers of antiquity, some of them even assuming the existence of two other *oekumenes* in the western hemisphere, one north and one south of the equator; and it is creditable to Isidore that he does not, like his predecessors, rave about the iniquity of imagining the existence of people on the opposite side of the earth.

Isidore also wrote a smaller treatise, *De rerum natura*, giving more details about some of the subjects touched on in the larger work, and here again we find him occupying a place midway between "the philosophers" and the bigoted patristic writers. The heaven is a sphere revolving once in a day and a night,[46] and though Ambrosius in his *Hexaemeron* says that philosophers make out that there are seven heavens of the seven planets, yet human temerity does not dare to say how many there may be. God the Creator tempered the nature of the heaven with water, lest the conflagration of the upper fire should kindle the lower elements. Therefore He named the circumference of the lower heaven the firmament, because it supports the upper waters.[47] The moon is much smaller than the sun,[48] and is nearest to us; the order of the planets is Moon, Mercury, Venus, Sun, &c., and they complete their circles in 8, 23, 9, 19, 15, 22, and 30 years![49] The stars (fixed stars) move with the world, it is not they which move while the world stands still.[50] A strange mixture of truth and error.

But though enlightened students like Philoponus and Isidore might accept some of the teaching of antiquity, the school of cosmographers of the Kosmas type continued to flourish. From about the seventh century we have a cosmography which goes under the name of one Æthicus of Istria and professes to be translated and abbreviated from a Greek original by a priest named Hierony-

---

44. For instance: "Nam philosophi dicunt coelum in sphæræ figuram undique esse convexam, omnibus partibus æquale, concludens terram in media mundi mole libratam. Hoc moveri dicunt, et cum motu ejus sidera in eo fixa ab oriente usque ad occidentem circuire." xiii. 5, compare iii. 31–32.

45. Liber. xiv. cap. 5, *De Libya*, § 17.

46. Cap. xii.

47. Cap. xiii.

48. Cap. xvi.

49. "Nam luna octo annis fertur explere circulum suum, Mercurius annis xxiii., Lucifer annis ix., Sol annis xix., *Pyrois annis* xv., Phaeton annis xxii., Saturnus annis xxx." Cap. xxiii., *Opera omnia*, Romæ, 1803, t. vii. p. 36. But on a figure showing concentric orbits, the figures 19, 20, 9, 19, 15, 12, 30, are marked (explet cursum annis 19, &c.). Of course Isidore has misunderstood the meaning of these periods, which are not periods of revolution but periods after which the planets occupy the same places among the stars.

50. Cap. xii.

mus; but nothing is known either of the alleged author or of the translator, who has very probably compiled the book himself. He has wonderful things to tell about Alexander the Great, Gog and Magog, centaurs and minotaurs, and dog-headed men; in fact the whole book reads like the ravings of a lunatic. But as he enjoyed a considerable reputation in the Middle Ages, he cannot be passed over in an account of the cosmical opinions of that time. The earth of course is flat, the sun likewise (it is spoken of as a table, *mensa solis*), and it passes through the gate of the east every morning to lighten up the world, and passes in the evening through the gate of the west to return during the night to its starting-point through the south(!), hidden in the meanwhile by a thick mist which screens it from us but allows some of its light to reach the moon and stars. Under the earth is the abyss of the great waters. The heaven is spread out over the earth like a skin and encloses the sun, moon, and stars, all of them moving freely and separated by it from the six upper heavens, the dwellings of the heavenly host.[51]

Another geographer from the end of the seventh century, the "anonymous geographer of Ravenna," whose work is chiefly statistical, views the world quite like the patristic writers. The world is bounded on the west by the ocean, on the east by a boundless desert, which even made Alexander the Great turn back. The sun illuminates the whole world at the same time. To the north, beyond the ocean, there are great mountains, placed there by God to make a screen, behind which sun and moon disappear. Some people indeed had denied the existence of these mountains, and asked if anyone had ever seen them, but it is clear that the Creator has made them inaccessible in order that mankind should know nothing about them.[52]

This is, however, the last writer of note who refuses obstinately to listen to common sense. No doubt there continued throughout the Middle Ages to be clerics to whom the sphericity of the earth was an abomination, and, even among those who acknowledged it, very few had the courage to confess openly that there was nothing impossible in assuming the existence of human beings on the other side of the sphere. But in the peaceful retreat of the monastery the study of the ancient Latin writers had long before the time of the Ravennese geographer taken root, and the geocentric system slowly but steadily began to resume its place among generally accepted facts. The next figure among medie-val writers after Isidore is the Venerable Bede, and he followed his predecessor in his opinions about the world. Born about 673 in the north of England, Bede spent most of his life in two monasteries in that neighbourhood, to which a considerable number of books had been brought from Rome by their founder, and he made good use of them in preparing his numerous writings. So great

51. *Cosmographiam Æthici Istrici* . . . edidit H. Wuttke, Leipzig, 1853, caps. 6, 8, 13, &c. The editor is very enthusiastic about his hero and greedily swallows all his marvels; he also willingly accepts the translator's identity with St. Jerome and assumes that the original was written before the time of Constantine!

52. Originally written in Greek, but only a Latin translation has come down to us. Edited by Pinder and Parthey, *Ravennatis Anonymi Cosmographia*, Berlin, 1860.

was the reputation which he acquired by these that long after his death (even four or five centuries later) spurious tracts were produced and palmed off on the reading world as products of the great English monk, to whom posterity had given the title of "Venerable" in token of its admiration. There is, however, no difficulty in separating these spurious treatises from the genuine ones, particularly as Bede four years before his death (which took place about 735) drew up a list of his writings and appended it to his famous *Ecclesiastical History*.

Among the undoubted writings of Bede is a treatise *De natura rerum*,[53] which in 51 paragraphs deals with the stars, the earth and its divisions, thunder, earthquakes, &c. The contents are taken from Pliny, often almost verbatim; and the spherical form of the earth, the order of the seven planets circling round it, the sun being much larger than the earth, and similar facts are plainly stated.[54] But the unlucky water around the heaven and the usual explanation of its existence could of course not be kept out of the book,[55] even though Pliny does not mention it, and Bede had stated that the heaven was a sphere. Another and much longer book by Bede deals with chronology (*De temporum ratione*) and shows a fair knowledge of the annual motion of the sun and the other principal celestial phenomena. When mentioning the zones of the earth[56] he says that only two of them are capable of being inhabited, while no assent can be given to the fables about antipodes, since nobody had ever heard or read of anyone having crossed the torrid zone and found human beings dwelling beyond it.

That it was worth while to be very cautious in speaking of antipodes appears from the ruin which threatened Fergil, an Irish ecclesiastic of the eighth century, better known as Virgilius of Salzburg. He was originally Abbot of Aghaboe (in the present Queen's County) and started for the Holy Land about 745, but he did not get further than Salzburg, where he became Abbot of St. Peter's. In 748 he came into collision with Boniface, the head of the missionary Churches of Germany, about the validity of a baptism administered by a priest ignorant of Latin, and when Boniface reported this to the Pope (Zacharias) he took the opportunity to complain that Virgil in his lectures had taught that there was "another world and other people under the earth." Zacharias replied that Boniface should call a council and expel Virgil from the Church, if he really had taught that. Whether any proceedings were taken against Virgil is not known,[57] but in any case he cannot have been condemned as guilty of heresy, since he became Bishop of Salzburg in 767 (when both Boniface and Zacharias were long dead) and ruled that see till his death in

---

53. *Venerabilis Bedæ Opera*, ed. Giles, Vol. vi. pp. 100–122 (London, 1848).
54. He even copies from Pliny (ii. 49) that the moon is larger than the earth (cap. xix.). Possibly Pliny misunderstood the papyrus of Eudoxus, col. xx. 15.
55. Cap. vii.
56. Cap. xxxiv.
57. In the *Thesaurus Monumentorum* of Canisius, iii. 2, p. 273 (Antwerp, 1725), it is said that Virgil did not obey the summons to Rome. The accusation is not mentioned in the *Monumenta Germaniæ* (Script. T. xi. p. 84 sq.).

784 or 785. No writings of his are extant, and nothing is known of his doctrines except the words quoted above from the Pope's reply, to which in one
edition is added that the other world underneath ours had its own sun and
moon.[58] But this is probably a marginal improvement made by some tran-
scriber to emphasize the shocking heresy of Virgil, and we cannot doubt that
Virgil merely taught the existence of antipodes.[59] And after all there is nothing
very remarkable in the fact that an Irish monk knew the earth to be a sphere.
Not only were many Irish monasteries centres of culture and learning, where
the fine arts and classical literature were studied at a time when thick night
covered most of the Continent and to a less extent England; but devoted missionaries had before the time of Virgil spread the light of Christianity as far
north as the Orkneys, while Adamnan, the biographer of St. Columba, had
had personal intercourse with Arculf, who had made a pilgrimage to the Holy
Land. That the sphericity of the earth, asserted by the Greek and Roman
writers, was an undoubted fact, must have been made clear by comparing
notes with these travellers, whose experience extended over 25° of latitude. In
the following century we find another Irishman of note, Dicuil, who finished
his geographical compilation, the *Liber de mensura orbis terræ*, in 825. Though
he says nothing about the figure of the earth, he tells us of Irish missionaries
who thirty years earlier visited Thule (which here undeniably means Iceland),
where they saw the sun barely hidden at midnight in midsummer, as if it
went behind a little hill, so that there was nearly as much light as in the
middle of the day, "and I believe that at the winter solstice and during the days
thereabouts the sun is visible for a very short time only in Thule, while it is
noon at the middle of the earth."[60] Dicuil must therefore have clearly understood the phenomena of the "oblique sphere."

However dangerous it might be to assert the existence of human beings in
what was thought to be an inaccessible part of the earth, beings who could
not be assumed to be descended from Adam or to have been redeemed by the
death of Christ, the idea that religion and secular learning were of necessity
opposed to each other was fast disappearing, and it had by this time become
quite a customary thing among men of learning to recognize that the earth is
a sphere. Still some people chose to say nothing about it, e.g. Hrabanus Maurus,
Abbot of Fulda and afterwards Archbishop of Mainz (d. 856), who, though he
did much to encourage classical studies, yet in his encyclopedic work *De*

---

58. "De perversa autem et iniqua doctrina, quam contra Deum et animam suam locutus est,
si clarificatum fuerit ita eum confiteri, quod alius mundus et alii homines sub terras sint; hunc
accito concilio, ab Ecclesia pelle sacerdotii honore privatum." So in Usher's works, ed. by
Elrington, IV. p. 464, and in S. Bonifacii *Opera*, ed. I. A. Giles, London, 1844, I. p. 173. But in
Sacrosancta Concilia, studio P. Labbæi et Gabr. Cossartii, Venice, 1729, t. VIII. p. 256, the words
"seu sol et luna" occur after "sub terra sint." In the Benedictine *Histoire littéraire de la France*,
t. IV. p. 26 (1738), it is said that Virgil discovered the antipodes "ou un autre monde qui a son
soleil, sa lune et ses saisons comme le nôtre." Virgil was canonized in 1233 by Pope Gregory IX.
on account of the miracles wrought by his bones after they had been found in 1171; see Canisius,
l.c. pp. 399 sq. and Riccioli, *Almag. nov.* II. p. 489.
59. So Maestlin understood him; Kepleri *Opera*, I. p. 58.
60. *Dicuilii liber de mensura orbis terræ*, a G. Parthey recognitus, Berlin, 1870, pp. 42–43.
Remains of the Irish settlements in Iceland were found by the Northmen on their arrival in 874.

Universo merely says that the earth is situated in the middle of the world.[61] The inhabited land, he says, is called orbis "from the rotundity of the circle, because it is like a wheel";[62] but he sees the necessity of assuming it to be a square, since Scripture speaks of its four corners, and he finds it awkward to explain why the horizon is a circle. But he refers to the fourth book of Euclid and seems to think that a square inscribed in a circle will save the situation. His statement that the heaven has two doors, east and west, through which the sun passes,[63] looks, however, as if his point of view was much the same as that of the patristic writers. But when an eminent mathematician like Gerbert ascended the papal throne as Sylvester II. (in 999, died 1003), the game was up for the followers of Lactantius. The example of Bede, who had openly taught the sphericity of the earth, had borne fruit, and so did doubtless that of a Pope who was familiar with the scientific writings of the ancients[64] and in his younger days had constructed celestial and terrestrial globes to assist his lectures on astronomy, and had been in the habit of exchanging them for MSS of Latin classics. And the horizon of mankind continued to be widened out by the spread of geographical knowledge through the intercourse with the Arabs in Spain on the one side, and the travels and adventures of the Northmen on the other. Adam of Bremen (about 1076), whose chronicle is of great importance for the study of the history of his time, has nothing in common with Kosmas or the geographer of Ravenna; he understands perfectly the cause of the inequality of the day and night in different latitudes and shows himself an apt student of Bede's writings. The maps of this period also mark a considerable advance. Beside the ordinary "wheel maps" or *T-O* maps, so called from the form, which resembles a *T* inscribed in a circle (Asia being above the horizontal stroke of the *T*, the vertical stroke of which is the Mediterranean), we find more elaborate maps. They are mostly founded on a design by Beatus, a Spanish priest who lived at the end of the eighth century; they represent Africa as not reaching to the equator, and though they do not show any sign of the antipodes, they contain nothing against the rotundity of the earth; and by degrees as the designers of maps became better acquainted with ancient works on geography, they made bolder attempts at depicting the earth.

From about the ninth century the rotundity of the earth and the geocentric system of planetary motions may be considered to have been reinstated in the places they had held among the philosophers of Greece from the days of Plato. The works of these philosophers were still unknown in the West, where Greek had been an unknown tongue after the fifth century; but the writings of Pliny, Chalcidius, Macrobius and Martianus Capella supplied a good deal

---

61. *De Universo*, xii. 1.—B. Rabani Mauri *Opera omnia*, ed. J. P. Migne, Paris, 1864, T. v. col. 331.

62. Ibid. xii. 2, col. 332–333.

63. Ibid. ix. 5, col. 265.

64. Among the sources of his Geometry Gerbert mentions Plato's *Timæus*, Chalcidius, Eratosthenes, &c. Cantor, *Gesch. d. Math.* i. p. 811.

of information to anyone who read them, and since the days of Charles the Great (768–814) Roman literature was rapidly becoming better known. We possess two works of unknown date which have been founded on these writers. They go under the name of Bede, but they are undoubtedly much later productions and have not been included in the modern edition of his works. One of them is entitled *De mundi cœlestis terrestrisque constitutione liber*,[65] and it is hardly possible that anyone can ever have believed it to have been written by Bede, as he is quoted in it and there are several references to the chronicles of Charles the Great, so that it must at any rate have been written after the year 814. The author has a fair knowledge of the general celestial phenomena such as could be gathered from the above-mentioned writers, but no more. He proves that the earth is a sphere by the different length of the day in different latitudes, and by the fact that the various phenomena in the heavens do not occur at the same time for different localities. He says that Plato followed the Egyptians in placing the solar orbit immediately outside that of the moon, but his own opinion seems to be that Venus and Mercury are sometimes above the sun and sometimes below it, as it is recorded in the history of Carolus that Mercury for nine days was visible as a spot on the sun, though clouds prevented both the ingress and the egress being seen. When they are below the sun they are visible in the middle of the day, and he refers to the star seen at the time of Cæsar's funeral, which he supposes to have been Venus.[66] The limits of the planets in latitude are also given.[67] The writer shows himself somewhat independent of his authorities by adding a fair sprinkling of astrology, and still more by giving the various theories current about the unavoidable "supercelestial waters."[68] One idea is that there are hollows in the outer surface of the heaven in which water may lie (as it does on the earth's surface), and notwithstanding the rapid rotation of the heaven it is not spilt, just as water will remain in a vessel swung rapidly round! Another idea is, that the water is only vapour like clouds; another that it is frozen owing to the great distance from the sun, the principal source of heat, and that Saturn is called the most frigid star because it is nearest the water. But the waters are simply held there by the power of God in order to cool the heaven, and above them are the spiritual heavens in which the angelic powers dwell.

The other work, which formerly was counted among the writings of Bede, is entitled "Περὶ διδάξεων *sive elementorum philosophiæ libri IV*."[69] It has been ascribed to William of Conches, a Norman of the first half of the 12th century, and in any case it cannot have been written much earlier, as it shows a freedom of thought which would have been impossible at the time of Bede.[70] This is particularly the case with regard to the question whether there is water

65. Ven. Bedæ *Opera*, Col. Agripp. 1612, T. i. cols. 323–344.
66. It was a comet.
67. Taken from Martianus Capella, but the sun's range is 2°, as in Pliny.
68. p. 332.
69. *Opera* (1612), T. ii. pp. 206–230.
70. There are two other editions differing very little from the Pseudo-Bede: *Philos. et astron. institutionum Guil. Hirsaugiensis libri* iii. Basle, 1531, and *De Philosophia Mundi; Honorii Opera, Max. Bibl. Pat.* T. xx.

above the ether.[71] Quoting the passage from Genesis about the water above the firmament, the writer says that it is *contra rationem*, for if congealed it would be heavy and the earth would then be the proper place for it, while the water above would be next the fire and either put it out or be dissolved by it, as we could not suppose that there is any boundary between them. The air is called the firmament because it strengthens and regulates the earthly things, and above it there is water suspended in the form of clouds, which are indeed different from the water below the air. "Although we think it was said more allegorically than literally." Turning to the planets, the writer is aware of the difference of opinion as to the position of the solar orbit. He dismisses the idea that the sun has been placed next after the moon in order that the heat and dryness of the sun might thus counteract the cold and humidity of the moon, which otherwise might become excessive owing to the proximity of the earth. Also the idea that the sun has to be next the moon because the latter is illuminated by it. But as the sun, Venus and Mercury move nearly in the same period round the zodiac, their circles must be nearly equal in size and are not contained within each other but intersect each other.[72] The sun is eight times as large as the earth. The air reaches to the moon; above that is ether or fire, which is so subtle that it cannot burn unless mixed with something humid and dense; while the sun and stars are not made of fire alone, but also of the other elements, though fire is the predominating material. In all this there is nothing new.

As regards the earth, the writer says that it is in the middle, as the yolk in the egg, and outside it is the water like the white round the yolk; around the water is the air like the skin round the white, and finally fire, corresponding to the eggshell. The two temperate zones are inhabitable, but we believe only one to be inhabited by men. "But because philosophers talk about the inhabitants of both, not because they are there, but because they may be there, we shall state what we believe there are, from our philosophical reading." The zone in which we live consists of two parts, of which we inhabit one and our antipodes the other, and similarly the other inhabitable zone consists of two parts, of which the upper is that of our *anthei* and the lower that of their antipodes. Thus we and our antipodes have summer or winter together, but when we have day they have night. In other words, the author adheres to the old idea of the four *oekumenes*, but he uses the word antipodes in a sense which is not the usual one, but which signifies people who live in our hemisphere but 180° distant from us in longitude.

From about the same time we have the *Imago Mundi* by Honorius of Autun, a kind of short encyclopædia from the first half of the twelfth century.[73] The cosmographical part is borrowed from Pliny, but with the necessary

---

71. p. 213.

72. The range in latitude of the planets is illustrated by a diagram, the figures resulting from this (not given in the text) being practically the same as in the book *De mundi constitutione*. But the diagram is perhaps a later addition, and a very absurd one.

73. *Mundi Synopsis sive De Imagine Mundi libri tres.* Ab Honorio Solitario Augustudunense. Spiræ, 1583. There are several other editions from the 15th and 16th centuries.

additions to suit the taste of medieval readers.[74] The two doors of heaven are duly mentioned, though they do not fit well in the geocentric system of the world. The upper heaven is called the firmament; it is spherical, adorned all over with stars which are round and fiery, and outside it are waters in the form of clouds, above which is the spiritual heaven, unknown to man, where the habitations of the angels are, arranged in nine orders.[75] Here is the Paradise of Paradises, where the souls of saints are received, and this is the heaven which was created in the beginning together with the earth.[76] In the centre of the earth is Hell, which is described in some detail. The writer does not seem to know where Purgatory is situated.

The work of Honorius found several imitators, both in prose and verse, among the latter being the *Image du monde,* written in 1245 by a certain Omons (otherwise unknown), who mentions Honorius and William of Conches among his authorities.[77] The ideas set forth are like theirs. Ptolemy, King of Egypt, invented clocks and various instruments and wrote several books, one of which is called the *Almagest.* There are two heavens, the crystalline and the empyrean; angels dwell in the latter, and from it the demons were expelled. Children, on account of their innocence, can hear the celestial music. The air of heaven is called ether, and the bodies of the angels are formed of it. The writer says nothing about the planetary system.

The taste for encyclopedic writing became strongly developed in the thirteenth century and was much influenced by the knowledge of Aristotle's works, which at last had begun to spread in the western countries. About the middle of the twelfth century Arabian translations of Aristotle began to be introduced into France from Spain, and with them came the commentaries of Alexander and Simplicius and works of other Greek philosophers. These had to be translated into Latin; and though the translations were not very accurate, having passed through Syriac and Arabic before putting on the Latin garb, still they opened up to a wondering world the treasures of Greek thought. At first the Church was hostile to this movement, a natural consequence of the mass of mystical, pseudo-neoplatonic and Arabian speculations, which had

74. Among the things borrowed from Pliny may be mentioned the greatest latitudes of the planets (i. 79) and the musical intervals of the planets (i. 81); one tone = 15,625 miliaria, which is the distance of the moon, from that to Mercury is $7812\frac{1}{2}$ miliaria, and so on, so that the distance of heaven (seven tones) is 109,375 miliaria.

75. Compare the *Libri Sententiarum* of Peter the Lombard, Bishop of Paris (d. 1164), where the nature of angels and their hierarchy are discussed in the second book. As to the waters above, he quotes the opinion of Bede that they form the solid heaven, as crystal is made of water, and that of Augustine that they are in the form of vapour much lighter than that which we see in clouds. "Anyhow we must not doubt that they are there." The hierarchy of angels was fixed by the Pseudo-Dionysius Areopagita and universally accepted during the Middle Ages; it is arranged as follows in the *Summa Theologiæ* of Thomas Aquinas (i. 108). Seraphim, Cherubim, Thrones form the uppermost, empyrean hierarchy; the Thrones pass on the commands of God to the first order of the second hierarchy, the Dominations, next to whom come the Virtues, who guide the motions of the stars and planets, and the Powers who remove anything which might hinder these motions. The third hierarchy, Principalities, Archangels, Angels rule the earthly affairs. Compare Dante, *Convito,* ii. 6.

76. Honorius, i. 87–90, 138–140.

77. *Notices et Extraits des manuscrits,* T. v. pp. 243–266.

been imported under the guise of Aristotelian treatises; and at a provincial council, held at Paris in 1209, it was decreed that neither Aristotle's books on Natural Philosophy nor commentaries on them should be read either in public or privately in Paris. In 1215 this prohibition was renewed in the statutes of the University of Paris. But by degrees the fears of the Church wore off, so that in 1254 official orders were issued, prescribing how many hours should be used in explaining the physical treatises of Aristotle; and the Aristotelian natural philosophy was from henceforth and for nearly four hundred years firmly established at the Paris University, and indeed at every seat of learning. Fresh translations had already earlier been made by order of the Emperor Frederic II.; others made directly from the Greek were soon provided at the instance of Albertus Magnus (1193–1280) and his disciple Thomas Aquinas (1227–1274), and soon Aristotle had become the recognized ally of the theologians. Both Albert and Thomas contributed by their writings greatly to the spread of knowledge of ancient science, a work in which the gigantic encyclopædia of Vincent of Beauvais (*Speculum Naturale,* completed in 1256) also had a great share.

The most representative writer among the scholastics is Thomas Aquinas, and among his works there is one in particular which must be mentioned here. It is a commentary on Aristotle's book on the heavens,[78] and the spirit in which it is written shows the vast strides from darkness towards light which had been recently made. Though Aquinas was deeply convinced that revelation is a more important source of knowledge than human reason, he considers both to be two distinct and separate ways of finding truth; and in expounding Aristotle he therefore never lets himself be disturbed by the difference between his doctrine and that of the Bible, but assumes both to be ultimately derived from the same source. His commentary is very interesting to read,[79] much clearer than that of Simplicius, with which he is well acquainted, and which he frequently quotes together with the works of Plato, Ptolemy, and others. Wherever necessary, he points out that philosophers after Aristotle have come to differ from him, as for instance in substituting epicycles for the homocentric spheres, or as regards the motion of the starry sphere, which Aristotle assumed to be the uppermost one, while later astronomers say that the sphere of the fixed stars has a certain proper motion (i.e. precession), for which reason they place another sphere above it, to which they attribute the first motion.[80] In speaking about the position of the earth at rest in the centre of the world, he quotes Ptolemy's arguments in its favour.[81]

Another and very much humbler writer may also be mentioned here, as his little book on the sphere remained the principal elementary text-book on

---

78. S. Thomæ Aquinatis *Opera omnia,* T. III., *Commentaria in libros Aristotelis de Coelo et Mundo* . . . Romæ, 1886, fol., a magnificent edition.
79. Especially if one reads it after wading through the patristic writers.
80. Lib. II. lect. IX. p. 153 a; compare XVII. p. 189 a, where Ptolemy's 1° in 100 years is mentioned.
81. Lib. II. lect. XXVI. p. 220 b.

astronomy for nearly four centuries. We know next to nothing of the life of Johannes de Sacro Bosco, or John of Holywood, except that he died at Paris in 1256. He quotes Ptolemy and Alfargani (the latter had been translated in the middle of the twelfth century) and describes the equants, deferents and epicycles, being the first European writer in the Middle Ages to give even a short sketch of the Ptolemaic system of planetary motions.[82] After the long and undisturbed reign of Pliny and Martianus Capella, Ptolemy at last began to come to the front again.

But to the great majority of scholastics there was no going beyond Aristotle, who was held to represent the last possibility of wisdom and learning. One man there was, however, who was not content to be a mere slave of Aristotle, any more than the Alexandrian thinkers had been. Roger Bacon (1214–1294) in his *Opus Majus* shows himself thoroughly acquainted with the literature of the Greeks and Arabians. But in opposition to the general tendency of the previous thousand years he does not think it enough to write wordy commentaries on the ancients: he is able to think for himself, and he lays stress on the importance of experiments as offering the only chance of helping science out of the state of infancy in which he is fully aware it still lies. The scholastic doctors also, after the manner of the ancients, talked finely about experience as the only safe guide in the visible world. But it began and ended in talk; they did not find a single new fact in natural philosophy, they did not determine a single value of any astronomical constant. Roger Bacon was a man of a different stamp, and had he lived under more favourable circumstances we cannot doubt that he would have opened a new era in the history of science, instead of being merely a voice crying in the wilderness, whose wonderful work had to lie in manuscript for nearly five hundred years before it was printed. His object was to effect a reform in natural philosophy by brushing aside the blind worship of authority and by setting forth the value of mathematical investigations. As he was only a poor persecuted student, he had not the means to carry out his ideas; but his treatise on perspective shows what he was capable of, and what he would have done, if he had been the petted son of the Church instead of being its prisoner. In his general ideas about the universe he followed Ptolemy, and we shall therefore here only allude to one or two points. He remarks that the earth is only an insignificant dot in the centre of the vast heaven; according to Alfargani, the smallest star is larger than the earth, a sixth magnitude star being 18 times as large, while a first magnitude star is 107 and the sun 170 times as large (i.e. in volume).[83] Ptolemy has shown that a star takes 36,000 years to travel round the heaven (i.e. by precession), while a man can walk round the earth in less than three

---

82. Sacrobosco's knowledge of the Ptolemaic system was evidently of a very elementary nature and only acquired second-hand, for he copies the mistake of Alfargani and Albattani, that the two points on the epicycle in which the planet is stationary are the points of contact of the two tangents from the earth.

83. *Opus. Majus*, ed. S. Jebb, London, 1733, p. 112. On p. 143 he gives the dimensions of the orbits in Roman miles according to Alfargani, the diameter of the starry sphere being 130,715,000 miliaria.

years. In the chapter on geography[84] it is interesting to see that he discusses at some length the question how large a part of the earth is covered by the sea, and, from the statements of Aristotle, Seneca, and Ptolemy, comes to the conclusion that the ocean between the east coast of Asia and Europe is not very broad. This part of Bacon's work was almost literally copied by Cardinal d'Ailly (Petrus de Alyaco) into his *Imago Mundi* (written in 1410, first printed in 1490) without any mention of Bacon; it was quoted by Columbus in his letter from Hispaniola to the Spanish monarchs in 1498, and it had evidently made a very strong impression on him.[85] It is pleasant to think that the persecuted English monk, then two hundred years in his grave, was able to lend a powerful hand in widening the horizon of mankind.

A reader of Roger Bacon cannot fail to be struck with the vast difference between him and the patristic writers. While they struggled hard to accept the most literal interpretation of every iota in Scripture, Roger Bacon fearlessly points out difficulties in various passages of the Old Testament, and urges that the only way to get over them is by making a thorough study of science, which the Fathers of the Church had failed to do. He mentions as examples the first chapter of Genesis, the sun standing still at the bidding of Joshua, and the shadow on the dial going back ten degrees.[86] Similarly the statement of St. Jerome (on Isaiah) that there are twenty-two stars in Orion, nine of which are of the third, nine of the fourth, and the rest of the fifth magnitude, which does not agree with the eighth book of the *Almagest*.

But though a few enlightened men like Thomas Aquinas and Roger Bacon knew the works of Ptolemy, they certainly remained quite unknown to the leading men of the thirteenth century. This fact is strongly illustrated by the cosmographical ideas of Dante, whose *Divina Commedia* represents the prevailing views of his time (around the year 1300) as to the structure of the world. In general it is a risky thing to draw conclusions from astronomical allusions in poetical works to the state of scientific knowledge of the time,[87] but in the case of Dante it is quite legitimate to do so, as he in the *Commedia* as well as in his other writings shows himself fully equipped with the learning then attainable. He was a pupil of Brunetto Latini, who during his residence in France, from 1260 to about 1267, became infected with the mania for encyclopedic writing prevailing in that country, and composed his celebrated work, *Li Livres dou Tresor*, in the North-French language.[88] Like all the other books of its kind, this is a mere compilation from classical and medieval sources, the astronomical part being very meagre. Though Dante had doubtless

84. pp. 181–236.

85. Humboldt, *Kritische Untersuchungen*, i. p. 71 sq.

86. Isaiah xxxviii. 8.

87. For instance, the novels of the 19th century would lead one to think that nothing was known at that time about the motion of the moon, since it is quite a common thing to read in them of a young moon rising in the evening, the full moon sailing high in the heavens in summer, &c.

88. The French original was not printed till 1863 (Paris, edited by P. Chabaille), but an Italian translation has been printed several times, the chapters on astronomy separately by B. Sorio, *Il Trattato della sfera di Ser Brunetto Latini*, Milano, 1858.

studied the structure of the world deeper than Brunetto had done, none of his writings show any familiarity with the *Syntaxis* of Ptolemy, while Aristotle (with the commentary of Thomas Aquinas), Pliny, and especially Alfargani, seem to have been the authors by the study of whom he had profited most.[89] He began writing an encyclopedic work, the *Convito,* or Banquet, intended to comprise fourteen books, of which, however, only four were written. In this work his cosmological ideas are set forth more systematically, with the addition of a good deal of astrology and other fancies.[90]

In Dante's majestic poem, Hell is a conical cavity reaching to the centre of the earth. Around the sloping sides the places of punishment are arranged in circles of gradually decreasing diameter, so that the worst sinners are placed nearest to the apex of the cone, where Lucifer dwells in the very centre of the earth. When Dante and his guide Virgil have passed to the bottom of the abyss and continue their journey straight on, Dante looks back and sees Lucifer upside down, whereupon his guide explains that they have now commenced their ascent to the other side of the earth.[91] Purgatory is a large, conical hill, rising out of the vast ocean at a point diametrically opposite to Jerusalem, the navel of the dry land. Having passed over the seven terraces of the mount, and reached the earthly paradise at the top, the poet is finally permitted to rise through the celestial spheres. These are, of course, ten in number, first that of the moon (to which the blue air reaches[92]), then the spheres of Mercury, Venus (to which the shadow of the earth reaches[93]), the sun, Mars, Jupiter, and Saturn. In each of these spheres spirits, though they have not their permanent abode there, appear to Dante, in order to illustrate to him the gradually increasing glory which they have been found worthy to enjoy, and to indicate their former earthly characters and temperaments, which had been chiefly influenced by one of the seven planets.[94] The eighth sphere is that of the fixed stars, the ninth is the Primum Mobile, the velocity of which is almost incomprehensible owing to the fervent desire of each part of it to be

89. At the end of his poem *Il Tesoretto,* Brunetto tells how he on Mount Olympus met Ptolemy, master of astronomy and philosophy, and asked him to explain about the four elements; upon which Ptolemy "rispose in questa guisa"—and there the poem ends abruptly!

90. The opinions set forth in the *Convito* differ in a few cases somewhat from the ideas of the *D. C.,* the most notable astronomical instance being the spots in the moon. In the *Conv.* II. 14, Dante say the spots are caused by the rarity of parts of the lunar body, which do not reflect the sun's rays well. In *Par.* II. Beatrice delivers a long lecture showing that this theory is erroneous (because those parts would be transparent and would show themselves to be so during solar eclipses); the moon shines by its own light, which differs in various places under the influence of the various angelic guides, just as the stars in the eighth sphere differ in brightness, owing to the different virtue communicated to them by the Cherubim who rule them.

91. *Inferno,* XXXIV. 87 seq.

92. *Purg.* I. 15.

93. *Par.* IX. 118.

94. In the *Convito* (II. 14-15) Dante explains that the first seven spheres correspond to the Trivium and Quadrivium of the seven liberal arts. For instance, Mercury, the smallest planet and the one most veiled in the sun's rays, corresponds to Dialectics, an art less in body and more veiled than any other, as it proceeds by more sophistic and uncertain arguments. The eighth or starry sphere corresponds to Physics and Metaphysics, the ninth to Moral Science, and the tenth or Empyrean heaven to Theology.

conjoined to the restful and most Divine Heaven, the tenth or Empyrean, the dwelling of the Deity.[95] The nine spheres are moved by the three triads of angelic intelligences, the Seraphim guiding the Primum Mobile, the Cherubim the fixed stars, the Thrones the sphere of Saturn, and so on down to the moon's sphere, which is in charge of the angels.[96] In the eleventh canto of *Purgatorio* (v. 108) there is a distinct allusion to the precession of the equinoxes or, as it was still assumed to be, of the sphere of the fixed stars: "che più tardi in cielo è torto." There is only one slight allusion to epicycles,[97] otherwise the planets are merely said to move in the ecliptic,[98] and it is curious to find the sun's motion stated to be along spirals,[99] just as Plato of old had said in the *Timæus*. Another old acquaintance meets us in the statement that the sphere of the moon has the slowest motion.[100]

Dante continued throughout his life to be deeply interested in cosmography. In 1320, the year before his death, he delivered a lecture "De Aqua et Terra" in order to refute the opinion occasionally promulgated in the Middle Ages, and even later, that the water- and land-surface of the earth do not form part of one and the same sphere, but that the earth consists of a land-sphere and a water-sphere, the centres of which do not coincide.[101]

We may here close our review of medieval cosmology. Dante died in the year 1321, almost exactly a thousand years after the Emperor Constantine had made the Christian faith the state religion of the Roman Empire. It had been a long and perfectly stationary period, at the end of which mankind occupied exactly the same place as regards culture as at the beginning; scarcely even that, as Greek science, philosophy and poetry were still very imperfectly known in the West, so that no serious attempt could be made to build further on the foundation they offered. For centuries men had feebly chewed the cud on the first chapter of Genesis; then compilers like Pliny and Martianus Capella had grudgingly obtained a hearing; finally Aristotle had been discovered, and had almost at once been accepted as the infallible guide.

But in the East the light once issuing from Greece had not been so long obscured. The flame had been kept alive by the very people who at first had

---

95. *Convito*, ii. 4.

96. *Convito*, ii. 6; ibid. ii. 5, the motive powers of the spheres are said to be known as angels among "la volgare gente."

97. *Par.* viii. 2, "Che la bella Ciprigna il folle amore
             Raggiasse, volta nel terzo epiciclo."
Compare *Convito*, ii. 4 (on the back of this circle in the heaven of Venus is a little sphere which turns by itself in this heaven, and the circle of which astronomers call epicycle), also ii. 6 near the end, where the same is said of the planets in general.

98. *Par.* x. 7,   "Leva dunque, Lettor, all' alte rote
               Meco la vista dritto a quella parte
               Dove l'un moto e l' altro si percote."

99. *Par.* x. 32.

100. *Par.* iii. 51. compare above, Chapter iii. pp. 70 and 81.

101. When Columbus in 1498 near the coast of South America noticed the steady current of water opposing his progress (coming from the Orinoco), he thought he was near the highest point of the sea, from which the water rushed down.

seemed destined to trample all civilisation under foot, as the Huns had once done in Europe; and from the Arabs came the first impulse which led to the awakening of the West.

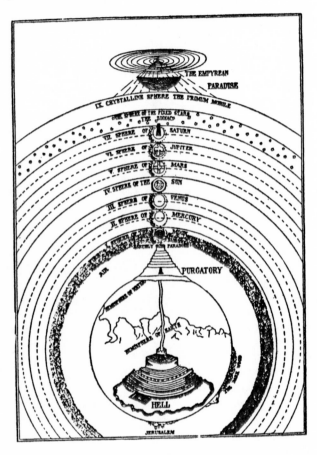

Dante's Conception of the Universe

(See *Studies in the History and Method of Science*, ed. Chas. Singer, 1917, Vol. I, Fig. 4.)

The Copernican Revolution and Its Aftermath

Introduction

CLASSICAL COSMOLOGY, as we have seen, rested upon at least three characteristic claims: that the earth is at the center of the universe, that the universe is spatially bounded, and finally, that the physical principles to be used for purposes of explanation are qualitative ones and involve a fundamental distinction between those to be used in interpreting terrestrial as distinguished from celestial phenomena. Each of these items of belief came under repeated and varied modes of attack beginning in the very late Middle Ages and with mounting effectiveness resulted, during the modern period, in the eventual destruction of the older cosmology. Looked at in a negative light, therefore, astronomy and physics as they came into their own in the modern era showed the need for removing the earth from its central position in the cosmos, for denying the classical conception of a universe bounded by the spherical shell of fixed stars, and finally for abandoning a qualitative, dualistic physics in interpreting astronomical phenomena. No single well-rounded and widely accepted cosmology emerges on the constructive side to take the place of the older scheme during the long period we are about to consider, a period that stretches, say, from the late fourteenth century to the turn of the present one. Yet important discoveries of an observational sort and conceptual schemes of increasing quantitative refinement begin to make their appearance and it is out of the ground prepared by these that the present-day surge forward in the field of scientific cosmology takes its start.

Among the many contributions, even before Copernicus, that helped in loosening the hold of the cosmology of a geocentric and finite universe, that of the fifteenth-century theologian Nicolas Cusanus is frequently recalled. In his book *Of Learned Ignorance,* Cusanus offers an account of his mystical or negative theology from whose principles he undertakes to derive a number of consequences relating to cosmology. In appealing to what he calls the doctrine of the "coincidence of contraries" Cusanus maintains that for the universe as a whole "the maximum and minimum coincide." This is to say that the universe has neither center nor circumference, or what amounts to the same thing, the center of the universe is everywhere, the circumference nowhere. The universe, while not actually infinite, has no boundaries. Neither the earth nor any other body is its absolute center. He maintains, moreover, that there can be no essential distinction between sublunary and celestial matter. Again, Cusanus argues that since everywhere throughout the universe there is motion, it would be illogical to insist that the earth alone is an exception. That we do not have a sensory experience of the earth's motion is an illustration of the familiar fact that only by reference to something fixed do we recognize motion. Just as a man on a boat in the middle of a river would not know he was moving if he could not see the fixed banks and were ignorant of the flow of the water, so men on the earth, for example, regard themselves as at rest, though actually in motion. Some of Cusanus's arguments, though often highly *aprioristic* and involving on occasion a good deal of obscurity as well as a confusion of empirical concepts with theologic ones, are nevertheless important. They are illustrations of the kind of thinking which, already a century before Copernicus, helped undermine bit by bit the foundations on which the orthodox cosmology rested.

The great contribution of Copernicus was neither to observational astronomy nor to the construction of a new and adequate cosmology as such, but rather to the theory of planetary motions. From the point of view of the over-all progress in cosmology, as distinct from planetary theory, the primary significance of Copernicus's work was thus in fact negative rather than constructive. It effectively removed the geocentrism of classical theory. Copernicus's substitution of the sun for the earth as the center of the universe was not in fact a step in the right direction as far as cosmology is concerned. For his heliocentrism was eventually to be discarded in favor of a view which recognized that our sun is no more privileged in its cosmic status than is our earth, and neither can be taken as the unique center of the universe, if indeed center there is. In taking the sun instead of the earth as his fixed center of co-ordinates, Copernicus demonstrated the greater mathematical simplicity to be attained in the account of planetary motions. In other respects, Copernicus was a conservative. He did not venture to abandon the use of patterns of circular motion in describing the paths of heavenly bodies and indeed continued to use the Ptolemaic device of epicycles. Although he assimilated the status of the earth to that of the other planets, thus preparing the ground for breaking down the essential dualism in Aristotelian physics between terrestrial and celestial phenomena, he himself did not challenge in any systematic way the traditional reverence for the Aristotelian physics. Again, although he admits that the question as to whether the universe is spatially infinite or finite cannot be properly settled by astronomy but must be left to the "natural philosophers," he did not definitely give up the view that the universe is finite. He believed, in fact, that it is bounded, as classical cosmology maintained, by the outermost shell of fixed stars. The effect, however, of Copernicus's challenge of the orthodox belief in the central and fixed position of the earth and his assignment to it of both a diurnal rotation on its axis as well as an annual revolution about the sun, was to make the other features of the traditional theory still retained by Copernicus himself more easily susceptible to attack. They were, all of them, indeed, given up in succeeding periods of astronomical and cosmological investigation. By showing through an effective and carefully worked out alternative how the geocentrism of classical theoretical astronomy can be avoided, Copernicus liberated the study of cosmology for fresh discoveries and perspectives.

Galileo's immortal work, *Dialogue Concerning the Two Chief World Systems*, was undoubtedly the greatest example of the efforts made to show to the public at large the superiority of the Copernican system to the older cosmology. With trenchant wit and brilliant style, Galileo undertakes through the mouth of Salviati in the *Dialogue* to ridicule the arguments of those Peripatetics, symbolized by Simplicio in the *Dialogue*, who in Galileo's day still slavishly adhered to the tenets of a moribund Aristotelianism. At the same time he cogently expounds the chief features of the newer cosmology, at the heart of which was the doctrine of the motion of the earth. Although in one portion of the *Dialogue* Galileo undertakes to present his own novel theory concerning the cause of the tides, it is as a restatement of the Copernican theory that his book remains a perennially readable classic.

When turning to Kepler's role in the development of cosmology, again we must see in him primarily an ardent defender and expositor of the main features of the Copernican cosmology. His own claim to immortal fame rests upon the improvement in planetary theory he made possible by the discovery of the famous three laws of planetary motion, among them the one which establishes that the planets move in elliptical orbits rather than in circles. Having taken advantage of the rich fund of laboriously obtained observational data bequeathed to him by Tycho Brahe, Kepler's work as a theoretical astronomer was kept in bounds and controlled by these empirical materials. When otherwise left free to speculate, as in the field of cosmology generally, Kepler proves to be far more of a mystic and an enthusiast. Here his penchant for working out an elaborate doctrine of the harmony of the spheres and of pointing up the central position of the sun in the entire scheme of things proved to be of less importance for the forward development of cosmology, whatever its intrinsic interest as one more illustration of the power of the Pythagorean way of thinking and the excesses to which it can lead.

In another direction the work of Bruno and Thomas Digges needs to be included in any account of the early modern development of cosmology for showing the possibilities in cosmology in wrenching attention away from the earth and other planets and turning instead to the stars as the "primary" units in terms of which one may construct a view of the universe. Bruno, whose Copernicanism was only tangential, and adopted in a form which is made to support a conception of an infinite universe and innumerable worlds, had his principal intellectual roots in Lucretius and Cusanus. It is to them that he is indebted for working out a philosophy which had at its core a combination of speculative materialistic atomism and pantheism. It was the combination of these ideas that served as the background for his conviction that our world is but one of an infinite multitude of similar "worlds."

In the case of Thomas Digges, as F. R. Johnson points out, we are dealing with a scientist who, taking the work of Copernicus as his primary starting point, argues that the new ideas allow for conceiving an *infinite* universe of *stars*. What is important in this claim is not that it could be warranted, which of course it could not be, but that it showed the possibility in terms of the new

Copernican astronomy for overthrowing another chief feature of classic cosmology, the belief that the universe is spatially bounded by the sphere of the fixed stars.

In Christiaan Huygens's *Cosmotheoros* we find still another example of the efforts made at trying to solve the question of the structure of the system of stars. Although Huygens restrains himself from venturing any definite opinion as to whether the system of stars is infinite or not, his reasoning in this essay concerning the status of the sun as one among the multitude of stars, and particularly his efforts at gauging the distances of the stars (a necessary first step in any attempt at a successful answer) are particularly noteworthy.

Henceforth the development of cosmology shifts its attention both on observational and theoretical grounds to the stars. The sun, now recognized to be no longer, as for Copernicus, Kepler and others, *the* center of the universe assumes its more modest position as one member of a vast multitude of stars. The principle question that now engages the attention of those interested in "the construction of the heavens" is the extent of the domain of stars, whether it is finite—an isolated "island" in space—and if so what its own spatial shape and structure is, whether there are other comparable systems of stars beyond the confines of this stellar system, or, finally, whether there is, on the contrary, only one limitless system of stars. The work, whether speculative, theoretical or observational, or some combination of these, of investigators like Newton, Thomas Wright, Kant, Lambert, William Herschel and others, belongs to this period in the development of the subject and constitutes the principal theme for inquiry from the eighteenth century up to the present epoch.

What Newton brought to the nascent subject of cosmology in its modern form were some brief suggestions showing the relevance of dynamical questions, specifically gravitational ones, to the construction of an acceptable view of the universe as a whole. The greatest achievements, of course, of his system of mechanics were to be found in the comprehensiveness, unity, and relative simplicity that his Laws of Motion and the Universal Law of Gravitation were able to introduce into the explanation and prediction of ordinary terrestrial phenomena as well as in the field of planetary motions. Beyond this familiar domain of classical physics, the data were simply lacking in the field of cosmology proper by which to undertake any precise quantitative analysis. Newton's sallies into this area (as in his *Letters to Bentley*), while rough and qualitative, are, nevertheless, important for being among the first efforts to apply the concepts of dynamics, as distinguished from kinematics, to understanding the distribution of matter and motion in the universe at large.

Meanwhile in another direction an important step forward in attempting to answer the question as to the structure of the universe of stars was taken by Thomas Wright of Durham in the middle of the eighteenth century. It is to him that astronomy owes a prophetically correct analysis of the structure of the Milky Way. Wright is the first to have surmised the general "disc-like" finite structure of our Galaxy of stars, of which the Milky Way is that fragmentary portion appearing to observers situated, as we are, roughly in the plane of our Galaxy. It is to Wright that Kant is indebted for the main clues in working out his own cosmology. In Kant's *Natural History and Theory of the Heavens,* one of the most interesting and important of his pre-Critical works, we find a carefully elaborated vision of an infinite universe built up of a limitless system of galaxies. It is a fascinating and brilliantly written exercise in world-building, one that seeks to combine a due regard for the ideas of Newtonian gravitational theory, atomistic materialism, and the suggestions of Wright concerning the Milky Way and other comparable systems. A similar speculative cosmology, making use of the notions of Newtonian mechanics and the idea of a hierarchy of celestial systems culminating in a single infinite universe, was worked out by Johann Lambert, a contemporary of Kant, in his *Cosmological Letters,* along lines that are in some respects strikingly similar to those of Kant.

While the efforts of speculative cosmologists like Kant and Lambert at the time they were propounded were incapable of being tested in a way demanded of a genuine scientific theory, a failure due primarily to the lack of instrumental resources and the extreme paucity of relevant observational data, a major advance in the direction of obtaining such data was soon to be initiated by the great astronomer William Herschel. By means of laborious and painstaking observations with his own specially devised telescopes, Herschel in effect opened the way to the later more refined inquires that reach their climax in present-day studies. From Herschel's day to the present, the story of observational astronomy in its relevance to the cosmologic problem is one of expanding horizons and increasing detail. The use of improved telescopes, together with the introduction of such important auxiliary instruments as the spectroscope and the photographic camera, opened up with an accelerating pace and on many fronts a domain that was only glimpsed by Herschel. That great astronomer, while primarily an observer, was also driven by a scientific curiosity to try to fathom the depths of the universe not only by his instruments but by his understanding. Throughout his long career we find him, on the basis provided by his own observational results, trying to achieve some acceptable theory of the "construction of the heavens." The paper by him reproduced below is an early example of his efforts in this direction.

NICOLAS CUSANUS:

Of Learned Ignorance

THE FACT that the ignorance which is learning has shown the truth of the foregoing doctrine will perhaps be a surprise to those who had not heard of such teaching before. By it we now know that the universe is a trinity; that there is not a being in the universe which is not a unity composed of potency, act and the movement connecting them and that none of these three is capable of absolute subsistence without the others, with the result that they are necessarily found in all things in the greatest diversity of degrees—in degrees so different that it is impossible to find in the universe two beings perfectly equal in all things. Consequently, once we have taken the different movements of the stars (orbium) into account, we see that it is impossible for the motor of the world to have the material earth, air, fire or anything else for a fixed, immovable centre. In movement there is no absolute minimum, like a fixed centre, since necessarily the minimum and the maximum are identical.

Therefore the centre and the circumference are identical. Now the world has no circumference. It would certainly have a circumference if it had a centre, in which case it would contain within itself its own beginning and end; and that would mean that there was some other thing which imposed a limit to the world—another being existing in space outside the world. All of these conclusions are false. Since, then, the world cannot be enclosed within a material circumference and centre, it is unintelligible without God as its centre and circumference. It is not infinite, yet it cannot be conceived as finite, since there are no limits within which it is enclosed.

The earth, which cannot be the centre, must in some way be in motion; in fact, its movement even must be such that it could be infinitely less. Just as the earth is not the centre of the world, so the circumference of the world is not the sphere of the fixed stars, despite the fact that by comparison the earth seems nearer the centre and heaven nearer the circumference. The earth, then, is not the centre of the eighth or any other sphere, and the appearance above the horizon of the six stars is no proof that the earth is at the

From Nicolas Cusanus, *Of Learned Ignorance*, translated by Fr. Germain Heron, Yale University Press, 1954, pp. 107–11. Reprinted by kind permission of the publishers, Yale University Press.

centre of the eighth sphere. If even at some distance from the centre it were revolving on its axis through the poles, in such a way that one part would be facing upwards towards one pole and the other part facing downwards towards the other pole, then, it is evident, that to men as distant from the poles as the horizon only half of the sphere would be visible. Further, the centre itself of the world is no more within than outside the earth; and this earth of ours has no centre nor has any other sphere a centre. Since the centre is a point equidistant from the circumference, and since it is impossible to have a sphere or circle so perfect that a more perfect one could not be given, it clearly follows that a centre could always be found that is truer and more exact than any given centre. Only in God are we able to find a centre which is with perfect precision equidistant from all points, for He alone is infinite equality. God, ever to be blessed, is, therefore, the centre of the world: He it is who is centre of the earth, of all spheres and of all things in the world; and at the same time He is the infinite circumference of all.

In addition, in the heavens there are no fixed, immovable poles, though the heaven of the fixed stars seems to move in describing circles smaller and smaller in magnitude—smaller than the equinoctial colures or the equinoctial minores: and so on for the intermediaries. Necessarily all parts of the heavens are in movement, though their movement is not uniform by comparison with the circles that the stars in their movement describe. That explains why certain stars seem to describe the maximum circle, whilst others seem to describe the minimum; but there is not a star which does not describe a circle. It is clear that a centre equidistant from the poles cannot be found, for the simple reason that there is no fixed pole on the sphere. In consequence, in the eighth sphere there is not a star which in its revolution describes the maximum circle, for that would necessarily mean that it was equidistant from the poles; but the poles do not exist. It follows also that there is no star which describes the minimum circle.

The poles, therefore, of the spheres and the centre coincide so that there is no centre but the pole, which is God ever to be blessed. It is only by reference to a fixed point—poles or centres—that we are able to detect movement, and we take such fixed points for granted in our measurements of movements. By reason of these assumptions which we make we find ourselves involved in error on all points, and, because we do not question the notions the ancients had about centres, poles and measurements, we are puzzled when we discover that the stars are not in the position indicated by their system.

It is evident from the foregoing that the earth is in movement. We have learned from the movement of a comet that the elements of air and fire are in movement and that the moon is moved less from east to west than Mercury or Venus or the sun; and so on by degrees. It follows, then, that the earth itself is moved least of all. In its movement, however, the earth does not, like a star, describe the minimum circle around a centre or pole; nor does the eighth sphere describe the maximum circle, as we have just proved.

As a keen observer, then, consider that just as the stars moved around imaginary poles on the eighth sphere, so, by imagining there is a pole where the centre is supposed to be, the earth, moon and planets move like stars around this pole at a distance from it, and with different movements. The earth, therefore, is in movement and, though as a star it may be nearer the central pole, it does not describe in its movement the minimum circle, as we have shown. In addition, though it may seem otherwise to us, neither the sun nor the moon nor the earth, nor any sphere can describe a true circle by its movement, since its movement is not on a fixed point. A given circle cannot be so true that a truer one cannot be found; and the movement of a sphere at one moment is never precisely equal to its movement at another, nor does it ever describe two circles similar and equal, even if from appearances the opposite may seem true.

If you really wish to understand something of what we have just said about the movement of the universe, you must regard the centre and the poles as coincident, using the help of your imagination as much as possible. Suppose one person were on the earth and under the arctic pole and that another were on the arctic pole; to him on the earth the pole would seem at the zenith, whereas to the person on the pole the centre would appear at the zenith. And just as the antipodes have the heavens above them as we have, so the earth would appear at the zenith to those on both poles; and no matter where a person were he would believe he was at the centre. Take, then, all these various images you have formed and merge them in one, so that the centre becomes the zenith and vice versa; and your intellect, which is aided so much by the ignorance that is learning, then sees the impossibility of comprehending the world, its movement and form, for it will appear as a wheel in a wheel, a sphere in a sphere without a centre or circumference anywhere, as has been said.

The ancient philosophers did not reach these truths we have just stated, because they lacked learned ignorance. It is now evident that this earth really moves though to us it seems stationary. In fact, it is only by reference to something fixed that we detect the movement of anything. How would a person know that a ship was in movement, if, from the ship in the middle of the river, the banks were invisible to him and he was ignorant of the fact that water flows? Therein we have the reason why every man, whether he be on earth, in the sun or on another planet, always has the impression that all other things are in movement whilst he himself is in a sort of immovable centre; he will certainly always choose poles which will vary accordingly as his place of existence is the sun, the earth, the moon, Mars, etc. In consequence, there will be a machina mundi whose centre, so to speak, is everywhere, whose circumference is nowhere, for God is its circumference and centre and He is everywhere and nowhere.

NICOLAUS COPERNICUS:

On the Revolutions
of the Heavenly Spheres

To the Most Holy Lord, Pope Paul III

The Preface of Nicolaus Copernicus to the Books of the Revolutions

I may well presume, most Holy Father, that certain people, as soon as they hear that in this book *On the Revolutions of the Spheres of the Universe* I ascribe movement to the earthly globe, will cry out that, holding such views, I should at once be hissed off the stage. For I am not so pleased with my own work that I should fail duly to weigh the judgment which others may pass thereon; and though I know that the speculations of a philosopher are far removed from the judgment of the multitude—for his aim is to seek truth in all things as far as God has permitted human reason so to do—yet I hold that opinions which are quite erroneous should be avoided.

Thinking therefore within myself that to ascribe movement to the Earth must indeed seem an absurd performance on my part to those who know that many centuries have consented to the establishment of the contrary judgment, namely that the Earth is placed immovably as the central point in the middle of the Universe, I hesitated long whether, on the one hand, I should give to the light these my Commentaries written to prove the Earth's motion, or whether, on the other hand, it were better to follow the example of the Pythagoreans and others who were wont to impart their philosophic mysteries only to intimates and friends, and then not in writing but by word of mouth, as the letter of Lysis to Hipparchus witnesses. In my judgment they did so not, as some would have it, through jealousy of sharing their doctrines, but as fearing lest these so noble and hardly won discoveries of the learned should be despised by such as either care not to study aught save for gain, or—if by the encouragement and example of others they are stimulated to philoso-

From Nicolaus Copernicus, *De Revolutionibus,* translated by John F. Dobson and Selig Brodetsky, Preface and Book I. Printed originally as *Occasional Notes Royal Astronomical Society,* No. 10, 1947. Reprinted with the kind permission of the Royal Astronomical Society.

phic liberal pursuits—yet by reason of the dulness of their wits are in the company of philosophers as drones among bees. Reflecting thus, the thought of the scorn which I had to fear on account of the novelty and incongruity of my theory, well-nigh induced me to abandon my project.

These misgivings and actual protests have been overcome by my friends. First among these was Nicolaus Schönberg, Cardinal of Capua, a man renowned in every department of learning. Next was one who loved me well, Tiedemann Giese, Bishop of Kulm, a devoted student of sacred and all other good literature, who often urged and even importuned me to publish this work which I had kept in store not for nine years only, but to a fourth period of nine years. The same request was made to me by many other eminent and learned men. They urged that I should not, on account of my fears, refuse any longer to contribute the fruits of my labours to the common advantage of those interested in mathematics. They insisted that, though my theory of the Earth's movement might at first seem strange, yet it would appear admirable and acceptable when the publication of my elucidatory comments should dispel the mists of paradox. Yielding then to their persuasion I at last permitted my friends to publish that work which they have so long demanded.

That I allow the publication of these my studies may surprise your Holiness the less in that, having been at such travail to attain them, I had already not scrupled to commit to writing my thoughts upon the motion of the Earth. How I came to dare to conceive such motion of the Earth, contrary to the received opinion of the Mathematicians and indeed contrary to the impression of the senses, is what your Holiness will rather expect to hear. So I should like your Holiness to know that I was induced to think of a method of computing the motions of the spheres by nothing else than the knowledge that the Mathematicians are inconsistent in these investigations.

For, first, the mathematicians are so unsure of the movements of the Sun and Moon that they cannot even explain or observe the constant length of the seasonal year. Secondly, in determining the motions of these and of the other five planets, they do not even use the same principles and hypotheses as in their proofs of seeming revolutions and motions. So some use only concentric circles, while others eccentrics and epicycles. Yet even by these means they do not completely attain their ends. Those who have relied on concentrics, though they have proven that some different motions can be compounded therefrom, have not thereby been able fully to establish a system which agrees with the phenomena. Those again who have devised eccentric systems, though they appear to have well-nigh established the seeming motions by calculations agreeable to their assumptions, have yet made many admissions which seem to violate the first principle of uniformity in motion. Nor have they been able thereby to discern or deduce the principal thing—namely the shape of the Universe and the unchangeable symmetry of its parts. With them it is as though an artist were to gather the hands, feet, head and other members for his images from divers models, each part excellently drawn, but not related to a single body, and since they in no way match each other, the result would be monster rather than man. So in the course of their exposition, which the mathematicians call their system (μέθοδος) we find that they have either omitted some indispensable detail or introduced something foreign and wholly irrelevant. This would of a surety not have been so had they followed fixed principles; for if their hypotheses were not misleading, all inferences based thereon might be surely verified. Though my present assertions are obscure, they will be made clear in due course.

I pondered long upon this uncertainty of mathematical tradition in establishing the motions of the system of the spheres. At last I began to chafe that philosophers could by no means agree on any one certain theory of the mechanism of the Universe, wrought for us by a supremely good and orderly Creator, though in other respects they investigated with meticulous care the minutest points relating to its orbits. I therefore took pains to read again the works of all the philosophers on whom I could lay hand to seek out whether any of them had ever supposed that the motions of the spheres were other than those demanded by the mathematical schools. I found first in Cicero that Hicetas had realized that the Earth moved. Afterwards I found in Plutarch that certain others had held the like opinion. I think fit here to add Plutarch's own words, to make them accessible to all:—

The rest hold the Earth to be stationary, but Philolaus the Pythagorean says that she moves around the (central) fire on an oblique circle like the Sun and Moon. Heraclides of Pontus and Ecphantus the Pythagorean also make the Earth to move, not indeed through space but by rotating round her own centre as a wheel on an axle from West to East.

Taking advantage of this I too began to think of the mobility of the Earth; and though the opinion seemed absurd, yet knowing now that others before me had been granted freedom to imagine such circles as they chose to explain the phenomena of the stars, I considered that I also might easily be allowed to try whether, by assuming some motion of the Earth, sounder explanations than theirs for the revolution of the celestial spheres might so be discovered.

Thus assuming motions, which in my work I ascribe to the Earth, by long and frequent observations I have at last discovered that, if the motions of the rest of the planets be brought into relation with the circulation of the Earth and be reckoned in proportion to the orbit of each planet, not only do their phenomena presently ensue, but the orders and magnitudes of all stars and spheres, nay the heavens themselves, become so bound together that nothing in any part thereof could be moved from its place without producing confusion of all the other parts and of the Universe as a whole.

In the course of the work the order which I have pursued is as here follows. In the first book I describe all positions of the spheres together with such movements as I ascribe to Earth; so that this book contains, as it were, the general system of the Universe. Afterwards, in the remaining books, I relate the motions of the other planets and all the spheres to the mobility of Earth, that we may gather thereby how far the motions and appearances of the rest of the planets and spheres may be preserved, if related to the motions of the Earth.

I doubt not that gifted and learned mathematicians will agree with me if they are willing to comprehend and appreciate, not superficially but thoroughly, according to the demands of this science, such reasoning as I bring to bear in support of my judgment. But that learned and unlearned alike may see that I shrink not from any man's criticism, it is to your Holiness rather than anyone else that I have chosen to dedicate these studies of mine, since in this remote corner of Earth in which I live you are regarded as the most eminent by virtue alike of the dignity of your Office and of your love of letters and science. You by your influence and judgment can readily hold the slanderers from biting, though the proverb hath it that there is no remedy against a sycophant's tooth. It may fall out, too, that idle babblers, ignorant of mathematics, may claim a right to pronounce a judgment on my work, by reason of a certain passage of Scripture basely twisted to suit their purpose. Should any such venture to criticize and carp at my project, I make no account of them; I consider their judgment rash, and utterly despise it. I well know that even Lactantius, a writer in other ways distinguished but in no sense a mathematician, discourses in a most childish fashion touching the shape of the Earth, ridiculing even those who have stated the Earth to be a sphere. Thus my supporters need not be amazed if some people of like sort ridicule me too.

Mathematics are for mathematicians, and they, if I be not wholly deceived, will hold that these my labours contribute somewhat even to the Commonwealth of the Church, of which your Holiness is now Prince. For not long since, under Leo X, the question of correcting the ecclesiastical calendar was debated in the Council of the Lateran. It was left undecided for the sole cause that the lengths of the years and months and the motions of the Sun and Moon were not held to have been yet determined with sufficient exactness. From that time on I have given thought to their more accurate observation, by the advice of that eminent man Paul, Lord Bishop of Sempronia, sometime in charge of that business of the calendar. What results I have achieved therein, I leave to the judgment of learned mathematicians and of your Holiness in particular. And now, not to seem to promise your Holiness more than I can perform with regard to the usefulness of the work, I pass to my appointed task.

Nicolai Copernici Revolutionum

1. That the Universe Is Spherical[1]

In the first place we must observe that the Universe is spherical. This is

---

1. This title, like that of many other chapters, is taken from the Almagest of Ptolemy. The work of Copernicus is so closely bound up with that of Ptolemy that it will be convenient here to review the history of the *Almagest*.

either because that figure is the most perfect,[2] as not being articulated[3] but whole and complete in itself; or because it is the most capacious and therefore best suited for that which is to contain and preserve all things; or again because all the perfect parts of it, namely, Sun, Moon and Stars, are so formed; or because all things tend to assume this shape, as is seen in the case of drops of water and liquid bodies in general if freely formed.[4] No one doubts that such a shape has been assigned to the heavenly bodies.

2. That the Earth Also Is Spherical[5]

The Earth also is spherical, since on all sides it inclines toward the centre. At first sight, the Earth does not appear absolutely spherical, because of the mountains and valleys; yet these make but little variation in its general roundness, as appears from what follows. As we pass from any point northward, the North Pole[6] of the daily rotation gradually rises, while the other pole sinks correspondingly and more stars near the North Pole cease to set, while certain stars in the South do not rise. Thus, *Canopus*, invisible in Italy, is visible in

The *Almagest* of Ptolemy was unknown in the earlier Middle Ages. Its first appearance in the West is in a translation made direct from the Greek in Sicily in the year 1160. Translation direct from the Greek was very unusual at the period. This translation was excessively rare and effectively without influence. About 1170 the Englishman Daniel of Morley was studying the Arabic text of Ptolemy at Toledo with the help of a native Arabic-speaking Christian, one Ibn Ghalib. Daniel tells us that he listened to Ibn Ghalib at work with the famous translator Gerard of Cremona (died 1187). Gerard's translation of the *Almagest* was completed about 1175 and was in use in the later Middle Ages. It was made from Arabic and not from Greek. The *Almagest* was again translated from Greek in the fifteenth century by George of Trebizond (1396–1486). In the same century an *Epitome*—also direct from the Greek—was commenced by George Purbach (1423–1461) and completed by his pupil Johann Müller of Königsberg, known as Regiomontanus (1436–1476). The earliest edition of this *Epitome* was printed at Venice in 1496. The first complete Latin edition is that of Liechtenstein, Venice 1515. Liechtenstein used the translation from the Arabic version. George of Trebizond's translation was printed at Venice in 1528 and the Greek text edited by Simon Grynaeus at Basel in 1538. There are many later editions of both Greek and Latin texts.

2. The conception of the sphere as the most perfect of all figures occurs in Plato's *Timaeus*. It became an Aristotelian commonplace which pervades the whole of subsequent Astronomy until Kepler. It is accepted as a matter of course by Copernicus. The *locus classicus* for the description of the spherical Universe is Aristotle's *De Coelo* I, §§ 5–12 and II, §§ 1, 4, 5, 6.

3. *Nulla indigens compagine.* The term *compago* is the usual mediaeval word for the fabric of the human body, *compagines membrorum*. The articulated human body, the *Microcosm*, is thus implicitly contrasted with the *Macrocosm*, the Universe, which is not thus articulated. The parallel between Macrocosm and Microcosm is the commonest basis of the mediaeval teaching concerning Man and the World. It is the key to mediaeval science in somewhat the same way that Evolution is the key to modern science. The work of Copernicus appeared in the same year as the great monograph on the structure of the human body by Vesalius. Thus in 1543 the axe was laid to the tree of Mediaeval Science from both sides. For Science, therefore, that year may be regarded as the opening of the modern period.

4. This idea, the comparison of the spherical world to drops of water, is taken from *Pliny* II, § 65. "As to whether there be *Antipodes* is in dispute between the learned and the vulgar. We maintain that there are men on every part of the earth. . . . If any should ask why those opposite to us do not fall off, we ask in return why those on the opposite side do not wonder that we do not fall off. . . . But what the vulgar most strenuously resist is the belief that water (which covers the surface of the earth) is forced into a rounded form. Yet nothing is more obvious. For we see everywhere that hanging drops assume the form of small globes . . . and are observed to be completely round."

5. *Almagest*, I, § 4.

6. The word *pole* in the writings of Copernicus and of his predecessors is used for the celestial pole, rather than for the pole of the Earth.

Egypt, while the last star of Eridanus, seen in Italy, is unknown in our colder zone. On the other hand, as we go southward, these stars appear higher, while those which are high for us appear lower. Further, the change in altitude of the pole is always proportional to the distance traversed on the Earth, which could not be save on a spherical figure.[7] Hence the Earth must be finite and spherical.

Furthermore, dwellers in the East do not see eclipses of the Sun and Moon which occur in the evening here, nor do they in the West see those which occur here in the morning. Yet mid-day eclipses here are seen later in the day by the eastward dwellers, earlier by the westerners. Sailors too have noted that the sea also assumes the same shape, since land invisible from the ship is often sighted from the mast-head. On the other hand, if some shining object on the mast-head be observed from the shore, it seems gradually to sink as the vessel leaves the land. It is also a sure fact that water free to flow always seeks a lower level, just as earth does, nor does the sea come higher up the shore than the convexity of the earth allows. It therefore follows that land, rising above the level of Ocean, is by so much further removed from the centre.

3. How Earth, with the Water on It, Forms One Sphere

The waters spread around the Earth form the seas and fill the lower declivities. The volume of the waters must be less than that of the Earth, else they would swallow up the land (since both, by their weight, press toward the same centre). Thus, for the safety of living things, stretches of the Earth are left uncovered, and also numerous islands widely scattered. Nay, what is a continent, and indeed the whole of the Mainland, but a vast island?

We must pass by certain Peripatetics who claim the volume of the waters to be ten times that of the earth.[8] They base themselves on a mere guess that in the transmutation of the elements, one part of earth is resolved into ten of water. They say, in fact, that the earth rises to a certain height above the water because, being full of cavities, it is not symmetrical as regards weight and therefore the centre of weight does not accord with the geometrical centre. Ignorance of geometry prevents them from seeing that the waters cannot be even seven times as great if some part of the earth is to be left dry, unless the earth, as being heavier, be quite removed from the centre of gravity to make room for the waters. For spheres are to each other as the cubes of their diameters. If, therefore, there had been seven parts of water to one of earth, the Earth's diameter could not be greater than the radius of the waters. Even less is it possible that the waters could be ten times as great as the Earth.[9]

---

7. The certainty of Copernicus that the form of the Earth is spherical and that the contour of its surface does not correspond to any other curve than that of a circle is of a piece with his general insistence on the sphere and the circle as characteristic of all cosmic form and movement.

8. This view is that of Alexander of Aphrodisias (c. 200 A.D.) whose works were very widely read in the North Italian Universities.

9. This argument requires little clarification. Assume that the land is in the form of a sphere of volume one-seventh that of the waters. Were one to plunge this sphere of land into the sphere of water and to restore the spherical form of the sphere of waters, the whole would be eight

There is, in fact, no difference between the Earth's centre of gravity and its geometric centre, since the height of the land above the Ocean does not increase continuously—for so it would utterly exclude the waters and there could be no great gulfs of seas between parts of the Mainland.[10] Further, the depth of Ocean would constantly increase from the shore outwards, and so neither island nor rock nor anything of the nature of land would be met by sailors, how far soever they ventured. Yet, we know that between the Egyptian Sea and the Arabian Gulf, well-nigh in the middle of the great land-mass, is a passage barely 15 stades wide. On the other hand, in his *Cosmography* Ptolemy would have it that the habitable land extends to the middle circle[11] with a *terra incognita* beyond where modern discovery has added Cathay and a very extensive region as far as 60° of longitude. Thus we know now that the Earth is inhabited to a greater longitude than is left for Ocean.

This will more evidently appear if we add the islands found in our own time under the Princes of Spain and Portugal, particularly America, a land named after the Captain who discovered it and, on account of its unexplored size, reckoned as another Mainland—besides many other islands hitherto unknown.[12] We thus wonder the less at the so-called Antipodes or Antichthones.[13] For geometrical argument demands that the Mainland of America on account of its position be diametrically opposite to the Ganges basin in India.

From such considerations then, it is clear that Land and Water have the same centre of gravity, which coincides with the centre of the Earth's volume. Yet since earth is the heavier, and its chasms filled with water, therefore the quantity of the water is but moderate as against earth, though, as to the surface, there may perhaps be more water. Moreover, the Earth, with the waters around

---

times as large as the sphere of land alone, and the sphere of land would in consequence *touch* the sphere of water only in the interior while the centre of the whole sphere would lie only on the circumference of the sphere of land. The sphere of land could thus no longer rise above the circumference of the sphere of water except by ceasing to touch the middle point of the whole water body.

10. The meaning is that there is not one single uniform land mass collected on one side of the Earth making it lop-sided.

11. By "middle circle" Copernicus means the 180th degree of longitude reckoning eastward from the "islands of the blessed" in the Western Ocean. The authority of Copernicus is Ptolemy. In his *Geography* VI, § 16, where the position of Serica is discussed, Ptolemy says "Serica is bounded on the east by the unknown land and is between 35 and 63 degrees broad on a meridian which has a geographical length of 180°." Ptolemy reckons geographical longitude from the "islands of the blessed" (the Canaries). Elsewhere (*Geography* I, § 12) Ptolemy says that "the length (that is measurement from West to East) of the known world from the meridian in the islands of the blessed to (the chief town of) Serica is 177¼°." The latitude of the "Metropolis of Sera" is fixed at 38° 36′ (VI, § 16).

12. The newly discovered land of America was of course first regarded as part of Cathay of which the West Indies were outlying islands. Early in the sixteenth century it became suspected that America was a separate continent. This was confirmed by Vasco Minez de Balboa (c. 1475–1517) when he first sighted the Pacific Ocean (1513), and brought out in many maps, e.g. that of Johan Schöner (1515) in which a clear differentiation is exhibited between Cathay and America. The point was proved by the voyage and circumnavigation (1519–1522) by Ferdinand Magellan (c. 1480–1521). The matter became common knowledge with the publication of Antonio Pigafelta's account of the journey in 1524.

13. The *Antichthon* is strictly speaking the *counter-earth* of the Pythagorean system of the Universe. Aristotle *De Coelo* II, § 13, 2, Cicero and later Latin writers, however, use *antichthones* as equivalent to inhabitants of the other hemisphere.

it, must have a shape conformable with its shadow. Now, at the Moon's eclipse we see a perfect arc of a circle; the Earth therefore is not flat as Empedocles and Anaxagoras would have had it, nor drum-shaped as Leucippus held, nor bowl-shaped as Heraclitus said, nor yet concave in some other way as Democritus believed; nor again cylindrical as Anaximander maintained, nor yet infinitely thick with roots extending below as Xenophanes represented; but perfectly round, as the Philosophers rightly hold.[14]

4. That the Motion of the Heavenly Bodies Is Uniform, Circular, and Perpetual, or Composed of Circular Motions

We now note that the motion of heavenly bodies is circular. Rotation is natural to a sphere and by that very act is its shape expressed. For here we deal with the simplest kind of body, wherein neither beginning nor end may be discerned nor, if it rotate ever in the same place, may the one be distinguished from the other.

Now in the multitude of heavenly bodies various motions occur. Most evident to sense is the diurnal rotation, the νυχθήμερον, as the Greeks call it, marking day and night. By this motion the whole Universe, save Earth alone, is thought to glide from East to West.[15] This is the common measure of all motions, since Time itself is numbered in days. Next we see other revolutions in contest, as it were, with this daily motion and opposing it from West to East. Such opposing motions are those of Sun and Moon and the five planets. Of these the Sun portions out the year, the Moon the month, the common measures of time. In like manner the five planets define each his own independent period.

But these bodies exhibit various differences in their motion. First their axes are not that of the diurnal rotation, but of the Zodiac, which is oblique thereto. Secondly, they do not move uniformly even in their own orbits; for are not Sun and Moon found now slower, now swifter in their courses? Further, at times the five planets become stationary at one point and another and even go backward. While the Sun ever goes forward unswerving on his own course, they wander in divers ways, straying now southward, now northward. For this reason they are named *Planets*.[16] Furthermore, sometimes they approach Earth, being then in *Perigee*, while at other times receding they are in *Apogee*.

Nevertheless, despite these irregularities, we must conclude that the motions of these bodies are ever circular or compounded of circles. For the irregularities themselves are subject to a definite law and recur at stated times, and this could not happen if the motions were not circular, for a circle alone

---

14. In mediaeval phraseology *the philosopher* is a synonym for Aristotle. The *philosophers* to whom Copernicus here refers are the followers of Aristotle, the Peripatetics.

15. *Almagest*, I, § 8.

16. Greek πλανήτης = wanderer.

can thus restore the place of a body as it was. So with the Sun which, by a compounding of circular motions, brings ever again the changing days and nights and the four seasons of the year. Now therein it must be that divers motions are conjoined, since a simple celestial body cannot move irregularly in a single orbit. For such irregularity must come of unevenness either in the moving force (whether inherent or acquired) or in the form of the revolving body. Both these alike the mind abhors regarding the most perfectly disposed bodies.[17]

It is then generally agreed that the motions of Sun, Moon and Planets do but seem irregular either by reason of the divers directions of their axes of revolution, or else by reason that Earth is not the centre of the circles in which they revolve, so that to us on Earth the displacements of these bodies when near seem greater than when they are more remote, as is shown in the *Optics*.[18] If then we consider equal arcs in the paths of the planets we find that they seem to describe differing distances in equal periods of time. It is therefore above all needful to observe carefully the relation of the Earth toward the Heavens, lest, searching out the things on high, we should pass by those nearer at hand, and mistakenly ascribe earthly qualities to heavenly bodies.

5. WHETHER CIRCULAR MOTION BELONGS TO THE EARTH; AND CONCERNING ITS POSITION

Since it has been shown that Earth is spherical, we now consider whether her motion is conformable to her shape and her position in the Universe. Without these we cannot construct a proper theory of the heavenly phenomena. Now authorities agree that Earth holds firm her place at the centre of the Universe, and they regard the contrary as unthinkable, nay as absurd. Yet if we examine more closely it will be seen that this question is not so settled, and needs wider consideration.

A seeming change of place may come of movement either of object or of observer, or again of unequal movements of the two (for between equal and parallel motions no movement is perceptible). Now it is Earth from which the rotation of the Heavens is seen. If then some motion of Earth be assumed it will be reproduced in external bodies, which will seem to move in the opposite direction.

Consider first the diurnal rotation. By it the whole Universe, save Earth alone and its contents, appears to move very swiftly. Yet grant that Earth revolves from West to East, and you will find, if you ponder it, that my con-

---

17. The passage is a remarkable illustration of the very firm hold that Aristotelian conceptions had taken. The incorruptible heavens, the necessity that all perfect movement must be in a circle, the eternal heavenly bodies as contrasted with this changeful, corruptible, temporal earth are ideas from which Copernicus, like all his contemporaries and predecessors, was quite unable to free himself.

18. For the book on *Optics* ascribed to Euclid and Theon's summary of it (edition by J. L. Heiberg, Leipzig, 1895, p. 41), see T. L. Heath, *Manual of Greek Mathematics*, p. 266, Oxford, 1931.

clusion is right. It is the vault of Heaven[19] that contains all things, and why should not motion be attributed rather to the contained than to the container, to the located than the locater? The latter view was certainly that of Heraclides[20] and Ecphantus the Pythagorean[21] and Hicetas of Syracuse (according to Cicero).[22] All of them made the Earth rotate in the midst of the Universe, believing that the Stars set owing to the Earth coming in the way, and rise again when it has passed on.

There is another difficulty, namely, the position of Earth. Nearly all have hitherto held that Earth is at the centre of the Universe. Now, grant that Earth is not at the exact centre but at a distance from it which, while small compared to the starry sphere, is yet considerable compared with the orbits of Sun and the other planets. Then calculate the consequent variations in their seeming motions, assuming these to be really uniform and about some centre other than the Earth's. One may then perhaps adduce a reasonable cause for these variable motions. And indeed since the Planets are seen at varying distances from the Earth, the centre of Earth is surely not the centre of their orbits. Nor is it certain whether the Planets move toward and away from Earth, or Earth toward and away from them. It is therefore justifiable to hold that the Earth has another motion in addition to the diurnal rotation. That the Earth, besides rotating, wanders with several motions and is indeed a Planet, is a view attributed to Philolaus the Pythagorean, no mean mathematician, and one whom Plato is said to have eagerly sought out in Italy.[23]

19. A pun is here involved which cannot be reproduced in English (caelum caelat).

20. Heracleides of Heracleia in Pontus was a pupil of Plato and his successor Speusippus and studied also with Aristotle and under the Pythagoreans. On Plato's death, being disappointed in not obtaining the headship of the Academy, he returned to his native town. He wrote many philosophical dialogues on ethical and physical topics which have now disappeared. It seems that he first taught the movement of the planets Venus and Mercury round the Sun and the movement of the Sun round the Earth. Later he showed how by the assumption of a real heliocentric system the celestial phenomena could be explained, but there is no evidence that he finally adopted this as the only possible explanation. Information about him has been collected by O. Voss, De Heracleidis vita et scriptis, Rostock, 1896.
As regards the works of Heracleides and the other Pythagoreans, the reader is referred to Sir T. L. Heath, Aristarchus of Samos, Oxford, 1913.

21. Ecphantus of Syracuse, a Pythagorean, was perhaps a pupil of Hicetas. The few references that survive concerning him are in Hippolytus and Aëtius (Diels, Vorsokratiker, I, p. 340, 1920), and all the little that is known of him has been put together by M. Wellmann in Pauly-Wissowa's Real-Enkyklopädie.

22. The reference to Cicero is to the Quaestiones Academicae, IV, § 29, where we read "Hicetas the Syracusan, so Theophrastus says, regarded the Heavens, the Sun, the Moon and the Stars, in fine all outside the Earth, as standing still, and that nothing in the world moves except the Earth which turns and revolves on its axis with great rapidity and produces exactly the same appearances as if the entire Heavens turned around an immobile Earth. Some think that Plato in his Timaeus expresses the same opinion but in more obscure terms."
This Hicetas was a Pythagorean the sole remains of whose works is this sentence and a reference to Aëtius (see Diels, Vorsokratiker, I, p. 340, 1922). All that we know of Hicetas has been put together by M. Wellmann in Pauly-Wissowa and amounts to hardly more than we have here.

23. The statement that Plato visited Philolaus in Italy rests on the unsupported statement of Diogenes Laertius. It is probable that Philolaus was an older contemporary of Socrates. Substantial fragments of his works have come down to us (see H. Diels, Vorsokratiker, I, p. 301, 1922). Copernicus had probably gained his knowledge of Philolaus from Plato and Stobaeus.

Many, however, have thought that Earth could be shown by geometry to be at the centre and like a mere point in the vast Heavens. They have thought too that Earth, as centre, ever remains unmoved, since if the whole system move the centre must remain at rest, and the parts nearest the centre must move most slowly.

6. Of the Vastness of the Heavens Compared with the Size of the Earth[24]

That the size of Earth is insignificant in comparison with the Heavens, may be inferred thus.

The bounding Circles (interpreting the Greek word *horizons*) bisect the Celestial Sphere. This could not be if the size of the Earth or its distance from the centre were considerable compared with the Heavens—for a circle to bisect a sphere must pass through its centre and be in fact a "great circle." Let the circle ABCD represent the celestial horizon, and E that point of the Earth from which we observe. The "horizon" or boundary line between bodies visible and bodies invisible has its centre at this point. Sup-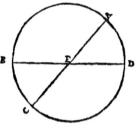pose that from point E we observe with Dioptra or Astrolabe or Chorobates[25] the first point of the sign Cancer rising at C and at the same moment the first point of Capricorn setting at A. AEC, since it is observed as a straight line through the Dioptra, is a diameter of the Ecliptic, for six Zodiacal Signs form a semicircle and its centre E coincides with that of the horizon. Next, suppose that after some time the first point of Capricorn rises at B; then Cancer will be seen setting at D, and BED will be a straight line, again a diameter of the ecliptic. Hence, it is clear that E, the point of intersection of the two lines, is the centre of the horizon. Therefore the horizon always bisects the ecliptic, which is a great circle on the sphere. But a circle that bisects a great circle must itself be a great circle. Therefore the horizon is a great circle and its centre is that of the ecliptic.

It is true that a line from the surface of Earth cannot coincide with the one from its centre. Yet owing to their immense length compared to the size of Earth these lines are practically parallel. Moreover, owing to the great distance of their meeting point they are practically one line—for the distance between them is immeasurably small in comparison with their length—as is shown in the *Optics*.[26] It therefore follows that the Heavens are immeasurable in com-

---

24. *Almagest*, I, § 6.
25. The Dioptra was known to Copernicus from works circulating at the time in the names of Euclid and Heron of Alexandria. By *Horoscopium* he means not the familiar plan of astrologers but the instrument or instruments used to obtain it, *i.e.*, an astrolabe or armillary sphere; both these instruments had been familiar since the earlier Middle Ages. The Chorobates is described in *Vitruvius*, VIII, § 5, whose work was accessible to Copernicus.
26. *Cf.* note 18.

parison with the Earth. Thus the Earth appears as a mere point compared to the Heavens, as a finite thing to the infinite.[27]

Yet it does not follow that the Earth must be at rest at the centre of the Universe. Should we not be more surprised if the vast Universe revolved in twenty-four hours, than that little Earth should do so? For the idea that the centre is at rest and the parts nearest it move least does not imply that Earth remains still. It is merely as one should say that the Heavens revolve, but the poles are still, and the parts nearest them move least (as *Cynosura* moves slower than *Aquila* or *Procyon* because, being nearer the pole, it describes a smaller circle). These all belong to the same sphere, whose motion becomes zero at the axis. Such motion does not admit that all the parts have the same rate of motion, since the revolution of the whole brings back each point to the original position in the same time, though the distances moved are unequal.

So too, it may be said, Earth, as part of the celestial sphere, shares in the motion thereof, though being at the centre she moves but little. Being herself a body and not a mere point, she will therefore move through the same angle as the Heavens but with a smaller radius in any given period of time. The falsity of this is clear, for if true it would always be mid-day in one place and midnight in another, and the daily phenomena of rising and setting could not occur, for the motion of the whole and the part are one and inseparable. A quite different theory is required to explain the various motions observed, namely that bodies moving in smaller paths revolve more quickly than those moving in larger paths. Thus Saturn, most distant of the Planets, revolves in 30 years, and Moon, nearest the Earth, compasses her circuit in a month. Lastly, then, the Earth must be taken to go round in the course of a day and a night, and so doubt is again cast on the diurnal rotation of the Heavens.

Besides we have not yet fixed the exact position of the Earth, which as shown above, is quite uncertain. For what was proved is only the vast size of the Heavens compared with the Earth, but how far this immensity extends is quite unknown.

7. Why the Ancients Believed that the Earth Is at Rest, like a Centre, in the Middle of the Universe[28]

The ancient Philosophers tried by divers other methods to prove Earth fixed in the midst of the Universe. The most powerful argument was drawn from the doctrine of the heavy and the light. For, they argue, Earth is the heaviest element, and all things of weight move towards it, tending to its centre. Hence since the Earth is spherical, and heavy things more vertically to it, they would all rush together to the centre if not stopped at the surface. Now those things which move towards the centre must, on reaching it, remain at rest. Much more then will the whole Earth remain at rest at the centre of

---

27. The passage resembles one in the work of Archimedes *The sand-reckoner.* See T. L. Heath, *Works of Archimedes,* p. 222, Cambridge, 1897.

28. *Almagest,* I, § 7.

the Universe. Receiving all falling bodies, it will remain immovable by its own weight.[29]

Another argument is based on the supposed nature of motion. Aristotle says that the motion of a single and simple body is simple. A simple motion may be either straight, or circular. Again a straight motion may be either up or down. So every simple motion must be either toward the centre, namely downward, or away from the centre, namely upward, or round the centre, namely circular. Now it is a property only of the heavy elements earth and water to move downward, that is to seek the centre. But the light elements air and fire move upward away from the centre. Therefore we must ascribe rectilinear motion to these four elements. The celestial bodies however have circular motion. So far Aristotle.[30]

If then, says Ptolemy, Earth moves at least with a diurnal rotation, the result must be the reverse of that described above. For the motion must be of excessive rapidity, since in 24 hours it must impart a complete rotation to the Earth. Now things rotating very rapidly resist cohesion or, if united, are apt to disperse, unless firmly held together. Ptolemy therefore says that Earth would have been dissipated long ago, and (which is the height of absurdity) would have destroyed the Heavens themselves; and certainly all living creatures and other heavy bodies free to move could not have remained on its surface, but must have been shaken off. Neither could falling objects reach their appointed place vertically beneath, since in the meantime the Earth would have moved swiftly from under them. Moreover clouds and everything in the air would continually move westward.[31]

8. The Insufficiency of These Arguments, and Their Refutation

For these and like reasons, they say that Earth surely rests at the centre of the Universe. Now if one should say that the Earth *moves,* that is as much as to say that the motion is natural, not forced; and things which happen according to nature produce the opposite effects to those due to force. Things subjected to any force, gradual or sudden, must be disintegrated, and cannot long exist. But natural processes being adapted to their purpose work smoothly.

Idle therefore is the fear of Ptolemy that Earth and all thereon would be disintegrated by a natural rotation, a thing far different from an artificial act. Should he not fear even more for the Universe, whose motion must be as much more rapid as the Heavens are greater than the Earth? Have the Heavens become so vast because of the centrifugal force of their violent motion, and would they collapse if they stood still? If this were so the Heavens must be of infinite size. For the more they expand by the centrifugal force of their motion, the more rapid will become the motion because of the ever increasing distance to be traversed in 24 hours. And in turn, as the motion waxes, must the im-

---

29. The argument is drawn from Aristotle *De Coelo,* II, § 14; 296 *b.*
30. Aristotle, *De Coelo,* I, §§ 2–3, III, §§ 3–5.
31. *Almagest,* I, § 5.

mensity of the Heavens wax. Thus velocity and size would increase each the
other to infinity—and as the infinite can neither be traversed nor moved, the
Heavens must stand still![32]

They say too that outside the Heavens is no body, no space, nay not even
void, in fact absolutely nothing, and therefore no room for the Heavens to
expand.[33] Yet surely it is strange that something can be held by nothing.
Perhaps indeed it will be easier to understand this nothingness outside the
Heavens if we assume them to be infinite, and bounded internally only by
their concavity, so that everything, however great, is contained in them, while
the Heavens remain immovable. For the fact that it moves is the principal
argument by which men have inferred that the Universe is finite.

Let us then leave to Physicists[34] the question whether the Universe be
finite or no, holding only to this that Earth is finite and spherical. Why then
hesitate to grant Earth that power of motion natural to its shape, rather than
suppose a gliding round of the whole Universe, whose limits are unknown
and unknowable? And why not grant that the diurnal rotation is only apparent in the Heavens but real in the Earth? It is but as the saying of Aeneas
in Virgil—"We sail forth from the harbour, and lands and cities retire."[35] As
the ship floats along in the calm, all external things seem to have the motion
that is really that of the ship, while those within the ship feel that they and
all its contents are at rest.

It may be asked what of the clouds and other objects suspended in the air,
or sinking and rising in it? Surely not only the Earth, with the water on it,
moves thus, but also a quantity of air and all things so associated with the
Earth. Perhaps the contiguous air contains an admixture of earthy or watery
matter and so follows the same natural law as the Earth, or perhaps the air
acquires motion from the perpetually rotating Earth by propinquity and
absence of resistance. So the Greeks thought that the higher regions of the air
follow the celestial motion, as suggested by those swiftly moving bodies, the
"Comets," or "Pogoniae" as they called them,[36] for whose origin they assign
this region, for these bodies rise and set just like other stars. We observe that
because of the great distance from the Earth that part of the air is deprived
of terrestrial motion, while the air nearest Earth, with the objects suspended in
it, will be stationary, unless disturbed by wind or other impulse which moves
them this way or that—for a wind in the air is as a current in the sea.

We must admit the possibility of a double motion of objects which fall and
rise in the Universe, namely the resultant of rectilinear and circular motion.

---

32. Aristotle, *Phys. Aus.* III, § 4. "First we must determine in how many ways the word
infinite is employed. The first meaning is 'that which cannot be traversed.'" See also *De Coelo,*
I, § 5, *Phys. Aus.* IV, § 4 and especially *De Coelo,* I, § 7.
33. Aristotle, *De Coelo,* I, § 9.
34. By the *Physicists* is meant the commentators on the *Physica* of Aristotle, a book very
widely read in the North Italian schools.
35. *Aeneid,* III, 72.
36. Pogoniae = bearded. Comets are spoken of as *bearded stars* in Aristotle's *Meteorologica,* I,
§ 7, 4 and elsewhere.

Thus heavy falling objects, being specially earthy, must doubtless retain the nature of the whole to which they belong. So also there are objects which by their fiery force are carried up into the higher regions. This terrestrial fire is nourished particularly by earthy matter, and flame is simply burning smoke. Now it is a property of fire to expand that which it attacks, and this so violently that it cannot in any wise be restrained from breaking its prison and fulfilling its end. The motion is one of extension from the centre outward, and consequently any earthy parts set on fire are carried to the upper region.[37]

That the motion of a simple body must be simple is true then primarily of circular motion, and only so long as the simple body rests in its own natural place and state. In that state no motion save circular is possible, for such motion is wholly self-contained and similar to being at rest. But if objects move or are moved from their natural place rectilinear motion supervenes. Now it is inconsistent with the whole order and form of the Universe that it should be outside its own place. Therefore there is no rectilinear motion save of objects out of their right place, nor is such motion natural to perfect objects, since they would be separated from the whole to which they belong and thus would destroy its unity. Moreover, even apart from circular motion, things moving up and down do not move simply and uniformly; for they cannot avoid the influence of their lightness or weight. Thus all things which fall begin by moving slowly, but their speed is accelerated as they go. On the other hand earthly fire (the only kind we can observe) when carried aloft loses energy, owing to the influence of the earthy matter.

A circular motion must be uniform for it has a never failing cause of motion; but other motions have always a retarding factor, so that bodies having reached their natural place cease to be either heavy or light, and their motion too ceases.

Circular motion then is of things as a whole, parts may possess rectilinear motion as well. Circular motion, therefore, may be combined with the rectilinear—just as a creature may be at once animal and horse. Aristotle's method of dividing simple motion into three classes, from the centre, to the centre, and round the centre, is thus merely abstract reasoning; just as we form separate conceptions of a line, a point, and a surface, though one cannot exist without another, and none can exist without substance.

Further, we conceive immobility to be nobler and more divine than change and inconstancy, which latter is thus more appropriate to Earth than to the Universe. Would it not then seem absurd to ascribe motion to that which contains or locates, and not rather to that contained and located, namely the Earth?

Lastly, since the planets approach and recede from the Earth, both their

37. Comets, falling stars, and certain other celestial phenomena have, according to Aristotle, a less orderly arrangement than the events in what he regarded as the more distant heavens. They thus partook of a terrestrial nature. Aristotle assumes the existence of exhalations from the Earth which become ignited in consequence of the motions of the upper regions of the Cosmos. *Meteorologica*, I, § 4–5.

motion round the centre, which is held to be the Earth, and also their motion
outward and inward are the motion of one body. Therefore we must accept
this motion round the centre in a more general sense, and must be satisfied
provided that every motion has a proper centre. From all these considerations
it is more probable that the Earth moves than that it remains at rest. This is
especially the case with the diurnal rotation, as being particularly a property
of the Earth.

9. Whether More Than One Motion Can Be Attributed to the Earth, and of the Centre of the Universe

Since then there is no reason why the Earth should not possess the power
of motion, we must consider whether in fact it has more motions than one, so
as to be reckoned as a Planet.

That Earth is not the centre of all revolutions is proved by the apparently
irregular motions of the planets and the variations in their distances from the
Earth. These would be unintelligible if they moved in circles concentric with
Earth. Since, therefore, there are more centres than one, we may discuss
whether the centre of the Universe is or is not the Earth's centre of gravity.

Now it seems to me gravity is but a natural inclination, bestowed on the
parts of bodies by the Creator so as to combine the parts in the form of a sphere
and thus contribute to their unity and integrity. And we may believe this
property present even in the Sun, Moon and Planets, so that thereby they
retain their spherical form notwithstanding their various paths.[38] If, therefore,
the Earth also has other motions, these must necessarily resemble the many
outside motions having a yearly period. For if we transfer the motion of the
Sun to the Earth, taking the Sun to be at rest, then morning and evening
risings and settings of Stars will be unaffected, while the stationary points,
retrogressions, and progressions of the Planets are due not to their own proper
motions, but to that of the Earth, which they reflect. Finally we shall place
the Sun himself at the centre of the Universe. All this is suggested by the
systematic procession of events and the harmony of the whole Universe, if only
we face the facts, as they say, "with both eyes open."

10. Of the Order of the Heavenly Bodies

No one doubts that the Sphere of the Fixed Stars is the most distant of
visible things. As for the planets, the early Philosophers were inclined to

---

38. On this striking passage, Alexander von Humboldt remarks that "even the idea of universal
gravitation or attraction toward the Sun as the centre of the world seems to have hovered before
the mind of this great man." Yet the analogy to the Newtonian view, if it exists, is very distant.
Copernicus presents us only with the activity of the parts of a single world body and has nothing
to say as to the relation of the separate bodies with one another on all sides. Nor is his gravity an
essential property of bodies but is present because they are not in the places to which they
naturally belong.

believe that they form a series in order of magnitude of their orbits. They adduce the fact that of objects moving with equal speed, those further distant seem to move more slowly (as is proved in Euclid's *Optics*).[39] They think that the Moon describes her path in the shortest time because, being nearest to the Earth, she revolves in the smallest circle. Furthest they place Saturn, who in the longest time describes the greatest orbit. Nearer than his is Jupiter, and then Mars.

Opinions differ as to Venus and Mercury which, unlike the others, do not altogether leave the Sun. Some place them beyond the Sun, as Plato in his *Timaeus;*[40] others nearer than the Sun, as Ptolemy[41] and many of the moderns.[42] Alpetragius makes Venus nearer and Mercury further than the Sun.[43] If we agree with Plato in thinking that the planets are themselves dark bodies that do but reflect light from the Sun, it must follow, that if nearer than the Sun, on account of their proximity to him they would appear as half or partial circles; for they would generally reflect such light as they receive, upwards, that is toward the Sun, as with the waxing or waning Moon. Some think that since no eclipse even proportional to their size is ever caused by these planets they can never be between us and the Sun.

On the other hand, those who place Venus and Mercury nearer than the Sun adduce in support the great distance which they posit between Sun and Moon. For the maximum distance of Moon from Earth, namely 64 1/6 times Earth's radius, they calculate as about 1/18 of the minimum distance of the Sun from Earth, which is 1160 times Earth's radius. So the distance between the Sun and the Moon is 1096 such units.[44] So vast a space must not remain empty. By calculating the widths of the paths of these planets from their greatest and least distances from the Earth they find that the sum of the widths is approximately the same as this whole distance. Thus the perigee of Mercury comes immediately beyond the apogee of the Moon and the apogee of Mercury is followed by the perigee of Venus, who, finally, at her apogee practically reaches

---

39. *Optics*, § 56.

40. The order given in the *Timaeus* is Moon, Sun, Venus, Mercury, Mars, Jupiter, Saturn.

41. *Almagest*, IX, § 1.

42. Among the most widely read "moderns" who took this view was the Arabian Astronomer Alfraganus (Ahmed ben Muhammed ben Ketu al Fagani (d. *c.* 880), whose works had been rendered into Latin by Gerard of Cremona (died 1187) by the Jew Johannes Hispalensis (Avendeath *c.* 1150) by Hugo Sanctallensis (XIVth century) and Bencivenni Zucchero (1313). The work was edited by Melanchthon from the literary remains of Regiomontanus at Nuremberg in 1537.

43. Alpetragius is the Spanish Arab Nured-din el Betrugi (fl. *c.* 1180) whose *Liber astronomiae* was translated in 1217 by Michael Scot (d. *c.* 1235) but never printed. It was translated into Hebrew by Moses ben Tibbon in 1259 and retranslated from Hebrew into Latin by the Jew Kolonymus ben David in 1529. This work, to which Copernicus here doubtless refers, was printed at Venice in 1531 along with Sacro Bosco with the following title *Alpetragii Arabis Theorica planetarum physicis comm. probata nuperrime ad latinos translata a Calo Calonymo hebraeo Neapolitano.*

44. This computation is attributed to Eratosthenes (276–194 B.C.) by Ptolemy *op. cit.* I, § 12, from whom Copernicus must have taken it, since the works of Eratosthenes were not available to him.

the perigee of the Sun.[45] For they estimate that the difference between the greatest and least distances of Mercury is nearly 177½ of the aforesaid units, and that the remaining space is very nearly filled up by the difference between the maximum and minimum distances of Venus, reckoned at 910 units.

They therefore deny that the planets are opaque like the Moon, but think that they either shine by their own light or that their bodies are completely pervaded by the light of the Sun. They also claim that the Sun is not obstructed by them for they are very rarely interposed between our eyes and the Sun since they usually differ from him in latitude. They are small, too, compared with the Sun. According to Albategni Aratensis[46] even Venus, which is greater than Mercury, can scarcely cover a hundredth part of the Sun. He estimates the Sun's diameter to be ten times that of Venus; and, therefore, so small a spot to be almost invisible in so powerful a light. Averroes indeed, in his Paraphrase of Ptolemy,[47] records that he saw a kind of black spot when in-

---

45. The idea was that the orbit of each planet was confined within two spheres, through the apogee and perigee respectively, and that the farther sphere of the planet coincided with the nearer sphere of the next one, thus:—

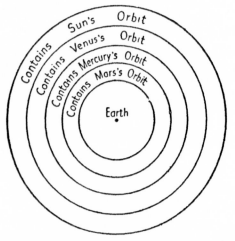

46. By Albategni Copernicus means Muhammed ben Sabir ben Sinan el Battani (died c. 919). Battan is a small place in the Hauran (Artensis) where he was born. Battani, Albattani, Albategni as he was variously known to the West, wrote a work *De motu stellarum* which was known in the Middle Ages by a Latin translation prepared by Plato of Tivoli about 1130. About the same date Robert of Retines translated into Latin the tables which accompanied this work. The translations were printed in 1537 at Nürnberg with the work of Regiomontanus on Alfraganus (see note 42) under the title *Mahometis Albatenii de scientias tellarum liber, cum aliquot additionibus Ioannis Regiomantani ex bibliotheca Vaticana transcriptus.*

47. Averroes is the mediaeval Latin form of the name of the heretical Spanish writer Muhammed ibn Ahmed ibn Muhammed ibn Roschd (1126–1198). Averroes takes a very important part in the history of mediaeval philosophy by reason of his commentaries on the works of Aristotle which profoundly influenced the North Italian schools and especially Padua. Averroes wrote little on Astronomy, among his few works on this subject being an Epitome of the *Almagest.* This work is not known in Latin but was translated into Hebrew at Naples by Jacob Anatoli in 1231. It is not clear how Copernicus obtained access to this document. It may be that he is not quoting Averroes direct but from another writer.

vestigating the numerical relations between the Sun and Mercury. This is the evidence that these two planets are nearer than the Sun.

But this reasoning is weak and uncertain. Whereas the least distance of the Moon is 38 times Earth's radius, according to Ptolemy, but, according to a truer estimate, more than 52 (as will be shown later) yet we are not aware of anything in all that space except air, and, if you will, the so called "fiery element." Besides, the diameter of the orbit of Venus, by which she passes to a distance of 45 degrees more or less on either side of the Sun, must be six times the distance from the Earth's centre to her perigee, as will also be shown later. What then will they say is contained in the whole of that space, which is so much bigger than that which could contain the Earth, the Air, the Aether, the Moon and Mercury, in addition to the space that the huge epicycle of Venus would occupy if it revolved round the resting Earth?

Unconvincing too is Ptolemy's proof that the Sun moves between those bodies that do and those that do not recede from him completely. Consideration of the case of the Moon, which does so recede, exposes its falseness. Again, what cause can be alleged, by those who place Venus nearer than the Sun, and Mercury next, or in some other order? Why should not these planets also follow separate paths, distinct from that of the Sun, as do the other planets? and this might be said even if their relative swiftness and slowness does not belie their alleged order. Either then the Earth cannot be the centre to which the order of the planets and their orbits is related, or certainly their relative order is not observed, nor does it appear why a higher position should be assigned to Saturn than to Jupiter, or any other planet.

Therefore I think we must seriously consider the ingenious view held by Martianus Capella the author of the *Encyclopaedia*[48] and certain other Latins, that Venus and Mercury do not go round the Earth like the other planets but run their courses with the Sun as centre, and so do not depart from him further than the size of their orbits allows. What else can they mean than that the centre of these orbits is near the Sun? So certainly the orbit of Mercury must be within that of Venus, which, it is agreed, is more than twice as great.

We may now extend this hypothesis to bring Saturn, Jupiter and Mars also into relation with this centre, making their orbits great enough to contain those of Venus and Mercury and the Earth; and their proportional motions according to the Table demonstrate this.[49] These outer planets are always nearer to the Earth about the time of their evening rising, that is, when they are in opposition to the Sun, and the Earth between them and the Sun. They are

---

48. Martianus Capella was a native of Madaura in Africa, who practised as a lawyer in Carthage about the beginning of the fifth century. His work *On the marriage of Philology and Mercury and on the Seven Liberal Arts* is a ridiculously strained and heavy allegory in difficult Latin treating of the nature and extent of human knowledge. The work was highly regarded in the Middle Ages during which the "Seven Liberal Arts" Grammar, Dialectic, Rhetoric, Geometry, Arithmetic, Astronomy and Music formed the basis of the Academic discipline. The work was frequently printed in the early sixteenth century beginning with 1499. The passage to which Copernicus refers is in Book VIII, Eyssenliant's edition, Leipzig, 1866, p. 317, line 14ff.

49. The Table is appended to Book V.

more distant from the Earth at the time of their evening setting, when they
are in conjunction with the Sun and the Sun between them and the Earth.
These indications prove that their centre pertains rather to the Sun than to the
Earth, and that this is the same centre as that to which the revolutions of Venus
and Mercury are related.

But since all these have one centre it is necessary that the space between
the orbit Venus and the orbit of Mars must also be viewed as a Sphere concentric with the others, capable of receiving the Earth with her satellite the
Moon and whatever is contained within the Sphere of the Moon—for we must
not separate the Moon from the Earth, the former being beyond all doubt
nearest to the latter, especially as in that space we find suitable and ample
room for the Moon.

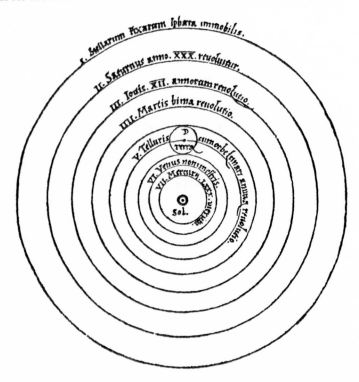

We therefore assert that the centre of the Earth, carrying the Moon's path,
passes in a great orbit among the other planets in an annual revolution round
the Sun; that near the Sun is the centre of the Universe; and that whereas the
Sun is at rest, any apparent motion of the Sun can be better explained by
motion of the Earth. Yet so great is the Universe that though the distance of
the Earth from the Sun is not insignificant compared with the size of any

other planetary path, in accordance with the ratios of their sizes, it is insignificant compared with the distance of the Sphere of the Fixed Stars.

I think it is easier to believe this than to confuse the issue by assuming a vast number of Spheres, which those who keep Earth at the centre must do. We thus rather follow Nature, who producing nothing vain or superfluous often prefers to endow one cause with many effects. Though these views are difficult, contrary to expectation, and certainly unusual, yet in the sequel we shall, God willing, make them abundantly clear at least to mathematicians.

Given the above view—and there is none more reasonable—that the periodic times are proportional to the sizes of the orbits, then the order of the Spheres, beginning from the most distant, is as follows. Most distant of all is the Sphere of the Fixed Stars, containing all things, and being therefore itself immovable. It represents that to which the motion and position of all the other bodies must be referred. Some hold that it too changes in some way,[50] but we shall assign another reason for this apparent change, as will appear in the account of the Earth's motion.[51] Next is the planet Saturn, revolving in 30 years. Next comes Jupiter, moving in a 12 year circuit: then Mars, who goes round in 2 years. The fourth place is held by the annual revolution in which the Earth is contained, together with the orbit of the Moon as on an epicycle. Venus, whose period is 9 months, is in the fifth place, and sixth is Mercury, who goes round in the space of 80 days.

In the middle of all sits Sun enthroned. In this most beautiful temple could we place this luminary in any better position from which he can illuminate the whole at once? He is rightly called the Lamp, the Mind, the Ruler of the Universe; Hermes Trismegistus names him the Visible God,[52] Sophocles' Electra calls him the All-seeing.[53] So the Sun sits as upon a royal throne ruling his children the planets which circle round him. The Earth has the Moon at her service. As Aristotle says, in his de Animalibus, the Moon has the closest relationship with the Earth.[54] Meanwhile the Earth conceives by the Sun, and becomes pregnant with an annual rebirth.

So we find underlying this ordination an admirable symmetry in the Universe, and a clear bond of harmony in the motion and magnitude of the orbits such as can be discovered in no other wise. For here we may observe why the progression and retrogression appear greater for Jupiter than Saturn, and less than for Mars, but again greater for Venus than for Mercury; and why such oscillation appears more frequently in Saturn than in Jupiter, but less frequently in Mars and Venus than in Mercury; moreover why Saturn, Jupiter and Mars are nearer to the Earth at opposition to the Sun than when they are

--- - -

50. This refers to the Precession of the Equinoxes.
51. See Book III.
52. Cf. Hermetica, Vol. I, Bk. V, 83, p. 159 (Ed. W. Scott, Oxford 1924). The Hermetic epistles were available to Copernicus, having been edited by Marsilius Ficinus, Treviso, 1472 and J. Schoeffer, 1503.
53. Electra, 826–832 is the nearest approach to the reference of Copernicus.
54. Copernicus perhaps means De generatione animalium, IV, § 10.

lost in or emerge from the Sun's rays.[55] Particularly Mars, when he shines all
night, appears to rival Jupiter in magnitude, being only distinguishable by his
ruddy colour; otherwise he is scarce equal to a star of the second magnitude,
and can be recognised only when his movements are carefully followed. All
these phenomena proceed from the same cause, namely Earth's motion.

That there are no such phenomena for the fixed stars proves their immeasurable distance, compared to which even the size of the Earth's orbit is
negligible and the parallactic effect unnoticeable. For every visible object has a
certain distance beyond which it can no more be seen (as is proved in the
Optics).[56] The twinkling of the stars, also, shows that there is still a vast distance between the furthest of the planets, Saturn, and the Sphere of the Fixed
Stars, and it is chiefly by this indication that they are distinguished from the
planets. Further, there must necessarily be a great difference between moving
and non-moving bodies. So great is this divine work of the Great and Noble
Creator!

11. Explanation of the Threefold Motion of the Earth

Since then planets agree in witnessing to the possibility that Earth moves,
we shall now briefly discuss the motion itself, in so far as the phenomena can
be explained by this hypothesis. This motion we must take to be threefold.
The first defines the Greek *nychthēmerinon*, the cycle of night and day. It is
produced by the rotation of the Earth on its axis from West to East, corresponding to the opposite motion by which the Universe appears to move round
the equinoctial circle, that is the equator, which some call the "equidial" circle,
translating the Greek expression *isēmerinos*. The second is the annual revolution of the centre of the Earth, together with all things on the Earth. This
describes the ecliptic round the Sun, also from West to East, that is, backwards,[57] between the orbits of Venus and Mars. So it comes about that the Sun
himself seems to traverse the ecliptic with a similar motion. For instance, when
the centre of the Earth passes over Capricorn, as seen from the Sun, the Sun
appears to pass over Cancer as seen from the Earth; but seen from Aquarius,
he would seem to pass over Leo, and so on. The equator and Earth's axis are
variably inclined[58] to this circle, which passes through the middle of the Zodiac,
and to its plane, since if they were fixed and followed simply the motion of the
Earth's centre there would be no inequality of days and nights.[59] Then there
is a third motion, of declination, which is also an annual revolution, but forwards, that is, tending in opposition to the motion of the Earth's centre; and
thus, as they are nearly equal and opposite, it comes about that the axis of the

---

55. *Acronyct* when the planet rises at sunset and sets at sunrise, being visible all night (Greek
"high night").
56. *Cf.* note 18.
57. Backwards, that is, compared to the diurnal rotation of the stars.
58. *Convertibilem habere inclinationem.*
59. The text here repeats itself by saying that the season would remain unchanged.

Earth, and its greatest parallel, the equator, point in an almost constant direction, as if they were fixed.[60] But meantime the Sun is seen to move along the oblique direction of the Ecliptic with that motion which is really due to the centre of the Earth (just as if the Earth were the centre of the Universe, remembering that we see the line joining Sun and Earth projected on the Sphere of the Fixed Stars).

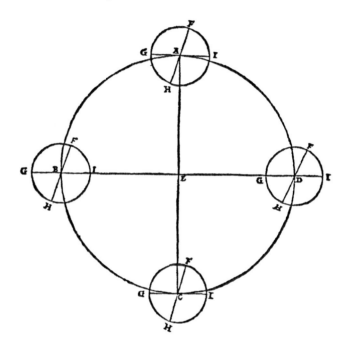

To express it graphically, draw a circle ABCD to represent the annual path of Earth's centre in the plane of the Ecliptic. Let E near its centre be the Sun. Divide this circle into four equal parts by the diameters AEC and BED. Let the first point of Cancer be at A, of Libra at B, of Capricorn at C and of Aries at D. Now let the centre of the Earth be first at A and round it draw the terrestrial Equator FGHI. This circle FGHI however is not in the same plane as the Ecliptic but its diameter GAI is the line of intersection with the ecliptic. Draw the diameter FAH, at right angles to GAI, and let F be the point of the greatest declination to the South, H to the North. This being so the inhabitants of the Earth will see the Sun near the centre E at its winter solstice in Capricorn, owing to the turning towards the Sun of the point of greatest Northern declination H. Hence in the diurnal rotation the inclination of the

---

60. The "motion of declination" as Copernicus calls it is his own discovery and a matter in which he had no forerunner. The idea follows on the conception of the motion of the Earth as related to the natural attraction of the Sun.

equator to AE makes the Sun move along the tropic of Capricorn, which is distant from the equator by an angle equal to EAH.

Now let the centre of the Earth travel forwards and let F, the point of greatest declination, move to the same extent backwards until both have completed quadrants of their circles at B. During this time the angle EAI remains always equal to the angle AEB, on account of the equality of the motions. The diameters FAH, FBH and GAI, GBI are also always parallel each to each, and the Equator remains parallel to itself. These parallel lines appear coincident in the immensity of the Heavens as has often been mentioned. Therefore, from the first point of Libra, E will appear to be in Aries, and the intersection of the planes will be the line GBIE, so that the diurnal rotation will give no declination, and all motion of the Sun will be lateral [in the plane of the Ecliptic]. The Sun is now at the vernal equinox. Further, suppose that the centre of the Earth continues its course. When it has completed a semi-circle at C, the Sun will appear to be entering Cancer. F, the point of greatest southern declination of the Equator, is now turned towards the Sun, and he will appear to be running along the Tropic of Cancer, distant from the Equator by an angle equal to ECF. Again, when F has turned through its third quadrant, the line of intersection GI will once more fall along the line ED, and from this position the Sun will be seen in Libra at the autumnal equinox. As the process continues and HF gradually turns towards the Sun, it will produce a return of the same phenomena as we observed at the starting-point.

Partes Boreæ.

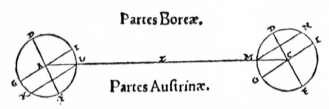

Partes Auſtrinæ.

We can explain it otherwise as follows. Take the diameter AEC in the plane of the paper. AEC is the line of intersection by this plane of a circle perpendicular to it. At points A and C, that is at Cancer and Capricorn respectively, describe in this plane a circle of longitude of the Earth DFGI. Let DF be the axis of the Earth, D the North Pole, F the South, and GI a diameter of the equator. Since then F turns towards the Sun at E, and the northern inclination of the Equator is the angle IAE, the rotation round the axis will describe a parallel south of the equator with diameter KL and at a distance from the equator equal to LI, the apparent distance from the equator of the Sun in Capricorn. Or better, by this rotation round the axis the line of sight AE describes a conical surface, with vertex at Earth's centre and as base a circle parallel to the equator. At the opposite point C the same phenomena occur, but conversely. Thus the contrary effects of the two motions, that of the centre and that of declination, constrain the axis of the Earth to remain in a constant direction, and produce all the phenomena of Solar motions.

We were saying that the annual revolution of the centre and of declination were *almost* equal. If they tallied exactly the equinoctial and solstitial points and the whole obliquity of the Ecliptic with reference to the Sphere of the Fixed Stars would be unchangeable. There is, however, a slight discrepancy, which has only become apparent as it accumulated in the course of ages. Between Ptolemy's time and ours it has reached nearly 21°, the amount by which the equinoxes have precessed. For this reason some have thought that the Sphere of the Fixed Stars also moves, and they have therefore postulated a ninth sphere. This being found insufficient, modern authorities now add a tenth. Yet they have still not attained the result which we hope to attain by the motion of the Earth. We shall assume this motion as a hypothesis and follow its consequences.[61]

---

61. In the original manuscript there follow on this chapter two and a half pages which have been heavily scored out. The translation of this section is as follows:

"Should we allow that the course of Sun and Moon could be diverted with Earth immovable it would yet be less allowable for the other planets. From these and similar causes Philolaus regarded Earth as movable. Some also say that Aristarchus of Samos though not moved by the reasoning that Aristotle advances and rejects (*De Coelo*, II, § 14) was of the same view. But since this cannot be understood, save by the bestowal of wit and industry, it has, as Plato would say, remained hidden from philosophers. There were thus but few who recognised the cause of motion of the stars. Yet if known to Philolaus or any other Pythagorean it is not probable that they would have published it. For it was not the habit of the Pythagoreans to vaunt their philosophical secrets in books nor indeed to reveal them at all but to entrust them to the faith of friends and intimates and so pass it on from hand to hand. In proof thereof is a letter from Lysis to Hipparchus."

Here Copernicus gives the text of the long and spurious letter of Lysis to Hipparchus which is to be found in *Iamblichus*, V, p. 75. It deals with the custom of the Pythagoreans and is without astronomical bearing.

GIORDANO BRUNO:

On the Infinite Universe and Worlds

First Dialogue

Speakers: {
Elpino
Philotheo (occasionally called Theophilo)
Fracastoro
Burchio
}

Elp. How is it possible that the universe can be infinite?

Phil. How is it possible that the universe can be finite?

Elp. Do you claim that you can demonstrate this infinitude?

Phil. Do you claim that you can demonstrate this finitude?

Elp. What is this spreading forth?

Phil. What is this limit?

Frac. To the point, to the point, if you please. Too long you have kept us in suspense.

Bur. Come quickly to argument, Philotheo, for I shall be vastly amused to hear this fable or fantasy.

Frac. More modestly, Burchio. What wilt thou say if truth doth ultimately convince thee?

Bur. Even if this be true I do not wish to believe it, for this INFINITE can neither be understood by my head nor brooked by my stomach. Although, to tell the truth, I could yet hope that Philotheo were right, so that if by ill luck I were to fall from this world I should always find myself on firm ground.

Elp. Certainly, O Theophilo, if we wish to judge by our senses, yielding suitable primacy to that which is the source of all our knowledge, perchance we shall not find it easier to reach the conclusion you expressed than to take the contrary view. Now be so kind as to begin my enlightenment.

Phil. No corporeal sense can perceive the infinite. None of our senses could be expected to furnish this conclusion; for the infinite cannot be the object

From Dorothea Waley Singer, *Giordano Bruno, His Life and Thought with Annotated Translation of His Work, On the Infinite Universe and Worlds,* New York, 1950, pp. 250–59, 302–4. Reprinted with the kind permission of the publishers, Abelard-Schuman, Inc.

of sense-perception; therefore he who demandeth to obtain this knowledge through the senses is like unto one who would desire to see with his eyes both substance and essence. And he who would deny the existence of a thing merely because it cannot be apprehended by the senses, nor is visible, would presently be led to the denial of his own substance and being. Wherefore there must be some measure in the demand for evidence from our sense-perception, for this we can accept only in regard to sensible objects, and even there it is not above all suspicion unless it cometh before the court aided by good judgement. It is the part of the intellect to judge, yielding due weight to factors absent and separated by distance of time and by space intervals. And in this matter our sense-perception doth suffice us and doth yield us adequate testimony, since it is unable to gainsay us; moreover it advertiseth and confesseth his own feebleness and inadequacy by the impression it giveth us of a finite horizon, an impression moreover which is ever changing. Since then we have experience that sense-perception deceiveth us concerning the surface of this globe on which we live, much more should we hold suspect the impression it giveth us of a limit to the starry sphere.

ELP. Of what use then are the senses to us? Tell me that.

PHIL. Solely to stimulate our reason, to accuse, to indicate, to testify in part; not to testify completely, still less to judge or to condemn. For our senses, however perfect, are never without some perturbation. Wherefore truth is in but very small degree derived from the senses as from a frail origin, and doth by no means reside in the senses.

ELP. Where then?

PHIL. In the sensible object as in a mirror. In reason, by process of argument and discussion. In the intellect, either through origin or by conclusion. In the mind, in its proper and vital form.

ELP. On, then, and give your reasons.

PHIL. I will do so. If the world is finite and if nothing lieth beyond, I ask you WHERE is the world? WHERE is the universe? Aristotle replieth, it is in itself.[1] The convex surface of the primal heaven is universal space, which being the primal container is by naught contained. For position in space is no other than the surfaces and limit of the containing body, so that he who hath no containing body hath no position in space.[2] What then dost thou mean, O Aristotle, by this phrase, that "space is within itself"? What will be thy conclusion concerning that which is beyond the world? If thou sayest, there is nothing, then the heaven[3] and the world will certainly not be anywhere.

FRAC. The world will then be nowhere. Everything will be nowhere.

PHIL. The world is something which is past finding out. If thou sayest (and it certainly appeareth to me that thou seekest to say something in order to escape Vacuum and Nullity), if thou sayest that beyond the world is a divine

1. Cf. *Physica*, IV, 3, 210a 29; 5, 212b 13, etc.
2. Cf. *Physica*, IV, 4, 211b 4; 212a 5–6.
3. *Cielo*.

intellect, so that God doth become the position in space of all things, why then thou thyself wilt be much embarrassed to explain to us how that which is incorporeal [yet] intelligible, and without dimension can be the very position in space occupied by a dimensional body; and if thou sayest that this incorporeal space containeth as it were a form, as the soul containeth the body, then thou dost not reply to the question of that which lieth beyond, nor to the enquiry concerning that which is outside the universe. And if thou wouldst excuse thyself by asserting that where naught is, and nothing existeth, there can be no question of position in space nor of beyond or outside, yet I shall in no wise be satisfied. For these are mere words and excuses, which cannot form part of our thought. For it is wholly impossible that in any sense or fantasy (even though there may be various senses and various fantasies), it is I say impossible that I can with any true meaning assert that there existeth such a surface, boundary or limit, beyond which is neither body, nor empty space, even though God be there. For divinity hath not as aim to fill space, nor therefore doth it by any means appertain to the nature of divinity that it should be the boundary of a body. For aught which can be termed a limiting body must either be the exterior shape or else a containing body. And by no description of this quality canst thou render it compatible with the dignity of divine and universal nature.[4]

BUR. Certainly I think that one must reply to this fellow that if a person would stretch out his hand beyond the convex sphere of heaven, the hand would occupy no position in space nor any place, and in consequence would not exist.

PHIL. I would add that no mind can fail to perceive the contradiction implicit in this saying of the Peripatetic. Aristotle defined position occupied by a body not as the containing body itself, nor as a certain [part of] space,[5] but as a surface of the containing body. Then he affirmeth that the prime, principal and greatest space is that to which such a definition least and by no means conformeth, namely, the convex surface of the first [outermost] heaven. This is the surface of a body of a particular sort, a body which containeth only, and is not contained. Now for the surface to be a position in space, it need not appertain to a contained body but it must appertain to a containing body. And if it be the surface of a containing body and yet be not joined to and continuous with the contained body, then it is a space without position, since the first [outermost] heaven cannot be a space except in virtue of the concave surface thereof, which is in contact with the convex surface of the next heaven. Thus we recognize that this definition is vain, confused and self-destructive, the confusion being caused by that incongruity which maintaineth that naught existeth beyond the firmament.

ELP. The Peripatetics would say that the outermost heaven is a containing

---

4. Lit., "And however thou mightest attempt to say this, thou wouldst be considered to detract from the dignity . . ."
5. For the following argument, cf. *Physica*, IV, 5, 212a–212b.

body in virtue of the concave and not of the convex surface thereof, and that in virtue of the concave surface it is a space.

FRAC. And I would add that therefore the surface of a containing body need not be a position in space.[6]

PHIL. In short then, to come straight to my proposition, it appeareth to me ridiculous to affirm that nothing is beyond the heaven, and that the heaven is contained in itself and is in place and hath position only by accident, that is, by means of the parts thereof. And however Aristotle's phrase *by accident* be interpreted, he cannot escape the difficulty that one cannot be transformed into two, for the container is eternally different from the contained,[7] so different, indeed, that according to Aristotle himself, the container is incorporeal while the contained is corporeal; the container is motionless while the contained hath motion; the container is a mathematical conception while the contained hath physical existence.[8]

Thus let this surface be what it will, I must always put the question, what is beyond? If the reply is NOTHING, then I call that the VOID or emptiness. And such a Void or Emptiness hath no measure and no outer limit, though it hath an inner; and this is harder to imagine than is an infinite or immense universe. For if we insist on a finite universe, we cannot escape the void. And let us now see whether there can be such a space in which is naught. In this infinite space is placed our universe (whether by chance, by necessity or by providence I do not now consider). I ask now whether this space which indeed containeth the world is better fitted to do so than is another space beyond?

FRAC. It certainly appeareth to me, not so. For where there is nothing, there can be no differentiation; where there is no differentiation there is no distinction of quality and perhaps there is even less of quality where there is naught whatsoever.

ELP. Neither also can there be then any lack of quality, and this more surely than the previous proposition.

PHIL. You say truly. Therefore I say that as the Void or Emptiness, which according to the Peripatetic view is necessary, hath no aptness to receive [i.e., no power of attracting the world], still less can it repel the world. But of these two faculties we see one in action, while the other we cannot wholly see except with the eye of reason. As therefore this world (called by the Platonists MATTER), lieth in this space which doth equal in size the whole of our world, so another world can be in that other space, and [other worlds] in innumerable spaces beyond of similar kind.[9]

FRAC. Certainly we may judge more confidently by analogy with what we

---

6. *Lit.*, "there existeth a surface of a containing body which surface is not a position in space."

7. *Physica*, IV, 5, 212b 13–14. Hardie and Gaye translate, "But other things are in place indirectly through something conjoined with them, as the soul and the heaven" (Oxford *Aristotle*, ed. W. D. Ross, Vol. II). The Latin edition (Venice, 1482) gives *secundum accidens*: "Alia uero secundum accidens ut anima et celestes partes non in loco quodammodo omnes sunt."

8. *Physica*, IV, 4, 212a 10–23.

9. *Equale a questo*. Cf. the same phrase with the same connotation, n. 16, below.

see and know than in opposition to what we see and know. Since then, on the evidence of our sight and experience, the universe hath no end nor is terminated in Void and Emptiness, about which indeed there is no information, therefore we should reasonably conclude as you do, since if all other reasonings were of equal weight, we should still see that our experience is opposed to a Void but not to a Plenum: therefore we shall always be justified in accepting the Plenum; but if we reject it, we shall not easily escape a thousand accusations and inconveniences. Continue, O Philotheo.

PHIL. As regards infinite space, we know for certain that this is apt for the reception of matter and we know naught else thereof; for me, however, it is enough that infinity is not repugnant to the reception of matter, if only because where there is naught, there at least is no outrage. It remaineth to see whether or not it is convenient that all space be filled? And here, if we consider no less what it may be than what it may do, we shall still find the Plenum not merely reasonable but inevitable. That this may be manifest I ask you whether it is well that this world[10] exist.

ELP. It is very well.

PHIL. Then it is well that this space equal in size to the world (I will call it Empty Space, like to and indistinguishable from the space which thou wouldst call the nullity beyond the convexity of the first heaven) that this space I say should similarly be filled.

ELP. Certainly.

PHIL. I ask thee further. Dost thou think that as in this our space there existeth this frame that we call the world, so the same could have existed or could exist in another space within this great Emptiness?

ELP. I will say yes, albeit I do not see how we can posit any distinction between one thing and another in mere nullity and empty space.

FRAC. I am sure that thou dost see, but thou art not anxious to declare it, for thou dost perceive whither this will lead thee.

ELP. Declare it indeed without hesitation.[11] For it behoveth us to declare and understand that our world[12] lieth in a space which without our world would be indistinguishable from that which is beyond your *primum mobile*.

FRAC. Continue.

PHIL. So just as this space can contain and hath contained this universal body, and is necessarily completed thereby as thou didst say, so also all the rest of space can be and hath been no less completed in this manner.

ELP. I admit it. What may be deduced therefrom? A thing can be or can have: therefore is it or hath it?

PHIL. I will expound so that, if thou wishest to make a frank confession, then wilt thou say that it can be, that it should be, that it is. For just as it would be ill were this our space not filled, that is, were our world[12] not to exist,

---

10. *Questo mondo*, i.e., our perceptible universe.
11. The beginning of Elpino's conversion.
12. *Mondo*.

then, since the spaces are indistinguishable, it would be no less ill if the whole of space were not filled. Thus we see that the universe[13] is of infinite size and the worlds[14] therein without number.

ELP. Wherefore then must they be so numerous rather than a single one?

PHIL. Because if it were ill that our world[15] should not exist, or that this Plenum should not be, then the same holdeth good of our space or space of similar kind.[16]

ELP. I say that 'twere ill as regards that which is in this our space, which might equally exist in another space of the same kind.[16]

PHIL. This, if thou considereth well, cometh all to the same. For the goodness of this corporeal being which is our space, or could be in another space similar to ours[16] doth explain and concern that goodness, suitability and perfection which may be in a space like to and as great as our own or in another similar[16] to ours, but doth not concern that goodness which may be in countless other spaces similar to our own.[17] This argument is the more cogent since, if it is reasonable to postulate a finite goodness, a bounded perfection, all the more reasonable is the conception of an infinite goodness. For whereas finite goodness appeareth to us reasonable and convenient, the infinite is an imperative necessity.

ELP. Infinite Good doth certainly exist, but is incorporeal.

PHIL. We are then at one concerning the incorporeal infinite; but what preventeth the similar acceptability of the good, corporeal and infinite being? And why should not that infinite which is implicit in the utterly simple and individual Prime Origin rather become explicit in his own infinite and boundless image able to contain innumerable worlds, than become explicit within such narrow bounds? So that it appeareth indeed shameful to refuse to credit that this world which seemeth to us so vast may not in the divine regard appear a mere point, even a nullity?

ELP. But since the greatness of God lieth not at all in corporeal size (not to mention that our world doth add nothing to him) so also we should not conceive the greatness of his image to consist in the greater or lesser extent of the size thereof.[18]

THEO. Well said. But you do not answer the pith of the argument. For I do not insist on infinite space, nor is Nature endowed with infinite space for the exaltation of size or of corporeal extent, but rather for the exaltation of corporeal natures and species, because infinite perfection is far better presented in innumerable individuals than in those which are numbered and finite. Needs must indeed that there should be an infinite image of the inaccessible divine countenance and that there should be in this image as infinite members thereof,

---

13. *Universo.*
14. *Mondi.*
15. *Mondo.*
16. *Equale a questo.* Cf. above, n. 9.
17. *Simili a questo.*
18. *Mole de' dimensioni.*

innumerable worlds, namely, these others that I postulate. But since innumerable grades of perfection must, through corporeal mode, unfold the divine incorporeal perfection, therefore there must be innumerable individuals, those great animals, whereof one is our earth, the divine mother who hath given birth to us, doth nourish us and moreover will receive us back;[19] and to contain these innumerable bodies there is needed an infinite space. Nevertheless it is well that there should be since there can be innumerable worlds similar to our own, even as our world hath achieved and doth achieve existence and it is well that it should exist.

ELP. We shall say that this finite world[20] with the finite stars embraceth the perfection of all things.

THEO. You may say so, but you cannot prove it. For the world[20] of this our finite space embraceth indeed the perfection of all those finite objects contained within our space, but not of those infinite potentialities of innumerable other spaces.

FRAC. Pray let us stop here and not act like those sophists who dispute merely for victory, and while they strive for their laurels prevent both themselves and others from comprehending the truth. For I believe there is none so pertinacious in perfidy and in slander withal as to deny that since space may contain infinity and in view of the goodness both individual and collective of the infinite number of worlds[21] which may be contained therein, therefore each of them, no less than this world which we know, may rationally and conveniently have his being. For infinite space is endowed with infinite quality and therein is lauded the infinite act of existence, whereby the infinite First Cause is not considered deficient, nor is the infinite quality thereof in vain. Let us then, O Elpino, be content to hear further arguments from Philotheo if they should occur to him.

ELP. To tell the truth, I see well that to pronounce the world (as you name the universe) boundless, carrieth no inconvenience and indeed freeth us from many difficulties in which the contrary opinion doth envelope us. In particular I recognize that, if we follow the Peripatetics, we must often assert that which hath no basis in our thought. For example, having denied the existence of empty space either without or within the universe,[22] when we seek to reply to the question "Where is the universe?" we must needs declare the universe to be within the very parts thereof, for fear of asserting that it is in no place whatsoever. As though we were to say *Nullibi, nusquam*. But it cannot be denied that by such arguments 'twere needful to declare that the parts occupy some position while the universe occupieth no position and is not in space.

---

19. *Lit.*, "moreover will not receive us back"; *non* is probably a printer's error.
20. *Mondo.*
21. *De infiniti mondi*, Cf. n. 8.
22. Cf. *Physica*, IV, 6–9. In *Metaphysica*, 6, 1048b 9–15, in the course of a discussion on potentiality, Aristotle groups the conceptions of the infinite and of the void: "The infinite and the void and all similar things are said to exist potentially and actually in a different sense from that in which many other things are said to exist . . . the infinite does not exist potenially in the sense that it will ever actually have separate existence."

And this (as all will recognize) is meaningless nonsense, and is clearly an obstinate flight in order to avoid confession of the truth, and to refuse admission either of the infinity of the world and of the universe, or of the infinity of space. From such attempts there followeth double confusion to whoever adopteth them. I therefore affirm that if the universe[23] be a single spherical body, and therefore hath form and limit, then it must terminate within infinite space. And if we would say that nothing is within infinite space, then we must admit a truly empty space, and if this exist, it is no less reasonable to conceive it of the whole than of this part which here we see capable of enclosing this world. But if vacant space doth not exist, then must [the whole of space] be a plenum, and consequently this universe must be infinite. And it were no less foolish to affirm that the world must have position after we have asserted that nothing lieth beyond it, or to maintain that it is within the very parts of itself, than if we were to say that Elpino must have position because his hand is on his arm, his eye on his face, his foot on his leg, his head on his body. But to come to a conclusion, not behaving like a sophist standing on manifest difficulties or spending my time in chatter, I declare that which I cannot deny, namely, that within infinite space either there may be an infinity of worlds similar to our own; or that this universe may have extended its capacity in order to contain many bodies such as those we name stars; or again that, whether these worlds be similar or dissimilar to one another, it may with no less reason be well that one than that another should exist. For the existence of one is no less reasonable than that of another; and the existence of many no less so than of one or of the other; and the existence of an infinity of them no less so than the existence of a large number. Wherefore, even as the abolition and non-existence of this world would be an evil, so would it be of innumerable others. . . .

Third Dialogue

Phil. [The whole universe] then is one, the heaven, the immensity of embosoming space, the universal envelope, the ethereal region through which the whole hath course and motion. Innumerable celestial bodies, stars, globes, suns and earths may be sensibly perceived therein by us and an infinite number of them may be inferred by our own reason. The universe, immense and infinite, is the complex of this [vast] space and of all the bodies contained therein.

Elp. So that there are no spheres with concave and convex surfaces nor deferent orbs; but all is one field, one universal envelope.

Phil. So it is.

Elp. The opinion of diverse heavens hath then been caused by diverse motions of the stars and by the appearance of a sky filled with stars revolving around the earth; nor can these luminaries by any means be seen to recede one from another; but, maintaining always the same distance and relation one

---

23. Il tutto.

to another, and a certain course, they [appear to] revolve around the earth, even as a wheel on which are nailed innumerable mirrors revolveth around his own axis. Thus it is considered obvious from the evidence of our eyes that these luminaries have no motion of their own; nor can they wander as birds through the air; but they move only by the revolution of the orbs to which they are fixed, whose motion is effected by the divine pulse of some [supreme] intelligence.

THEO. Such is the common opinion. But once the motion is understood of our own mundane star which is fixed to no orb, but impelled by her own intrinsic principle, soul and nature, taketh her course around the sun through the vastness of universal space, and spinneth around her own centre, then this opinion will be dispelled. Then will be opened the gate of understanding of the true principles of nature, and we shall be enabled to advance with great strides along the path of truth which hath been hidden by the veil of sordid and bestial illusions and hath remained secret until to-day, through the injury of time and the vicissitudes of things, ever since there succeeded to the daylight of the ancient sages the murky night of the foolhardy sophists.

> Naught standeth still, but all things swirl and whirl
> As far as in heaven and beneath is seen.
> All things move, now up, now down,
> Whether on a long or a short course,
> Whether heavy or light;
> Perchance thou too goest the same path
> And to a like goal.
> For all things move till overtaken,
> As the wave[24] swirleth through the water,
> So that the same part
> Moveth now from above downward
> And now from below upward,
> And the same hurly-burly
> Imparteth to all the same successive fate.[25]

ELP. Indubitable that the whole fantasy of spheres bearing stars and fires, of the axes, the deferents, the functions of the epicycles, and other such chimeras, is based solely on the belief that this world occupieth as she seemeth to do the very centre of the universe, so that she alone being immobile and fixed, the whole universe revolveth around her.

PHIL. This is precisely what those see who dwell on the moon and on the other stars in this same space, whether they be earths or suns.

ELP. Suppose then for the moment that the motion of our earth causeth the appearance of daily world motion, and that by her own diverse motions the earth causeth all those motions which seem to appertain to the innumerable stars, we should still say that the moon, which is another earth, moveth by her own force through the air around the sun. Similarly, Venus, Mercury and the

---

24. *Il buglo*, wave or bubble.
25. i.e., involves all things in all possible movements.

Due back
Feb 16, 04

The Power of One

Virtue: Individuality—discovering w
difference.
Memory Verse: I praise you becau
Psalm 139:14, CEV
Bible Story: The Power of One • *A*
Bible Byte: Always doing good. *Ac*
Bottom Line: You can make a diffe

Power Up: Engage the Heart (Large

PREVIEW
Power Up is designed to engage children i
prayer, and an innovative approach to the
interactive time.

GET READY
Gather all the supplies needed for large gr

others which are all earths, pursue their courses around the same father of life.

PHIL. It is so.

ELP. The proper motions of each of these are those of their apparent motions which are not due to our so-called world motion; and the proper motions of the bodies known as fixed stars (though both their apparent fixity and the world motion should be referred to our earth) are more diverse and more numerous than the celestial bodies themselves. For if we could observe the motion of each one of them, we should find that no two stars ever hold the same course at the same speed; it is but their great distance from us which preventeth us from detecting the variations. However much these stars circulate around the solar flame or spin round their own centres in order to participate in the vital heat [of a sun], it is impossible for us to detect their diverse approach toward and retreat from us.

PHIL. That is so.

ELP. There are then innumerable suns, and an infinite number of earths revolve around those suns, just as the seven we can observe revolve around this sun which is close to us.

PHIL. So it is.

ELP. Why then do we not see the other bright bodies which are earths circling around the bright bodies which are suns? For beyond these we can detect no motion whatever; and why do all other mundane bodies (except those known as comets) appear always in the same order and at the same distance?

PHIL. The reason is that we discern only the largest suns, immense bodies. But we do not discern the earths because, being much smaller, they are invisible to us. Similarly it is not impossible that other earths revolve around our sun and are invisible to us on account either of greater distance or of smaller size, or because they have but little watery surface, or because such watery surface is not turned toward us and opposed to the sun, whereby it would be made visible as a crystal mirror which receiveth luminous rays; whence we perceive that it is not marvellous or contrary to nature that often we hear that the sun hath been partially eclipsed though the moon hath not been interpolated between him and our sight. . . .

FRANCIS R. JOHNSON:

Thomas Digges
and the Infinity of the Universe

IN HIS BOOK on the new star, the *Alae,* Thomas Digges had not only won for himself a place beside John Dee as one of the two most eminent astronomers and mathematicians in England; he had also done great service to English science by his insistence on the experimental method. At the same time, he had become the recognized leader of the English supporters of the Copernican theory, and no one in Europe had shown greater eagerness to put the new heliocentric system to the test of observations and experiments which should either confirm it or necessitate some further modifications in Copernicus' hypothesis. The *Alae,* however, had been written in Latin for an international audience. Digges's next work was designed to make a knowledge of the essential features of the new system of the universe available to his less learned countrymen, and particularly to the skilled artisans and mechanics whose intelligent co-operation was so necessary to successful research in the sciences.

When, in 1576, the printer, Thomas Marshe, prepared to issue a new edition of Leonard Digges's *Prognostication euerlasting,* Thomas Digges took advantage of this opportunity to revise his father's work and to make some important additions. The most significant of these was a supplement describing the Copernican universe, entitled *A Perfit Description of the Caelestiall Orbes according to the most aunciente doctrine of the Pythagoreans, latelye reuiued by Copernicus and by Geometricall Demonstrations approued.* Besides a diagram of the heliocentric system, it gave an English translation of the principal sections of Book I of the *De revolutionibus,* in which the author had set forth the essential features of his new system and the chief arguments in its favor.

Digges clearly states, in his preface to the reader, the reasons which prompted him to make this addition to his father's book.[1] He was unwilling

From Francis R. Johnson, *Astronomical Thought in Renaissance England,* Baltimore, 1937, Chapter VI, pp. 161–69. Reprinted with the kind permission of the publishers, The Johns Hopkins Press.

1. *A Prognostication euerlastinge of righte good effecte . . . Published by Leonard Digges Gentleman. Lately corrected and augmented by Thomas Digges his sonne* (London: Thomas Marsh, 1576), sig. M1^r. Digges's supplement on the Copernican system is reprinted in Johnson and Larkey, "Thomas Digges, the Copernican System, and the Idea of the Infinity of the Universe in 1576."

to allow the description of the universe according to the Ptolemaic theory to go unchallenged, since the new system of Copernicus was so much more logical and mathematically harmonious. He therefore "thought it conuenient together with the olde Theorick also to publish this, to the ende such noble English minds (as delight to reache aboue the baser sort of men) might not be altogether defrauded of so noble a part of Philosophy."

The preface to Thomas Digges's treatise on the Copernican theory also makes clear that the author was in no way deceived by the common assertion, based on Osiander's spurious preface to the *De revolutionibus,* that Copernicus had put forward his hypothesis merely to simplify mathematical calculations and had not himself believed it to be physically true. Digges says:

> And to the ende it might manifestly appeare that *Copernicus* mente not as some haue fondly excused him to deliuer these grounds of the Earthes mobility onely as Mathematicall principles, fayned & not as Philosophicall truly auerred. I haue also from him deliuered both the Philosophicall reasons by *Aristotle* and others produced to maintaine the Earthes stability, and also their solutions and insufficiency, wherein I cannot a litle commende the modestie of that graue Philosopher *Aristotle,* who seing (no doubt) the insufficiency of his owne reasons in seeking to confute the Earthes motion, vseth these words. *De his explicatum est ea qua potuimus facultate* howbeit his disciples haue not with like sobriety maintayned the same.[2]

The attitude toward Aristotle which Digges expresses in this passage is the one which we have already noted as typical of the ablest English scientists of the period. It is interesting, moreover, that, in his attack upon the Aristotelians who rigidly maintained their master's infallibility in scientific matters, he quotes from the well-known work of Palingenius. Digges's familiarity with the *Zodiacus vitae* was commented upon by Gabriel Harvey, who tells us that "M. Digges hath the whole Aquarius of Palingenius bie hart: & takes mutch delight to repeate it often."[3] This knowledge is quite evident throughout the preface to Digges's work. He makes a special point of Palingenius' allusion to the early ideas of Anaxagoras and Democritus that every star was a world, and reprints the poet's lines in the following form:

> Singula nonnulli credunt quoque sydera posse
> Dici Orbes, TERRAMque appellant sydus opacu
> Cui minimus Diuûm praesit &c.

The word "Terram" is printed in extremely large capitals, with the obvious design of calling attention to some of the startling implications of the new theory by means of the adroit use of a quotation from an author known and admired by nearly all of Digges's fellow countrymen.

In translating the principal chapters of the first book of Copernicus' great work, Digges followed the usual Elizabethan practice of working phrase by phrase, rather than word by word, so that the general result is an excellent piece of Elizabethan prose, which, although not an absolutely literal translation, is consistently faithful to Copernicus' meaning. When Digges varies from

---

2. *A Perfit Description of the Caelestiall Orbes,* sigs. M1^r-M1^v.
3. G. C. Moore Smith, *Gabriel Harvey's Marginalia,* p. 161.

Copernicus, it is usually on the side of giving greater emphasis to the implied criticisms of some current cosmological ideas.[4] He also adds further evidence to support Copernicus' refutation of the Aristotelian contention that, if the earth rotated, objects dropped from a high tower would be left behind and finally hit the ground a considerable distance to the west of the tower. Digges brings forward the very pertinent experiment, which he had probably made himself, of dropping an object from the mast of a moving ship and noting that it appeared to fall to the deck in a straight line parallel to the mast. He says:

And of thinges ascedinge and descendinge in respect of the worlde we must confesse them to haue a mixt motion of right & circulare, albeit it seeme to vs right & streight, No otherwise then if in a shippe vnder sayle a man should softly let a plumet downe from the toppe alonge by the maste euen to the decke: This plummet passing alwayes by the streight maste, seemeth also too fall in a righte line, but beinge by discours of reason wayed his Motion is found mixt of right and circulare.

By far the most important addition which Digges made to Copernicus, however, was his assertion that the heliocentric universe should be conceived as infinite, with the fixed stars located at varying distances throughout infinite space. Copernicus, as previously noted, although bringing the question of the infinity of the universe into his refutation of his Aristotelian opponents, had refused to commit himself on this subject. Instead, he had retained in his diagram of the universe the finite sphere of fixed stars, merely making this sphere sufficiently large to account for the absence of any perceptible stellar parallax. Digges, however, clearly perceived that, the moment the rotation of the earth was conceded, there was no longer any necessity for picturing the stars as attached to a huge, rotating sphere at a definite distance from the earth. At the same time, Aristotle's oft-quoted mathematical proofs of the finiteness of the universe were automatically invalidated. Therefore, Digges had the courage to break completely with the older cosmologies by shattering the finite outer wall of the universe. He was the first modern astronomer of note to portray an infinite, heliocentric universe, with the stars scattered at varying distances throughout infinite space.

Digges's diagram of the universe, printed in the 1576 and in all later editions of the *Prognastication euerlasting,* was the representation of the new Copernican system most familiar to the average Englishman of the Renaissance.[5] It followed the plan printed in the *De revolutionibus* so far as the planets were concerned, but, instead of representing the sphere of the fixed stars by merely the customary circle standing for the eighth sphere, it scattered

---

4. For example, in speaking (sig. N1ᵛ) of the space between the earth and the moon, in which we know "nothinge but the aire, or fiery Orbe if any sutch be," Digges makes more emphatic the doubt cast upon the existence of the sphere of fire, just below the moon. Copernicus' words were (*De revolutionibus,* fol. 6ʳ) "nihil tamen aliud in tanto spacio nouimus contineri quàm aërem. & si placet etiam, quod igneum vocant elementum."

5. See Fig. L.

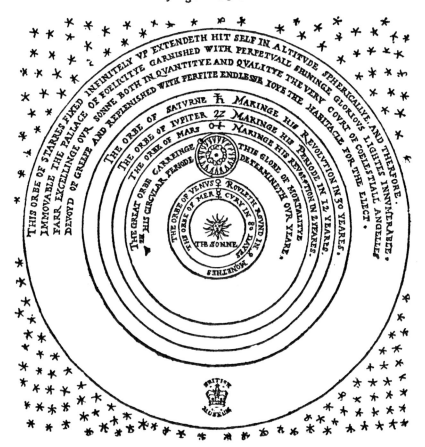

Fig. 1.—Thomas Digges's Diagram of the Infinite Copernican Universe

From *A Perfit Description of the Caelestiall Orbes* (1576)

the stars out to the borders of the diagram, and inserted the legend: "This orbe of starres fixed infinitely vp extendeth hit self in altitude sphericallye, and therfore immouable; the pallace of foelicitye garnished with perpetuall shininge glorious lightes innumerable, farr excellinge our sonne both in quantitye and qualitye . . ."

In the text of his translation of Copernicus, Digges inserted the following paragraph of his own, emphasizing the same idea:

Heerein can wee neuer sufficiently admire thys wonderfull & incomprehensible huge frame of goddes woorke proponed to our senses, seinge fyrst thys baull of ye earth wherein we moue, to the common sorte seemeth greate, and yet in respecte of the Moones Orbe is very small, but compared with *Orbis magnus* wherein it is caried, it scarcely retayneth any sensible proportion, so meruceilously is that Orbe of Annuall motion greater than this litle darcke starre wherein we liue. But that *Orbus magnus* beinge as is before declared but as a poynct in respect of the immesity of that immoueable heauen, we may easily consider what litle portion of gods frame, our Elementare corruptible worlde is, but neuer sufficiently be able to admire the immensity of the Rest. Especially of that fixed Orbe garnished with lightes innumerable and reachinge vp in *Sphaericall altitude* without ende. Of whiche lightes Celestiall it is to bee thoughte that we onely behoulde sutch as are in the inferioure partes of the same Orbe, and as they are hygher, so seeme they of lesse and lesser quantity, euen tyll our sighte beinge not able farder to reache or conceyue, the greatest part rest by reason of their wonderfull distance inuisible vnto vs. And this may wel be thought of vs to be the gloriouse court of ye great god, whose vnsercheable worcks inuisible we may partly by these his visible coiecture, to whose infinit power and maiesty such an infinit place surmountinge all other both in quantity and quality only is conueniente. But because the world hath so longe a tyme bin carryed with an opinion of the earths stabilitye, as the contrary cannot but be nowe very imperswasible, I haue thought good out of *Copernicus* also to geue a taste of the reasons philosophicall alledged for the earthes stabilitye, and their solutions, that sutch as are not able with *Geometricall* eyes to beehoulde the secrete perfection of *Copernicus Theoricke,* maye yet by these familiar, naturall reasons be induced to serche farther, and not rashly to condempne for phantasticall, so auncient doctrine reuiued, and by *Copernicus* so demonstratiuely approued.[6]

It is worth noting that Digges fully realized the huge size of the stars which followed as a mathematical consequence of the Copernican hypothesis, for his legend asserts that these stars far excel the sun both in quantity and quality.[7] Furthermore, in the paragraph he added to Copernicus, he made a point of the fact that the essential merits of the new system could be fully appreciated only by the mathematicians, yet proceeded to give, out of Copernicus, the refutations of the usual arguments of the Aristotelians against the new system, even inserting, as we have observed, some additional evidence of his own against the current physical theories.

The influence of Digges's treatise on contemporary astronomical thought can hardly be overestimated. The book in which it appeared was, as already

6. *A Perfit Description of the Caelestiall Orbes,* sigs. N3^v-N4^r.

7. According to the traditional figures of Alfraganus, the first-magnitude stars were only slightly smaller than the sun, being 107 times the size of the earth, as against 166 in the case of the sun.

indicated, one of the most popular works of the period. There are extant no less than seven editions containing Digges's *Perfit Description of the Caelestiall Orbes,* dated 1576, 1578, 1583, 1585, 1592, 1596, and 1605. For each edition the volume was completely reprinted; there are no cases of the reissue of old sheets with a new title-page. There may well have been several other editions of which no copy has survived. In Digges's book the average Englishman, who lacked the opportunity or the learning to read and understand Copernicus' great work in the original, got his first authoritative exposition of the new heliocentric theory and the arguments in its favor. Digges's reputation as one of the leading mathematicians and scientists in England lent added weight to his support of the Copernican system. The manner in which he set forth his idea of the infinite universe, moreover, apparently led the majority of his countrymen to believe that the notion that the fixed stars were located at varying distances throughout infinite space was an integral part of the heliocentric hypothesis.

The close association, in the English mind, of the new Copernican system and the conception of the infinity of the universe has usually been attributed to the influence of Giordano Bruno. Bruno was in England from 1583 to 1585, and the first works in which he published his speculations concerning the infinite universe were printed in England during that period. These books, however, were written in Italian, and Bruno himself spoke no English. During his sojourn in England he seems to have been known only to a limited circle, and references to his cosmological speculations in English books do not begin to appear until several years afterward.[8] Then they are based upon his later Latin works printed abroad, rather than on his Italian works printed in London.

It was Digges's treatise proclaiming the idea of an infinite universe in conjunction with the Copernican system, and not Bruno's speculations on this subject, that influenced the thought of sixteenth-century England. Digges, moreover, was an eminent scientist, and was eager to verify his ideas by experimental methods. Bruno, on the other hand, had arrived at his notions entirely through metaphysical speculations;[9] the Copernican theory was seized upon and incorporated in his philosophy merely because it could be made to give seeming proof, in the physical realm, of ideas already arrived at by methods in which mathematical science played no part. It is entirely possible that Digges's brief treatise on the Copernican system first suggested to Bruno's mind the thought of using the new heliocentric theory as a physical proof of his highly speculative notions concerning the infinity of the material world. The *Perfit Description of the Caelestiall Orbes* had passed through at least two editions before Bruno's arrival in England and was reprinted twice during his sojourn there, so that he had ample opportunity to know of the work and may even have met Digges himself.

8. See Oliver Elton, "Giordano Bruno in England," *Modern Studies* (London, 1907), pp. 1–36.
9. Cf. the recently published work by A. O. Lovejoy, *The Great Chain of Being* (Cambridge, Mass., 1936), pp. 116–21.

GALILEO GALILEI:

Dialogue Concerning
the Two Chief World Systems

SALVIATI. We shall next consider the annual movement generally attributed to the sun, but then, first by Aristarchus of Samos and later by Copernicus, removed from the sun and transferred to the earth. Against this position I know that Simplicio comes strongly armed, in particular with the sword and buckler of his booklet of theses or mathematical disquisitions. It will be good to commence by producing the objections from this booklet.

SIMPLICIO. If you don't mind, I am going to leave those for the last, since they were the most recently discovered.

SALVIATI. Then you had better take up in order, in accordance with our previous procedure, the contrary arguments by Aristotle and the other ancients. I also shall do so, in order that nothing shall be left out or escape careful consideration and examination. Likewise Sagredo, with his quick wit, shall interpose his thoughts as the spirit moves him.

SAGREDO. I shall do so with my customary lack of tact; and since you have asked for this, you will be obliged to pardon it.

SALVIATI. This favor will oblige me to thank and not to pardon you. But now let Simplicio begin to set forth those objections which restrain him from believing that the earth, like the other planets, may revolve about a fixed center.

SIMPLICIO. The first and greatest difficulty is the repugnance and incompatibility between being at the center and being distant from it. For if the terrestrial globe must move in a year around the circumference of a circle— that is, around the zodiac—it is impossible for it at the same time to be in the center of the zodiac. But the earth is at that center, as is proved in many ways by Aristotle, Ptolemy, and others.

SALVIATI. Very well argued. There can be no doubt that anyone who wants to have the earth move along the circumference of a circle must first prove

From Galileo Galilei, *Dialogue Concerning the Two Chief World Systems—Ptolemaic and Copernican*, translated by Stillman Drake, with a foreword by Albert Einstein, University of California Press, Berkeley and Los Angeles, 1953, pp. 318–28 ("The Third Day"). Reprinted with the kind permission of the publishers, The University of California Press.

that it is not at the center of that circle. The next thing is for us to see whether the earth is or is not at that center around which I say it turns, and in which you say it is situated. And prior to this, it is necessary that we declare ourselves as to whether or not you and I have the same concept of this center. Therefore tell me what and where this center is that you mean.

SIMPLICIO. I mean by "center," that of the universe; that of the world; that of the stellar sphere; that of the heavens.

SALVIATI. I might very reasonably dispute whether there is in nature such a center, seeing that neither you nor anyone else has so far proved whether the universe is finite and has a shape, or whether it is in infinite and unbounded. Still, conceding to you for the moment that it is finite and of bounded spherical shape, and therefore has its center, it remains to be seen how credible it is that the earth rather than some other body is to be found at that center.

SIMPLICIO. Aristotle gives a hundred proofs that the universe is finite, bounded, and spherical.

SALVIATI. Which are later all reduced to one, and that one to none at all. For if I deny him his assumption that the universe is movable all his proofs fall to the ground, since he proves it to be finite and bounded only if the universe is movable. But in order not to multiply our disputes, I shall concede to you for the time being that the universe is finite, spherical, and has a center. And since such a shape and center are deduced from mobility, it will be the more reasonable for us to proceed from this same circular motion of world bodies to a detailed investigation of the proper position of the center. Even Aristotle himself reasoned about and decided this in the same way, making that point the center of the universe about which all the celestial spheres revolve, and at which he believed the terrestrial globe to be situated. Now tell me, Simplicio: if Aristotle had found himself forced by the most palpable experiences to rearrange in part this order and disposition of the universe, and to confess himself to have been mistaken about one of these two propositions—that is, mistaken either about putting the earth in the center, or about saying that the celestial spheres move around such a center—which of these admissions do you think that he would choose?

SIMPLICIO. I think that if that should happen, the Peripatetics . . .

SALVIATI. I am not asking the Peripatetics; I am asking Aristotle himself. As for the former, I know very well what they would reply. They, as most reverent and most humble slaves of Aristotle, would deny all the experiences and observations in the world, and would even refuse to look at them in order not to have to admit them, and they would say that the universe remains just as Aristotle has written; not as nature would have it. For take away the prop of his authority, and with what would you have them appear in the field? So now tell me what you think Aristotle himself would do.

SIMPLICIO. Really, I cannot make up my mind which of these two difficulties he would have regarded as the lesser.

SALVIATI. Please, do not apply this term "difficulty" to something that may necessarily be so; wishing to put the earth in the center of the celestial revolutions was a "difficulty." But since you do not know to which side he would have leaned, and considering him as I do a man of brilliant intellect, let us set about examining which of the two choices is the more reasonable, and let us take that as the one which Aristotle would have embraced. So, resuming our reasoning once more from the beginning, let us assume out of respect for Aristotle that the universe (of the magnitude of which we have no sensible information beyond the fixed stars), like anything that is spherical in shape and moves circularly, has necessarily a center for its shape and for its motion. Being certain, moreover, that within the stellar sphere there are many orbs one inside another, with their stars which also move circularly, our question is this: Which is it more reasonable to believe and to say; that these included orbs move around the same center as the universe does, or around some other one which is removed from that? Now you, Simplicio, say what you think about this matter.

SIMPLICIO. If we could stop with this one assumption and were sure of not running into something else that would disturb us, I should think it would be much more reasonable to say that the container and the things it contained all moved around one common center rather than different ones.

SALVIATI. Now if it is true that the center of the universe is that point around which all the orbs and world bodies (that is, the planets) move, it is quite certain that not the earth, but the sun, is to be found at the center of the universe. Hence, as for this first general conception, the central place is the sun's, and the earth is to be found as far away from the center as it is from the sun.

SIMPLICIO. How do you deduce that it is not the earth, but the sun, which is at the center of the revolutions of the planets?

SALVIATI. This is deduced from most obvious and therefore most powerfully convincing observations. The most palpable of these, which excludes the earth from the center and places the sun there, is that we find all the planets closer to the earth at one time and farther from it at another. The differences are so great that Venus, for example, is six times as distant from us at its farthest as at its closest, and Mars soars nearly eight times as high in the one state as in the other. You may thus see whether Aristotle was not some trifle deceived in believing that they were always equally distant from us.

SIMPLICIO. But what are the signs that they move around the sun?

SALVIATI. This is reasoned out from finding the three outer planets—Mars, Jupiter, and Saturn—always quite close to the earth when they are in opposition to the sun, and very distant when they are in conjunction with it. This approach and recession is of such moment that Mars when close looks sixty times as large as when it is most distant. Next, it is certain that Venus and Mercury must revolve around the sun, because of their never moving far away from it, and because of their being seen now beyond it and now on this side of it, as Venus's changes of shape conclusively prove. As to the moon, it is true that this can never separate from the earth in any way, for reasons that will be set forth more specifically as we proceed.

SAGREDO. I have hopes of hearing still more remarkable things arising from this annual motion of the earth than were those which depended upon its diurnal rotation.

SALVIATI. You will not be disappointed, for as to the action of the diurnal motion upon celestial bodies, it was not and could not be anything different from what would appear if the universe were to rush speedily in the opposite direction. But this annual motion, mixing with the individual motions of all the planets, produces a great many oddities which in the past have baffled all the greatest men in the world.

Now returning to these first general conceptions, I repeat that the center of the celestial rotation for the five planets, Saturn, Jupiter, Mars, Venus, and Mercury, is the sun; this will hold for the earth too, if we are successful in placing that in the heavens. Then as to the moon, it has a circular motion around the earth, from which as I have already said it cannot be separated; but this does not keep it from going around the sun along with the earth in its annual movement.

SIMPLICIO. I am not yet convinced of this arrangement at all. Perhaps I should understand it better from the drawing of a diagram, which might make it easier to discuss.

SALVIATI. That shall be done. But for your greater satisfaction and your astonishment, too, I want you to draw it yourself. You will see that however firmly you may believe yourself not to understand it, you do so perfectly, and just by answering my questions you will describe it exactly. So take a sheet of paper and the compasses; let this page be the enormous expanse of the universe, in which you have to distribute and arrange its parts as reason shall direct you. And first, since you are sure without my telling you that the earth is located in this universe, mark some point at your pleasure where you intend this to be located, and designate it by means of some letter.

SIMPLICIO. Let this be the place of the terrestrial globe, marked A.

SALVIATI. Very well. I know in the second place that you are aware that this earth is not inside the body of the sun, nor even contiguous to it, but is distant from it by a certain space. Therefore assign to the sun some other place of your choosing, as far from the earth as you like, and designate that also.

SIMPLICIO. Here I have done it; let this be the sun's position, marked O.

SALVIATI. These two established, I want you to think about placing Venus in such a way that its position and movement can conform to what sensible experience shows us about it. Hence you must call to mind, either from past discussions or from your own observations, what you know happens with this star. Then assign it whatever place seems suitable for it to you.

SIMPLICIO. I shall assume that those appearances are correct which you have related and which I have read also in the booklet of theses; that is, that this star never recedes from the sun beyond a certain definite interval of forty degrees or so; hence it not only never reaches opposition to the sun, but not even quadrature, nor so much as a sextile aspect. Moreover, I shall assume that it displays itself to us about forty times as large at one time than at another; greater when, being retrograde, it is approaching evening conjunction with the sun, and very small when it is moving forward toward morning conjunction, and furthermore that when it appears very large, it reveals itself in a horned shape, and when it looks very small it appears perfectly round.

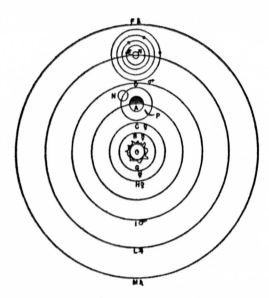

These appearances being correct, I say, I do not see how to escape affirming that this star revolves in a circle around the sun, in such a way that this circle cannot possibly be said to embrace and contain within itself the earth, nor to be beneath the sun (that is, between the sun and the earth), nor yet beyond the sun. Such a circle cannot embrace the earth because then Venus would sometimes be in opposition to the sun; it cannot be beneath the sun, for then Venus would appear sickle-shaped at both conjunctions; and it cannot be beyond the sun, since then it would always look round and never horned. Therefore for its lodging I shall draw the circle CH around the sun, without having this include the earth.

SALVIATI. Venus provided for, it is fitting to consider Mercury, which, as you know, keeping itself always around the sun, recedes therefrom much less than Venus. Therefore consider what place you should assign to it.

SIMPLICIO. There is no doubt that, imitating Venus as it does, the most appropriate place for it will be a smaller circle, within this one of Venus and also described about the sun. A reason for this, and especially for its proximity to the sun, is the vividness of Mercury's splendor surpassing that of Venus and all the other planets. Hence on this basis we may draw its circle here and mark it with the letters BG.

SALVIATI. Next, where shall we put Mars?

SIMPLICIO. Mars, since it does come into opposition with the sun, must embrace the earth with its circle. And I see that it must also embrace the sun; for, coming into conjunction with the sun, if it did not pass beyond it but fell short of it, it would appear horned as Venus and the moon do. But it always looks round; therefore its circle must include the sun as well as the earth. And since I remember your having said that when it is in opposition to the sun it looks sixty times as large as when in conjunction, it seems to me that this phenomenon will be well provided for by a circle around the sun embracing the earth, which I draw here and mark DI. When Mars is at the point D, it is very near the earth and in opposition to the sun, but when it is at the point I, it is in conjunction with the sun and very distant from the earth.

And since the same appearances are observed with regard to Jupiter and Saturn (although with less variation in Jupiter than in Mars, and with still less in Saturn than in Jupiter), it seems clear to me that we can also accommodate these two planets very neatly with two circles, still around the sun. This first one, for Jupiter, I mark EL; the other, higher, for Saturn, is called FM.

SALVIATI. So far you have comported yourself uncommonly well. And since, as you see, the approach and recession of the three outer planets is measured by double the distance between the earth and the sun, this makes a greater variation in Mars than in Jupiter because the circle DI of Mars is smaller than the circle EL of Jupiter. Similarly, EL here is smaller than the circle FM of Saturn, so the variation is still less in Saturn than in Jupiter, and this corresponds exactly to the appearances. It now remains for you to think about a place for the moon.

SIMPLICIO. Following the same method (which seems to me very convincing), since we see the moon come into conjunction and opposition with the sun, it must be admitted that its circle embraces the earth. But it must not embrace the sun also, or else when it was in conjunction it would not look horned but always round and full of light. Besides, it would never cause an eclipse of the sun for us, as it frequently does, by getting in between us and the sun. Thus one must assign to it a circle around the earth, which shall be this one, NP, in such a way that when at P it appears to us here on the earth A as in conjunction with the sun, which sometimes it will eclipse in this position. Placed at N, it is seen in opposition to the sun, and in that position it may fall under the earth's shadow and be eclipsed.

SALVIATI. Now what shall we do, Simplicio, with the fixed stars? Do we want to sprinkle them through the immense abyss of the universe, at various distances from any predetermined point, or place them on a spherical surface extending around a center of their own so that each of them will be the same distance from that center?

SIMPLICIO. I had rather take a middle course, and assign to them an orb described around a definite center and included between two spherical surfaces—a very distant concave one, and another closer and convex, between which are placed at various altitudes the innumerable host of stars. This might be called the universal sphere, containing within it the spheres of the planets which we have already designated.

SALVIATI. Well, Simplicio, what we have been doing all this while is arranging the world bodies according to the Copernican distribution, and this has now been done by your own hand. Moreover, you have assigned their proper movements to them all except the sun, the earth, and the stellar sphere. To Mercury and Venus you have attributed a circular motion around the sun without embracing the earth. Around the same sun you have caused the three outer planets, Mars, Jupiter, and Saturn, to move, embracing the earth within their circles. Next, the moon cannot move in any way except around the earth and without embracing the sun. And in all these movements you likewise agree with Copernicus himself. It now remains to apportion three things among the sun, the earth, and the stellar sphere: the state of rest, which appears to belong to the earth; the annual motion through the zodiac, which appears to belong to the sun; and the diurnal movement, which appears to belong to the stellar sphere, with all the rest of the universe sharing in it except the earth. And since it is true that all the planetary orbs (I mean Mercury, Venus, Mars, Jupiter, and Saturn) move around the sun as a center, it seems most reasonable for the state of rest to belong to the sun rather than to the earth—just as it does for the center of any movable sphere to remain fixed, rather than some other point of it remote from the center.

Next as to the earth, which is placed in the midst of moving objects—I mean between Venus and Mars, one of which makes its revolution in nine months and the other in two years—a motion requiring one year may be attributed to it much more elegantly than a state of rest, leaving the latter for the sun. And such being the case, it necessarily follows that the diurnal motion, too, belongs to the earth. For if the sun stood still, and the earth did not revolve upon itself but merely had the annual movement around the sun, our year would consist of no more than one day and one night; that is, six months of day and six months of night, as was remarked once previously.

See, then, how neatly the precipitous motion of each twenty-four hours is taken away from the universe, and how the fixed stars (which are so many suns) agree with our sun in enjoying perpetual rest. See also what great simplicity is to be found in this rough sketch, yielding the reasons for so many weighty phenomena in the heavenly bodies.

SAGREDO. I see this very well indeed. But just as you deduce from this simplicity a large probability of truth in this system, others may on the contrary make the opposite deduction from it. If this very ancient arrangement of the Pythagoreans is so well accommodated to the appearances, they may ask (and not unreasonably) why it has found so few followers in the course of centuries; why it has been refuted by Aristotle himself, and why even Copernicus is not having any better luck with it in these latter days.

SALVIATI. Sagredo, if you had suffered even a few times, as I have so often, from hearing the sort of follies that are designed to make the common people contumacious and unwilling to listen to this innovation (let alone assent to it), then I think your astonishment at finding so few men holding this opinion would dwindle a good deal. It seems to me that we can have little regard for imbeciles who take it as a conclusive proof in confirmation of the earth's motionlessness, holding them firmly in this belief, when they observe that they cannot dine today at Constantinople and sup in Japan, or for those who are positive that the earth is too heavy to climb up over the sun and then fall headlong back down again. There is no need to bother about such men as these, whose name is legion, or to take notice of their fooleries. Neither need we try to convert men who define by generalizing and cannot make room for distinctions, just in order to have such fellows for our company in very subtle and delicate doctrines. Besides, with all the proofs in the world what would you expect to accomplish in the minds of people who are too stupid to recognize their own limitations?

No, Sagredo, my surprise is very different from yours. You wonder that there are so few followers of the Pythagorean opinion, whereas I am astonished that there have been any up to this day who have embraced and followed it. Nor can I ever sufficiently admire the outstanding acumen of those who have taken hold of this opinion and accepted it as true; they have through sheer force of intellect done such violence to their own senses as to prefer what reason told them over that which sensible experience plainly showed them to the contrary. For the arguments against the whirling of the earth which we have already examined are very plausible, as we have seen; and the fact that the Ptolemiacs and Aristotelians and all their disciples took them to be conclusive is indeed a strong argument of their effectiveness. But the experiences which overtly contradict the annual movement are indeed so much greater in their apparent force that, I repeat, there is no limit to my astonishment when I reflect that Aristarchus and Copernicus were able to make reason so conquer sense that, in defiance of the latter, the former became mistress of their belief.

JOHANNES KEPLER:

On the Principal Parts of the World

What do you judge to be the lay-out of the principal parts of the world?

The Philosophy of Copernicus reckons up the principal parts of the world by dividing the figure of the world into regions. For in the sphere, which is the image of God the Creator and the Archetype of the world—as was proved in Book I—there are three regions, symbols of the three persons of the Holy Trinity—the centre, a symbol of the Father; the surface, of the Son; and the intermediate space, of the Holy Ghost. So, too, just as many principal parts of the world have been made—the different parts in the different regions of the sphere: the sun in the centre, the sphere of the fixed stars on the surface, and lastly the planetary system in the region intermediate between the sun and the fixed stars.

I thought the principal parts of the world are reckoned to be the heavens and the earth?

Of course, our uncultivated eyesight from the Earth cannot show us any other more notable parts . . . since we tread upon the one with our feet and are roofed over by the other, and since both parts seem to be commingled and cemented together in the common limbo of the horizon—like a globe in which the stars, clouds, birds, man, and the various kinds of terrestrial animals are enclosed.

But we are practised in the discipline which discloses the causes of things, shakes off the deceptions of eyesight, and carries the mind higher and farther, outside of the boundaries of eyesight. Hence it should not be surprising to anyone that eyesight should learn from reason, that the pupil should learn something new from his master which he did not know before—namely, that the Earth, considered alone and by itself, should not be reckoned among the primary parts of the great world but should be added to one of the primary parts, *i.e.*, to the planetary region, the movable world, and that the Earth has the proportionality of a beginning in that part; and that the sun in turn should

From Johannes Kepler, *Epitome of Copernican Astronomy* IV and V, translated by Charles Glenn Wallis, *Great Books of the Western World*, Volume 16, pp. 853–57. Reprinted with the kind permission of the publishers, Encyclopedia Britannica, Inc.

be separated from the number of stars and set up as one of the principal parts of the whole universe. But I am speaking now of the Earth in so far as it is a part of the edifice of the world, and not of the dignity of the governing creatures which inhabit it.

By what properties do you distinguish these members of the great world from one another?

The perfection of the world consists in light, heat, movement, and the harmony of movements. These are analogous to the faculties of the soul: light, to the sensitive; heat to the vital and the natural; movement, to the animal; harmony, to the rational. And indeed the adornment [*ornatus*] of the world consists in light; its life and growth in heat; and, so to speak, its action, in movement; and its contemplation—wherein Aristotle places blessedness—in harmonies. Now since three things necessarily come together for every affection, namely the cause *a qua*, the subject *in quo*, and the form *sub qua*—therefore, in respect to all the aforesaid affections of the world, the sun exercises the function of the efficient cause; the region of the fixed stars that of the thing forming, containing, and terminating; and the intermediate space, that of the subject—in accordance with the nature of each affection. Accordingly, in all these ways the sun is the principal body of the whole world.

For as regards light; since the sun is very beautiful with light and is as if the eye of the world, like a source of light or very brilliant torch, the sun illuminates, paints, and adorns the bodies of the rest of the world; the intermediate space is not itself light-giving, but light-filled and transparent and the channel through which light is conducted from its source, and there exist in this region the globes and the creatures upon which the light of the sun is poured and which make use of this light. The sphere of the fixed stars plays the role of the river-bed in which this river of light runs and is as it were an opaque and illuminated wall, reflecting and doubling the light of the sun: you have very properly likened it to a latern, which shuts out the winds.

Thus in animals the cerebrum, the seat of the sensitive faculty imparts to the whole animal all its senses, and by the act of common sense causes the presence of all those senses as if arousing them and ordering them to keep watch. And in another way, in this simile, the sun is the image of common sense; the globes in the intermediate space of the sense-organs; and the sphere of the fixed stars of the sensible objects.

As regards heat: the sun is the fireplace [*focus*] of the world; the globes in the intermediate space warm themselves at this fireplace, and the sphere of the fixed stars keeps the heat from flowing out, like the wall of the world, or a skin or garment—to use the metaphor of the Psalm of David. The sun is fire, as the Pythagoreans said, or a red-hot stone or mass, as Democritus said—and the sphere of the fixed stars is ice, or a crystalline sphere, comparatively speaking. But if there is a certain vegetative faculty not only in terrestrial creatures but also in the whole ether throughout the universal amplitude of the world—and both the manifest energy of the sun in warming and physical considerations concerning the origin of comets lead us to draw this inference— it is believable that this faculty is rooted in the sun as in the heart of the world, and that thence by the oarage of light and heat it spreads out into this most wide space of the world—in the way that in animals the seat of heat and of the vital faculty is in the heart and the seat of the vegetative faculty in the liver, whence these faculties by the intermingling of the spirits spread out into the remaining members of the body. The sphere of the fixed stars, situated diametrically opposite on every side, helps this vegetative faculty by concentrating heat, as they say; as it were a kind of skin of the world.

As regards movement: the sun is the first cause of the movement of the planets and the first mover of the universe, even by reason of its own body. In the intermediate space the movables, i.e., the globes of the planets, are laid out. The region of the fixed stars supplies the movables with a place and a base upon which the movables are, as it were, supported; and movement is understood as taking place relative to its absolute immobility. So in animals the cerebellum is the seat of the motor faculty, and the body and its members are that which is moved. The Earth is the base of an animal body; the body, the base of the arm or head; and the arm, the base of the finger. And the movement of each part takes place upon this base as upon something immovable.

Finally, as regards the harmony of the movements: the sun occupies that place in which alone the movements of the planets give the appearance of magnitudes harmonically proportioned [contemperatarum]. The planets themselves, moving in the intermediate space, exhibit the subject or terms, wherein the harmonies are found; the sphere of the fixed stars, or the circle of the zodiac, exhibits the measures whereby the magnitude of the apparent movements is known. So too in man there is the intellect, which abstracts universals and forms numbers and proportions, as things which are not outside of intellect; but individuals [individua], received inwardly through the senses are the foundation of universals; and indivisible [individuae] and discrete unities, of numbers; and real terms of proportions. Finally, memory, divided as it were into compartments of quantities and times, like the sphere of the fixed stars, is the storehouse and repository of sensations. And further, there is never judgment of sensations except in the cerebrum; and the effect of joy never arises from a sense-perception except in the heart.

Accordingly, the aforesaid vegetating corresponds to the nutritive faculty of animals and plants; heating corresponds to the vital faculty; movement, to the animal faculty; light, to the sensitive; and harmony, to the rational. Wherefore most rightly is the sun held to be the heart of the world and the seat of reason and life, and the principal one among three primary members of the world; and these praises are true in the philosophic sense, since the poets honor the sun as the king of the stars, but the Sidonians, Chaldees, and Persians— by an idiom of language observed in German too—as the queen of the heavens, and the Platonists, as the king of intellectual fire.

These three members of the world do not seem to correspond with sufficient neatness to the three regions of a sphere: for the centre is a point, but the sun is a body; and the outer surface is understood to be continuous, yet the region of the fixed stars does not shine as a totality, but is everywhere sown with shining points discrete from one another; and finally, the intermediate part in a sphere fills the whole expanse, but in the world the space between the sun and the fixed stars is not seen to be set in motion as a whole.

As a matter of fact the question indicates the neatest answer concerning the three parts of the world. For since a point could not be clothed or expressed except by some body—and thus the body which is in the centre would fail of the indivisibility of the centre—it was proper that the sphere of the fixed stars should fail of the continuity of a spherical surface, and should burst open in the very minute points of the innumerable fixed stars; and that finally the middle space should not be wholly occupied by movement and the other affections, not be completely transparent, but slightly more dense, since it could not be altogether empty but had to be filled by some body.

Are there solid spheres [orbes] whereon the planets are carried? And are there empty spaces between the spheres?

Tycho Brahe disproved the solidity of the spheres by three reasons: the first from the movement of comets; the second from the fact that light is not refracted; the third from the ratio of the spheres.

For if spheres were solid, the comets would not be seen to cross from one sphere into another, for they would be prevented by the solidity; but they cross from one sphere into another, as Brahe shows.

From light thus: since the spheres are eccentric, and since the Earth and its surface—where the eye is—are not situated at the center of each sphere; therefore if the spheres were solid, that is to say far more dense than that very limpid ether, then the rays of the stars would be refracted before they reached our air, as optics teaches; and so the planet would appear irregularly and in places far different from those which could be predicted by the astronomer.

The third reason comes from the principles of Brahe himself; for they bear witness, as do the Copernican, that Mars is sometimes nearer the Earth than the sun is. But Brahe could not believe this interchange to be possible if the spheres were solid, since the sphere of Mars would have to intersect the sphere of the sun.

ISAAC NEWTON:

Fundamental Principles
of Natural Philosophy
AND
Four Letters to Richard Bentley

Scholium

I DO NOT define time, space, place, and motion, as being well known to all. Only I must observe that the common people conceive those quantities under no other notions but from the relation they bear to sensible objects. And thence arise certain prejudices, for the removing of which it will be convenient to distinguish them into absolute and relative, true and apparent, mathematical and common.

1. Absolute, true, and mathematical time, of itself and from its own nature, flows equably without relation to anything external, and by another name is called 'duration'; relative, apparent, and common time is some sensible and external (whether accurate or unequable) measure of duration by the means of motion, which is commonly used instead of true time, such as an hour, a day, a month, a year.

2. Absolute space, in its own nature, without relation to any thing external, remains always similar and immovable. Relative space is some movable dimension or measure of the absolute spaces, which our senses determine by its position to bodies and which is commonly taken for immovable space; such is the dimension of a subterraneous, an aerial, or celestial space, determined by its position in respect of the earth. Absolute and relative space are the same in figure and magnitude, but they do not remain always numerically the same. For if the earth, for instance, moves, a space of our air, which relatively and in respect of the earth remains always the same, will at one time be one part

From Isaac Newton, "Scholium" (to Definitions) and "General Scholium," from *Philosophiae Naturalis Principia Mathematica*, Andrew Motte's translation revised by Florian Cajori, University of California Press, 1946. Reprinted with the kind permission of the publishers, The University of California Press.
"Four Letters to Richard Bentley" from *Opera Omnia*, IV, pp. 429–42.

of the absolute space into which the air passes; at another time it will be another part of the same, and so, absolutely understood, it will be continually changed.

3. Place is a part of space which a body takes up and is, according to the space, either absolute or relative. I say, a part of space; not the situation nor the external surface of the body. For the places of equal solids are always equal; but their surfaces, by reason of their dissimilar figures, are often unequal. Positions properly have no quantity; nor are they so much the places themselves as the properties of places. The motion of the whole is the same with the sum of the motions of the parts; that is, the translation of the whole, out of its place, is the same thing with the sum of the translations of the parts out of their places; and therefore the place of the whole is the same as the sum of the places of the parts, and for that reason it is internal and in the whole body.

4. Absolute motion is the translation of a body from one absolute place into another, and relative motion the translation from one relative place into another. Thus in a ship under sail the relative place of a body is that part of the ship which the body possesses, or that part of the cavity which the body fills and which therefore moves together with the ship, and relative rest is the continuance of the body in the same part of the ship or of its cavity. But real, absolute rest is the continuance of the body in the same part of that immovable space in which the ship itself, its cavity, and all that it contains is moved. Wherefore, if the earth is really at rest, the body, which relatively rests in the ship, will really and absolutely move with the same velocity which the ship has on the earth. But if the earth also moves, the true and absolute motion of the body will arise, partly from the true motion of the earth in immovable space, partly from the relative motion of the ship on the earth; and if the body moves also relatively in the ship, its true motion will arise, partly from the true motion of the earth in immovable space and partly from the relative motions as well of the ship on the earth as of the body in the ship; and from these relative motions will arise the relative motion of the body on the earth. As if that part of the earth where the ship is was truly moved toward the east with a velocity of 10,010 parts, while the ship itself, with a fresh gale and full sails, is carried toward the west with a velocity expressed by 10 of those parts, but a sailor walks in the ship toward the east with 1 part of the said velocity; then the sailor will be moved truly in immovable space toward the east, with a velocity of 10,001 parts, and relatively on the earth toward the west, with a velocity of 9 of those parts.

Absolute time, in astronomy, is distinguished from relative by the equation or correction of the apparent time. For the natural days are truly unequal, though they are commonly considered as equal and used for a measure of time; astronomers correct this inequality that they may measure the celestial motions by a more accurate time. It may be that there is no such thing as an equable motion whereby time may be accurately measured. All motions may be accelerated and retarded, but the flowing of absolute time is not liable to any change. The duration or perseverance of the existence of things remains the same, whether the motions are swift or slow, or none at all; and therefore this duration ought to be distinguished from what are only sensible measures thereof and from which we deduce it, by means of the astronomical equation. The necessity of this equation, for determining the times of a phenomenon, is evinced as well from the experiments of the pendulum clock as by eclipses of the satellites of Jupiter.

As the order of the parts of time is immutable, so also is the order of the parts of space. Suppose those parts to be moved out of their places, and they will be moved (if the expression may be allowed) out of themselves. For times and spaces are, as it were, the places as well of themselves as of all other things. All things are placed in time as to order of succession and in space as to order of situation. It is from their essence or nature that they are places, and that the primary places of things should be movable is absurd. These are therefore the absolute places, and translations out of those places are the only absolute motions.

But because the parts of space cannot be seen or distinguished from one another by our senses, therefore in their stead we use sensible measures of them. For from the positions and distances of things from any body considered as immovable we define all places; and then, with respect to such places, we estimate all motions, considering bodies as transferred from some of those places into others. And so, instead of absolute places and motions, we use relative ones, and that without any inconvenience in common affairs; but in philosophical disquisitions, we ought to abstract from our senses and consider things themselves, distinct from what are only sensible measures of them. For it may be that there is no body really at rest to which the places and motions of others may be referred.

But we may distinguish rest and motion, absolute and relative, one from the other by their properties, causes, and effects. It is a property of rest that bodies really at rest do rest in respect to one another. And therefore, as it is possible that in the remote regions of the fixed stars, or perhaps far beyond them, there may be some body absolutely at rest, but impossible to know from the position of bodies to one another in our regions whether any of these do keep the same position to that remote body, it follows that absolute rest cannot be determined from the position of bodies in our regions.

It is a property of motion that the parts which retain given positions to their wholes do partake of the motions of those wholes. For all the parts of revolving bodies endeavor to recede from the axis of motion, and the impetus of bodies moving forward arises from the joint impetus of all the parts. Therefore, if surrounding bodies are moved, those that are relatively at rest within them will partake of their motion. Upon which account the true and absolute motion of a body cannot be determined by the translation of it from those which only seem to rest; for the external bodies ought not only to appear at rest, but to be really at rest. For otherwise all included bodies, besides their translation from near the surrounding ones, partake likewise of their true motions; and though that translation were not made, they would not be really at rest, but only seem to be so. For the surrounding bodies stand in the like relation to the surrounded as the exterior part of a whole does to the interior, or as the shell does to the kernel; but if the shell moves, the kernel will also move, as being part of the whole, without any removal from near the shell.

A property near akin to the preceding is this, that if a place is moved, whatever is placed therein moves along with it; and therefore a body which is moved from a place in motion partakes also of the motion of its place. Upon which account all motions, from places in motion, are no other than parts of entire and absolute motions; and every entire motion is composed of the motion of the body out of its first place and the motion of this place out of its place; and so on, until we come to some immovable place, as in the before-mentioned example of the sailor. Wherefore entire and absolute motions cannot be otherwise determined than by immovable places; and for that reason I did before refer those absolute motions to immovable places, but relative ones to movable places. Now no other places are immovable but those that, from infinity to infinity, do all retain the same given position one to another, and upon this account must ever remain unmoved and do thereby constitute immovable space.

The causes by which true and relative motions are distinguished, one from the other, are the forces impressed upon bodies to generate motion. True motion is neither generated nor altered but by some force impressed upon the body moved, but relative motion may be generated or altered without any force impressed upon the body. For it is sufficient only to impress some force on other bodies with which the former is compared that, by their giving way, that relation may be changed in which the relative rest or motion of this other body did consist. Again, true motion suffers always some change from any force impressed upon the moving body, but relative motion does not necessarily undergo any change by such forces. For if the same forces are likewise impressed on those other bodies with which the comparison is made, that the relative position may be preserved, then that condition will be preserved in which the relative motion consists. And therefore any relative motion may be changed when the true motion remains unaltered, and the relative may be preserved when the true suffers some change. Thus, true motion by no means consists in such relations.

The effects which distinguish absolute from relative motion are the forces of receding from the axis of circular motion. For there are no such forces in a circular motion purely relative, but in a true and absolute circular motion they are greater or less, according to the quantity of the motion. If a vessel, hung by a long cord, is so often turned about that the cord is strongly twisted, then filled with water and held at rest together with the water, thereupon by the sudden action of another force it is whirled about the contrary way, and while the cord is untwisting itself the vessel continues for some time in this

motion, the surface of the water will at first be plain, as before the vessel began
to move; but after that the vessel, by gradually communicating its motion
to the water, will make it begin sensibly to revolve and recede by little and
little from the middle, and ascend to the sides of the vessel, forming itself into
a concave figure (as I have experienced); and the swifter the motion becomes,
the higher will the water rise, till at last, performing its revolutions in the
same times with the vessel, it becomes relatively at rest in it. This ascent of the
water shows its endeavor to recede from the axis of its motion; and the true
and absolute circular motion of the water, which is here directly contrary to
the relative, becomes known and may be measured by this endeavor. At first,
when the relative motion of the water in the vessel was greatest, it produced
no endeavor to recede from the axis; the water showed no tendency to the
circumference, nor any ascent toward the sides of the vessel, but remained of
a plain surface, and therefore its true circular motion had not yet begun. But
afterward, when the relative motion of the water had decreased, the ascent
thereof toward the sides of the vessel proved its endeavor to recede from the
axis; and this endeavor showed the real circular motion of the water continually increasing, till it had acquired its greatest quantity, when the water
rested relatively in the vessel. And therefore this endeavor does not depend
upon any translation of the water in respect of the ambient bodies; nor can
true circular motion be defined by such translation. There is only one real
circular motion of any one revolving body, corresponding to only one power
of endeavoring to recede from its axis of motion, as its proper and adequate
effect; but relative motions, in one and the same body, are innumerable,
according to the various relations it bears to external bodies, and, like other
relations, are altogether destitute of any real effect, any otherwise than they
may perhaps partake of that one only true motion. And therefore in their
system who suppose that our heavens, revolving below the sphere of the fixed
stars, carry the planets along with them, the several parts of those heavens
and the planets, which are indeed relatively at rest in their heavens, do yet
really move. For they change their position one to another (which never
happens to bodies truly at rest) and, being carried together with their heavens,
partake of their motions and, as parts of revolving wholes, endeavor to recede
from the axis of their motions.

Wherefore relative quantities are not the quantities themselves whose
names they bear, but those sensible measures of them (either accurate or
inaccurate) which are commonly used instead of the measured quantities
themselves. And if the meaning of words is to be determined by their use,
then by the names 'time,' 'space,' 'place,' and 'motion' their [sensible] measures
are properly to be understood; and the expression will be unusual, and purely
mathematical, if the measured quantities themselves are meant. On this
account, those violate the accuracy of language, which ought to be kept
precise, who interpret these words for the measured quantities. Nor do those

less defile the purity of mathematical and philosophical truths who confound real quantities with their relations and sensible measures.

It is indeed a matter of great difficulty to discover and effectually to distinguish the true motions of particular bodies from the apparent, because the parts of that immovable space in which those motions are performed do by no means come under the observation of our senses. Yet the thing is not altogether desperate; for we have some arguments to guide us, partly from the apparent motions, which are the differences of the true motions; partly from the forces, which are the causes and effects of the true motions. For instance, if two globes, kept at a given distance one from the other by means of a cord that connects them, were revolved about their common center of gravity, we might, from the tension of the cord, discover the endeavor of the globes to recede from the axis of their motion, and from thence we might compute the quantity of their circular motions. And then if any equal forces should be impressed at once on the alternate faces of the globes to augment or diminish their circular motions, from the increase or decrease of the tension of the cord we might infer the increment or decrement of their motions, and thence would be found on what faces those forces ought to be impressed that the motions of the globes might be most augmented; that is, we might discover their hindmost faces, or those which, in the circular motion, do follow. But the faces which follow being known, and consequently the opposite ones that precede, we should likewise know the determination of their motions. And thus we might find both the quantity and the determination of this circular motion, even in an immense vacuum, where there was nothing external or sensible with which the globes could be compared. But now, if in that space some remote bodies were placed that kept always a given position one to another, as the fixed stars do in our regions, we could not indeed determine from the relative translation of the globes among those bodies whether the motion did belong to the globes or to the bodies. But if we observed the cord and found that its tension was that very tension which the motions of the globes required, we might conclude the motion to be in the globes and the bodies to be at rest; and then, lastly, from the translation of the globes among the bodies, we should find the determination of their motions. But how we are to obtain the true motions from their causes, effects, and apparent differences, and the converse, shall be explained more at large in the following treatise. For to this end it was that I composed it.

General Scholium

The hypothesis of vortices is pressed with many difficulties. That every planet by a radius drawn to the sun may describe areas proportional to the times of description, the periodic times of the several parts of the vortices

208 ISAAC NEWTON

should observe the square of their distances from the sun; but that the periodic times of the planets may obtain the 3/2th power of their distances from the sun, the periodic times of the parts of the vortex ought to be as the 3/2th power of their distances. That the smaller vortices may maintain their lesser revolutions about Saturn, Jupiter, and other planets, and swim quietly and undisturbed in the greater vortex of the sun, the periodic times of the parts of the sun's vortex should be equal; but the rotation of the sun and planets about their axes, which ought to correspond with the motions of their vortices, recede far from all these proportions. The motions of the comets are exceedingly regular, are governed by the same laws with the motions of the planets, and can by no means be accounted for by the hypothesis of vortices; for comets are carried with very eccentric motions through all parts of the heavens indifferently, with a freedom that is incompatible with the notion of a vortex.

Bodies projected in our air suffer no resistance but from the air. Withdraw the air, as is done in Mr. Boyle's vacuum, and the resistance ceases; for in this void a bit of fine down and a piece of solid gold descend with equal velocity. And the same argument must apply to the celestial spaces above the earth's atmosphere; in these spaces, where there is no air to resist their motions, all bodies will move with the greatest freedom; and the planets and comets will constantly pursue their revolutions in orbits given in kind and position, according to the laws above explained; but though these bodies may, indeed, continue in their orbits by the mere laws of gravity, yet they could by no means have at first derived the regular position of the orbits themselves from those laws.

The six primary planets are revolved about the sun in circles concentric with the sun, and with motions directed toward the same parts and almost in the same plane. Ten moons are revolved about the earth, Jupiter, and Saturn, in circles concentric with them, with the same direction of motion, and nearly in the planes of the orbits of those planets; but it is not to be conceived that mere mechanical causes could give birth to so many regular motions, since the comets range over all parts of the heavens in very eccentric orbits; for by that kind of motion they pass easily through the orbs of the planets, and with great rapidity; and in their aphelions, where they move the slowest and are detained the longest, they recede to the greatest distances from each other, and hence suffer the least disturbance from their mutual attractions. This most beautiful system of the sun, planets, and comets could only proceed from the counsel and dominion of an intelligent and powerful Being. And if the fixed stars are the centers of other like systems, these, being formed by the like wise counsel, must be all subject to the dominion of One, especially since the light of the fixed stars is of the same nature with the light of the sun and from every system light passes into all the other systems; and lest the systems of the fixed stars should, by their gravity, fall on each other, he hath placed those systems at immense distances from one another.

This Being governs all things, not as the soul of the world, but as Lord over all; and on account of his dominion he is wont to be called "Lord God" παντοκράτωρ, or "Universal Ruler"; for 'God' is a relative word and has a respect to servants; and Deity is the dominion of God, not over his own body, as those imagine who fancy God to be the soul of the world, but over servants. The Supreme God is a Being eternal, infinite, absolutely perfect, but a being, however perfect, without dominion, cannot be said to be "Lord God"; for we say "my God," "your God," "the God of Israel," "the God of Gods," and "Lord of Lords," but we do not say "my Eternal," "your Eternal," "the Eternal of Israel," "the Eternal of Gods"; we do not say "my Infinite," or "my Perfect": these are titles which have no respect to servants. The word 'God'[1] usually signifies 'Lord,' but every lord is not a God. It is the dominion of a spiritual being which constitutes a God; a true, supreme, or imaginary dominion makes a true, supreme, or imaginary God. And from his true dominion it follows that the true God is a living, intelligent, and powerful Being; and, from his other perfections, that he is supreme or most perfect. He is eternal and infinite, omnipotent and omniscient; that is, his duration reaches from eternity to eternity; his presence from infinity to infinity; he governs all things and knows all things that are or can be done. He is not eternity and infinity, but eternal and infinite; he is not duration or space, but he endures and is present. He endures forever and is everywhere present; and, by existing always and everywhere, he constitutes duration and space. Since every particle of space is *always*, and every indivisible moment of duration is *everywhere*, certainly the Maker and Lord of all things cannot be *never* and *nowhere*. Every soul that has perception is, though in different times and in different organs of sense and motion, still the same indivisible person. There are given successive parts in duration, coexistent parts in space, but neither the one nor the other in the person of a man or his thinking principle; and much less can they be found in the thinking substance of God. Every man, so far as he is a thing that has perception, is one and the same man during his whole life, in all and each of his organs of sense. God is the same God, always and everywhere. He is omnipresent not *virtually* only but also *substantially;* for virtue cannot subsist without substance. In him[2] are all things contained and moved, yet neither affects the other; God suffers nothing from the motion of bodies, bodies find no resistance from the omnipresence of God. It is allowed by all that the Supreme God exists necessarily, and by the same necessity he exists *always* and

---

1. Dr. Pocock derives the Latin word *'Deus'* from the Arabic *'du'* (in the oblique case *'di'*), which signifies 'Lord.' And in this sense princes are called "gods," Psalm lxxxii. ver. 6; and John x. ver. 35. And Moses is called a "god" to his brother Aaron, and a "god" to Pharaoh (Exodus iv. ver. 16; and vii. ver. 1). And in the same sense the souls of dead princes were formerly, by the Heathens, called "gods," but falsely, because of their want of dominion.

2. This was the opinion of the Ancients. So Pythagoras, in *Cicer. de Nat. Deor.* lib. i. Thales, Anaxagoras, Virgil, *Georg.* lib. iv. ver. 220; and *Aeneid,* lib. vi. ver. 721. *Philo Allegor,* at the beginning of lib. i. Aratus, in his *Phaenom,* at the beginning. So also the sacred writers: as St. Paul, Acts xvii. ver. 27, 28. St. John's Gosp. chap. xiv. ver. 2. Moses, in Deuteronomy iv. ver. 39; and x. ver. 14. David, Psalm cxxxix, ver. 7, 8, 9. Solomon 1 Kings viii. ver. 27. Job xxii. ver. 12, 13, 14. Jeremiah, xxiii. ver. 23, 24. The Idolaters supposed the sun, moon, and stars, the souls of men, and other parts of the world to be parts of the Supreme God, and therefore to be worshiped; but erroneously.

everywhere. Whence also he is all similar, all eye, all ear, all brain, all arm, all power to perceive, to understand, and to act; but in a manner not at all human, in a manner not at all corporeal, in a manner utterly unknown to us. As a blind man has no idea of colors, so have we no idea of the manner by which the all-wise God perceives and understands all things. He is utterly void of all body and bodily figure, and can therefore neither be seen nor heard nor touched; nor ought he to be worshiped under the representation of any corporeal thing. We have ideas of his attributes, but what the real substance of anything is we know not. In bodies we see only their figures and colors, we hear only the sounds, we touch only their outward surfaces, we smell only the smells and taste the savors, but their inward substances are not to be known either by our senses or by any reflex act of our minds; much less, then, have we any idea of the substance of God. We know him only by his most wise and excellent contrivances of things and final causes; we admire him for his perfections, but we reverence and adore him on account of his dominion, for we adore him as his servants; and a god without dominion, providence, and final causes is nothing else but Fate and Nature. Blind metaphysical necessity, which is certainly the same always and everywhere, could produce no variety of things. All that diversity of natural things which we find suited to different times and places could arise from nothing but the ideas and will of a Being necessarily existing. But, by way of allegory, God is said to see, to speak, to laugh, to love, to hate, to desire, to give, to receive, to rejoice, to be angry, to fight, to frame, to work, to build; for all our notions of God are taken from the ways of mankind by a certain similitude, which, though not perfect, has some likeness, however. And thus much concerning God, to discourse of whom from the appearances of things does certainly belong to natural philosophy.

Hitherto we have explained the phenomena of the heavens and of our sea by the power of gravity, but have not yet assigned the cause of this power. This is certain, that it must proceed from a cause that penetrates to the very centers of the sun and planets, without suffering the least diminution of its force; that operates not according to the quantity of the surfaces of the particles upon which it acts (as mechanical causes used to do), but according to the quantity of the solid matter which they contain, and propagates its virtue on all sides to immense distances, decreasing always as the inverse square of the distances. Gravitation toward the sun is made up out of the gravitations toward the several particles of which the body of the sun is composed, and in receding from the sun decreases accurately as the inverse square of the distances as far as the orbit of Saturn, as evidently appears from the quiescence of the aphelion of the planets; nay, and even to the remotest aphelion of the comets, if those aphelions are also quiescent. But hitherto I have not been able to discover the cause of those properties of gravity from phenomena, and I frame no hypotheses; for whatever is not deduced from the phenomena is to be called a hypothesis, and hypotheses, whether metaphysical or physical, whether of occult qualities or mechanical, have no place in experimental philosophy. In this philosophy particular propositions are inferred from the phenomena and afterward rendered general by induction. Thus it was that the impenetrability, the mobility, and the impulsive force of bodies, and the laws of motion and of gravitation, were discovered. And to us it is enough that gravity does really exist and act according to the laws which we have explained, and abundantly serves to account for all the motions of the celestial bodies and of our sea.

And now we might add something concerning a certain most subtle spirit which pervades and lies hid in all gross bodies, by the force and action of which spirit the particles of bodies attract one another at near distances and cohere, if contiguous; and electric bodies operate to greater distances, as well repelling as attracting the neighboring corpuscles; and light is emitted, reflected, refracted, inflected, and heats bodies; and all sensation is excited, and the members of animal bodies move at the command of the will, namely, by the vibrations of this spirit, mutually propagated along the solid filaments of the nerves, from the outward organs of sense to the brain and from the brain into the muscles. But these are things that cannot be explained in few words; nor are we furnished with that sufficiency of experiment which is required to an accurate determination and demonstration of the laws by which this electric and elastic spirit operates.

Four Letters to Richard Bentley

I

To the Reverend Dr. Richard Bentley, at the Bishop of Worcester's House, in Park Street, Westminster

Sir,

When I wrote my treatise about our system, I had an eye upon such principles as might work with considering men for the belief of a Deity; and nothing can rejoice me more than to find it useful for that purpose. But if I have done the public any service this way, it is due to nothing but industry and patient thought.

As to your first query, it seems to me that if the matter of our sun and planets and all the matter of the universe were evenly scattered throughout all the heavens, and every particle had an innate gravity toward all the rest, and the whole space throughout which this matter was scattered was but finite, the matter on the outside of this space would, by its gravity, tend toward all the matter on the inside and, by consequence, fall down into the middle of the whole space and there compose one great spherical mass. But if the matter was evenly disposed throughout an infinite space, it could never convene into one mass; but some of it would convene into one mass and some into another, so as to make an infinite number of great masses, scattered at great distances from one to another throughout all that infinite space. And thus might the sun and fixed stars be formed, supposing the matter were of a lucid nature. But how the matter should divide itself into two sorts, and that part of it which is fit to compose a shining body should fall down into one mass and make a sun and the rest which is fit to compose an opaque body should coalesce, not into one great body, like the shining matter, but into many little ones; or if the sun at first were an opaque body like the planets or the planets lucid bodies like the sun, how he alone should be changed into a shining body whilst all they continue opaque, or all they be changed into opaque ones whilst he remains unchanged, I do not think explicable by mere natural causes, but am forced to ascribe it to the counsel and contrivance of a voluntary Agent.

The same Power, whether natural or supernatural, which placed the sun in the center of the six primary planets, placed Saturn in the center of the orbs of his five secondary planets and Jupiter in the center of his four secondary planets, and the earth in the center of the moon's orb; and therefore, had this cause been a blind one, without contrivance or design, the sun would have been a body of the same kind with Saturn, Jupiter, and the earth, that is, without light and heat. Why there is one body in our system qualified to give light and heat to all the rest, I know no reason but because the Author of the system thought it convenient; and why there is but one body of this kind, I know no reason but because one was sufficient to warm and enlighten all the rest. For the Cartesian hypothesis of suns losing their light and then turning into comets, and comets into planets, can have no place in my system and is plainly erroneous; because it is certain that, as often as they appear to us, they descend into the system of our planets, lower than the orb of Jupiter and sometimes lower than the orbs of Venus and Mercury, and yet never stay here, but always return from the sun with the same degrees of motion by which they approached him.

To your second query, I answer that the motions which the planets now have could not spring from any natural cause alone, but were impressed by an intelligent Agent. For since comets descend into the region of our planets and here move all manner of ways, going sometimes the same way with the planets, sometimes the contrary way, and sometimes in crossways, in planes inclined to the plane of the ecliptic and at all kinds of angles, it is plain that there is no natural cause which could determine all the planets, both primary and secondary, to move the same way and in the same plane, without any considerable variation; this must have been the effect of counsel. Nor is there any natural cause which could give the planets those just degrees of velocity, in proportion to their distances from the sun and other central bodies, which were requisite to make them move in such concentric orbs about those bodies. Had the planets been as swift as comets, in proportion to their distances from the sun (as they would have been had their motion been caused by their gravity, whereby the matter, at the first formation of the planets, might fall from the remotest regions toward the sun), they would not move in concentric orbs, but in such eccentric ones as the comets move in. Were all the planets as swift as Mercury or as slow as Saturn or his satellites, or were their several velocities otherwise much greater or less than they are, as they might have been had they arose from any other cause than their gravities, or had the distances from the centers about which they move been greater or less than they are, with the same velocities, or had the quantity of matter in the sun or in Saturn, Jupiter, and the earth, and by consequence their gravitating power, been greater or less than it is, the primary planets could not have revolved about the sun nor the secondary onces about Saturn, Jupiter, and the earth, in concentric circles, as they do, but would have moved in hyperbolas or parabolas or in ellipses very eccentric. To make this system, therefore, with all its motions, required a cause which understood and compared together the quantities of matter in the several bodies of the sun and planets and the gravitating powers resulting from thence, the several distances of the primary planets from the sun and of the secondary ones from Saturn, Jupiter, and the earth, and the velocities with which these planets could revolve about those quantities of matter in the central bodies; and to compare and adjust all these things together, in so great a variety of bodies, argues that cause to be, not blind and fortuitous, but very well skilled in mechanics and geometry.

To your third query, I answer that it may be represented that the sun may, by heating those planets most of which are nearest to him, cause them to be better concocted and more condensed by that concoction. But when I consider that our earth is much more heated in its bowels below the upper crust by subterraneous fermentations of mineral bodies than by the sun, I see not why the interior parts of Jupiter and Saturn might not be as much heated, concocted, and coagulated by those fermentations as our earth is; and therefore this various density should have some other cause than the various distances of the planets from the sun. And I am confirmed in this opinion by considering that the planets of Jupiter and Saturn, as they are rarer than the rest, so they are vastly greater and contain a far greater quantity of matter, and have many satellites about them; which qualifications surely arose, not from their being placed at so great a distance from the sun, but were rather the cause why the Creator placed them at great distance. For, by their gravitating powers, they disturb one another's motions very sensibly, as I find by some late observations of Mr. Flamsteed; and had they been placed much nearer to the sun and to one another, they would, by the same powers, have caused a considerable disturbance in the whole system.

To your fourth query, I answer that, in the hypothesis of vortices, the inclination of the axis of the earth might, in my opinion, be ascribed to the situation of the earth's vortex before it was absorbed by the neighboring vortices and the earth turned from a sun to a comet; but this inclination ought to decrease constantly in compliance with the motion of the earth's vortex, whose axis is much less inclined to the ecliptic, as appears by the motion of the moon carried about therein. If the sun by his rays could carry about the planets, yet I do not see how he could thereby effect their diurnal motions.

Lastly, I see nothing extraordinary in the inclination of the earth's axis for proving a Deity, unless you will urge it as a contrivance for winter and summer, and for making the earth habitable toward the poles; and that the diurnal rotations of the sun and planets, as they could hardly arise from any cause purely mechanical, so by being determined all the same way with the annual and menstrual motions they seem to make up that harmony in the system which, as I explained above, was the effect of choice rather than chance.

There is yet another argument for a Deity, which I take to be a very strong one; but till the principles on which it is grounded are better received, I think it more advisable to let it sleep.

I am your most humble servant to command,

Is. NEWTON

Cambridge, December 10, 1692

II

FOR MR. BENTLEY, AT THE PALACE AT WORCESTER

Sir,

I agree with you that if matter evenly diffused through a finite space, not spherical, should fall into a solid mass, this mass would affect the figure of the whole space, provided it were not soft, like the old chaos, but so hard and solid from the beginning that the weight of its protuberant parts could not make it yield to their pressure; yet, by earthquakes loosening the parts of this solid, the protuberances might sometimes sink a little by their weight, and thereby the mass might by degrees approach a spherical figure.

The reason why matter evenly scattered through a finite space would convene in the midst you conceive the same with me, but that there should be a central particle so accurately placed in the middle as to be always equally attracted on all sides, and thereby continue without motion, seems to me a supposition fully as hard as to make the sharpest needle stand upright on its point upon a looking glass. For if the very mathematical center of the central particle be not accurately in the very mathematical center of the attractive power of the whole mass, the particle will not be attracted equally on all sides. And much harder it is to suppose all the particles in an infinite space should be so accurately poised one among another as to stand still in a perfect equilibrium. For I reckon this as hard as to make, not one needle only, but an infinite number of them (so many as there are particles in an infinite space) stand accurately poised upon their points. Yet I grant it possible, at least by a divine power; and if they were once to be placed, I agree with you that they would continue in that posture without motion forever, unless put into new motion by the same power. When, therefore, I said that matter evenly spread through all space would convene by its gravity into one or more great masses, I understand it of matter not resting in an accurate poise.

But you argue, in the next paragraph of your letter, that every particle of matter in an infinite space has an infinite quantity of matter on all sides and, by consequence, an infinite attraction every way, and therefore must rest *in equilibrio,* because all infinites are equal. Yet you suspect a paralogism in this argument, and I conceive the paralogism lies in the position that all infinites are equal. The generality of mankind consider infinites no other ways than indefinitely; and in this sense they say all infinites are equal, though they would speak more truly if they should say they are neither equal nor unequal, nor have any certain difference or proportion one to another. In this sense, therefore, no conclusions can be drawn from them about the equality, proportions, or differences of things; and they that attempt to do it usually fall into paralogisms. So when men argue against the infinite divisibility of magnitude by saying that if an inch may be divided into an infinite number of parts the sum of those parts will be an inch; and if a foot may be divided into an infinite number of parts the sum of those parts must be a foot; and therefore, since all infinites are equal, those sums must be equal, that is, an inch equal to a foot.

The falseness of the conclusion shows an error in the premises, and the error lies in the position that all infinites are equal. There is, therefore, another way of considering infinites used by mathematicians, and that is, under certain definite restrictions and limitations, whereby infinites are determined to have certain differences or proportions to one another. Thus Dr. Wallis considers them in his *Arithmetica Infinitorum,* where, by the various proportions of infinite sums, he gathers the various proportions of infinite magnitudes, which way of arguing is generally allowed by mathematicians and yet would not be good were all infinites equal. According to the same way of considering infinites, a mathematician would tell you that, though there be an infinite number of infinite little parts in an inch, yet there is twelve times that number of such parts in a foot; that is, the infinite number of those parts in a foot is not equal to but twelve times bigger than the infinite number of them in an inch. And so a mathematician will tell you that if a body stood *in equilibrio* between any two equal and contrary attracting infinite forces, and if to either of these forces you add any new finite attracting force, that new force, howsoever little, will destroy their equilibrium and put the body into the same motion into which it would put it were those two contrary equal forces but finite or even none at all; so that in this case the two equal infinites, by the addition of a finite to either of them, become unequal in our ways of reckoning; and after these ways we must reckon, if from the considerations of infinites we would always draw true conclusions.

To the last part of your letter, I answer, first, that if the earth (without the moon) were placed anywhere with its center in the *orbis magnus* and stood still there without any gravitation or projection, and there at once were infused into it both a gravitating energy toward the sun and a transverse impulse of a just quantity moving it directly in a tangent to the *orbis magnus,* the compounds of this attraction and projection would, according to my notion, cause a circular revolution of the earth about the sun. But the transverse impulse must be a just quantity; for if it be too big or too little, it will cause the earth to move in some other line. Secondly, I do not know any power in nature which would cause this transverse motion without the divine arm. Blondel tells us somewhere in his book of Bombs that Plato affirms that the motion of the planets is such as if they had all of them been created by God in some region very remote from our system and let fall from thence toward the sun, and so soon as they arrived at their several orbs their motion of falling turned aside into a transverse one. And this is true, supposing the gravitating power of the sun was double at that moment of time in which they all arrive at their several orbs; but then the divine power is here required in a double respect, namely, to turn the descending motions of the falling planets into a side motion and, at the same time, to double the attractive power of the sun. So, then, gravity may put the planets into motion, but without the divine power it could never put them into such a circulating motion as they have about the sun; and therefore, for this as well as other reasons, I am compelled to ascribe the frame of this system to an intelligent Agent.

You sometimes speak of gravity as essential and inherent to matter. Pray do not ascribe that notion to me, for the cause of gravity is what I do not pretend to know and therefore would take more time to consider of it.

I fear what I have said of infinites will seem obscure to you; but it is enough if you understand that infinites, when considered absolutely without any restriction or limitation, are neither equal nor unequal, nor have any certain proportion one to another, and therefore the principle that all infinites are equal is a precarious one.

Sir, I am your most humble servant,

Is. NEWTON

Trinity College, January 17, 1692/3

III

For Mr. Bentley, at the Palace at Worcester

Sir,

Because you desire speed, I will answer your letter with what brevity I can. In the six positions you lay down in the beginning of your letter, I agree with you. Your assuming the *orbis magnus* 7,000 diameters of the earth wide implies the sun's horizontal parallax to be half a minute. Flamsteed and Cassini have of late observed it to be about 10 minutes, and thus the *orbis magnus* must be 21,000, or, in a round number, 20,000 diameters of the earth wide. Either computation, I think, will do well; and I think it not worth while to alter your numbers.

In the next part of your letter you lay down four other positions, founded upon the six first. The first of these four seems very evident, supposing you take attraction so generally as by it to understand any force by which distant bodies endeavor to come together without mechanical impulse. The second seems not so clear, for it may be said that there might be other systems of worlds before the present ones and others before those, and so on to all past eternity, and by consequence that gravity may be coeternal to matter and have the same effect from all eternity as at present, unless you have somewhere proved that old systems cannot gradually pass into new ones or that this system had not its original from the exhaling matter of former decaying systems but from a chaos of matter evenly dispersed throughout all space; for something of this kind, I think you say, was the subject of your Sixth Sermon, and the growth of new systems out of old ones, without the mediation of a divine power, seems to me apparently absurd.

The last clause of the second position I like very well. It is inconceivable that inanimate brute matter should, without the mediation of something else which is not material, operate upon and affect other matter without mutual contact, as it must be if gravitation, in the sense of Epicurus, be essential and inherent in it. And this is one reason why I desired you would not ascribe innate gravity to me. That gravity should be innate, inherent, and essential to matter, so that one body may act upon another at a distance through a *vacuum,* without the mediation of anything else, by and through which their action and force may be conveyed from one to another, is to me so great an absurdity that I believe no man who has in philosophical matters a competent faculty of thinking can ever fall into it. Gravity must be caused by an agent acting constantly according to certain laws, but whether this agent be material or immaterial I have left to the consideration of my readers.

Your fourth assertion, that the world could not be formed by innate gravity alone, you confirm by three arguments. But in your first argument you seem to make a *petitio principii;* for whereas many ancient philosophers and others, as well theists as atheists, have all allowed that there may be worlds and parcels of matter innumerable or infinite, you deny this by representing it as absurd as that there should be positively an infinite arithmetical sum or number, which is a contradiction *in terminis,* but you do not prove it as absurd. Neither do you prove that what men mean by an infinite sum or number is a contradiction in nature, for a contradiction *in terminis* implies no more than an impropriety of speech. Those things which men underderstand by improper and contradictious phrases may be sometimes really in nature without any contradiction at all: a silver inkhorn, a paper lantern, an iron whetstone, are absurd phrases, yet the things signified thereby are really in nature. If any man should say that a number and a sum, to speak properly, is that which may be numbered and summed, but things infinite are numberless or, as we usually speak, innumerable and sumless or insummable, and therefore ought not to be called a number or sum, he will speak properly enough, and your argument against him will, I fear, lose its force. And yet if any man shall take the words 'number' and 'sum' in a larger sense, so as to understand thereby things which, in the proper way of speaking, are numberless and sumless (as you seem to do when you allow an infinite number of points in a line), I could readily allow him the use of the contradictious phrases of 'innumerable number' or 'sumless sum,' without inferring from thence any absurdity in the thing he means by those phrases. However, if by this or any other argument you have proved the finiteness of the universe, it follows that all matter would fall down from the outsides and convene in the middle. Yet the matter in falling might concrete into many round masses, like the bodies of the planets, and these, by attracting one another, might acquire an obliquity of descent by means of which they might fall, not upon the great central body, but upon the side of it, and fetch a compass about and then ascend again by the same steps and degrees of motion and velocity with which they descended before, much after the manner that comets revolve about the sun; but a circular motion in concentric orbs about the sun they could never acquire by gravity alone.

And though all the matter were divided at first into several systems, and every system by a divine power constituted like ours, yet would the outside systems descend toward the middlemost; so that this frame of things could not always subsist without a divine power to conserve it, which is the second argument; and to your third I fully assent.

As for the passage of Plato, there is no common place from whence all the planets, being let fall and descending with uniform and equal gravities (as Galileo supposes), would, at their arrival to their several orbs, acquire their several velocities with which they now revolve in them. If we suppose the gravity of all the planets toward the sun to be of such a quantity as it really is, and that the motions of the planets are turned upward, every planet will ascend to twice its height from the sun. Saturn will ascend till he be twice as high from the sun as he is at present, and no higher; Jupiter will ascend as high again as at present, that is, a little above the orb of Saturn; Mercury will ascend to twice his present height, that is, to the orb of Venus; and so of the rest; and then, by falling down again from the places to which they ascended, they will arrive again at their several orbs with the same velocities they had at first and with which they now revolve.

But if, so soon as their motions by which they revolve are turned upward, the gravitating power of the sun, by which their ascent is perpetually retarded, be diminished by one half, they will now ascend perpetually, and all of them at equal distances from the sun will be equally swift. Mercury, when he arrives at the orb of Venus, will be as swift as Venus; and he and Venus, when they arrive at the orb of the earth, will be as swift as the earth; and so of the rest. If they begin all of them to ascend at once and ascend in the same line, they will constantly, in ascending, become nearer and nearer together, and their motions will constantly approach to an equality and become at length slower than any motion assignable. Suppose, therefore, that they ascended till they were almost contiguous and their motions inconsiderably little, and that all their motions were at the same moment of time turned back again or, which comes almost to the same thing, that they were only deprived of their motions and let fall at that time; they would all at once arrive at their several orbs, each with the velocity it had at first, and if their motions were then turned sideways and at the same time the gravitating power of the sun doubled, that it might be strong enough to retain them in their orbs, they would revolve in them as before their ascent. But if the gravitating power of the sun was not doubled, they would go away from their orbs into the highest heavens in parabolical lines. These things follow from my *Principia Mathematica*, Book I, Propositions XXXIII, XXXIV, XXXVI, XXXVII.

I thank you very kindly for your designed present, and rest

Your most humble servant to command,

Is. NEWTON

Cambridge, February 25, 1692/3

IV

FOR MR. BENTLEY, AT THE PALACE AT WORCESTER

Sir,

The hypothesis of deriving the frame of the world by mechanical principles from matter evenly spread through the heavens being inconsistent with my system, I had considered it very little before your letters put me upon it, and therefore trouble you with a line or two more about it, if this comes not too late for your use.

In my former I represented that the diurnal rotations of the planets could not be derived from gravity, but required a divine arm to impress them. And though gravity might give the planets a motion of descent toward the sun, either directly or with some little obliquity, yet the transverse motions by which they revolve in their several orbs required the divine arm to impress them according to the tangents of their orbs. I would now add that the hypothesis of matters being at first evenly spread through the heavens is, in my opinion, inconsistent with the hypothesis of innate gravity, without a supernatural power to reconcile them; and therefore it infers a Deity. For if there be innate gravity, it is impossible now for the matter of the earth and all the planets and stars to fly up from them, and become evenly spread throughout all the heavens, without a supernatural power; and certainly that which can never be hereafter without a supernatural power could never be heretofore without the same power.

You queried whether matter evenly spread throughout a finite space, of some other figure than spherical, would not, in falling down toward a central body, cause that body to be of the same figure with the whole space, and I answered yes. But in my answer it is to be supposed that the matter descends directly downward to that body and that that body has no diurnal rotation. This, sir, is all I would add to my former letters.

I am your most humble servant,

Is. NEWTON

Cambridge, February 11, 1693

CHRISTIAAN HUYGENS:

Cosmotheoros

BEFORE the invention of telescopes, it seemed to contradict Copernicus's opinion to make the Sun one of the fixed stars. For the stars of the first magnitude being esteemed to be about three minutes diameter; and Copernicus (observing that though the Earth changed its place, they always kept the same distance from us) having ventured to say that the *magnus orbis* was but a point in respect of the sphere in which they were placed, it was a plain consequence that every one of them that appeared any thing bright, must be larger than the path or orbit of the Earth: which is very absurd. This is the principal argument that Tycho Brahe set up against Copernicus. But when the telescopes took away those rays of the stars which appear when we look upon them with our naked eye, (which they do best when the eyeglass is blacked with smoke) they seemed just like little shining points, and then that difficulty vanished, and the stars may yet be so many suns. Which is the more probable, because their light is certainly their own: for it's impossible that ever the Sun should send, or they reflect it at such a vast distance. This is the opinion that commonly goes along with Copernicus's system. And the patrons of it do also with reason suppose, that all these stars are not in the same sphere, as well because there is no argument for it, as that the Sun, which is one of them, cannot be brought to this rule. But it is more likely they are scattered and dispersed all over the immense spaces of the heaven, and are as far distant perhaps from one another, as the nearest of them are from the Sun. Here again too I know Kepler is of another opinion in his epitome of Copernicus's system. . . . For though he agrees with us, that the stars are diffused through all the vast expanse of the heavens, yet he cannot allow that they have as large an empty space about them as our sun has. For then it was his opinion, we should see but very few, and those of very different magnitudes: "For, seeing the largest of all appear so small to us, that we can scarce observe or measure them with our best instruments; how must those appear that are three or four times farther from us? Why, supposing them

From an English translation of *Cosmotheoros* under the title, *The Celestial Worlds Discovered or Conjectures Concerning the Inhabitants, Plants and Productions of the Worlds in the Planets*, London, 1698. The following selection is taken from an edition printed for Robert Urie, Glasgow, 17—(?), pp. 110 ff.

no larger than these, they must seem three or four times less, and so on till a little farther they will not be to be seen at all: thus we shall have the sight of but very few stars, and those different one from another"; whereas we have above a thousand, and those not considerably bigger or less than one another. But this by no means proves what he would have it; and his mistake was chiefly, that he did not consider the nature of fire and flame which may be seen at such distances, and at such small angles as all other bodies would totally disappear under. A thing we need go no farther than the lamps set along the streets to prove. For although they are a hundred foot from one another, yet you may count twenty of them in a continued row with your eyes, and yet the twentieth part of them scarce makes an angle of six seconds. Certainly then the glorious light of the stars do much more than this; so that it is no wonder we should see a thousand or two of them with our bare eyes, and with a telescope discover twenty times that number. But Kepler had a private design in making the sun thus superior to all the other stars, and planting it in the middle of the world, attended with the planets; for his aim was hereby to strengthen his cosmographical mystery, that the distances of the planets from the sun are in a certain proportion to the diameters of the spheres that are inscribed within, and circumscribed about Euclid's regular bodies. Which could never be so much as probable, except there were but one chorus of planets moving round the sun, and so the sun were the only one of his kind.

But that whole mystery is nothing but an idle dream taken from Pythagoras or Plato's philosophy. And the author himself acknowledges that the proportions do not agree so well as they should, and is fain to invent two or three very silly excuses for it. And he uses yet poorer arguments to prove that the universe is of a spherical figure, and that the number of stars must necessarily be finite, because the magnitude of each of them is so. But what is worst of all is that he settles the space between the sun and the concavity of the sphere of the fixed stars, to be six hundred thousand of the earth's diameters: for this reason, which he has no foundation for, that as the diameter of the sun is to that of the orbit of Saturn, which he makes to be 1 to 2000, so is this diameter to that of the sphere of the fixed stars. I cannot but wonder how such things as these could fall from so ingenious a man, and so great an astronomer. But I must be of the same opinion with all the greatest philosophers of our age, that the sun is of the same nature with the fixed stars. And this will give us a greater idea of the world, than all those other opinions. For then why may not every one of these stars or suns have as great a retinue as our sun, of planets, with their moons, to wait upon them? Nay, there is a manifest reason why they should. For if we imagine ourselves placed at an equal distance from the sun and fixed stars, we should then perceive no difference between them. For, as for all the planets that we now see attend the sun, we should not have the least glimpse of them, either because their light would be too weak to affect us, or that all the orbs in which they move would make up one lucid point with the sun. In this station we should have not occasion to imagine any difference between the stars, and should make no doubt if we had but the sight, and knew the nature of one of them, to make that the standard of all the rest. We are then placed near one of them, namely, our sun, and so near as to discover six other globes moving round him, some of them having others performing them the same office. Why then may not we make use of the same judgment that we would in that case; and conclude that our star has no better attendance than the others? So that what we allowed the planets, upon the account of our enjoying it, we must likewise grant to all those planets that surround that prodigious number of suns. They must have their plants and animals, nay and their rational creatures too, and those as great admirers, and as diligent observers of the heavens as ourselves; and must consequently enjoy whatsoever is subservient to, and requisite for such knowledge.

What a wonderful and amazing scheme have we here of the magnificent vastness of the universe! So many suns, so many earths, and every one of them stocked with so many herbs, trees and animals, and adorned with so many seas and mountains! And how must our wonder and admiration be encreased when we consider the prodigious distance and multitude of the stars? That their distance is so immense, that the space between the earth and sun (which is no less than twelve thousand of the earth's diameters) is almost nothing when compared to it, has more proofs than one to confirm it. And this among the rest. If you observe two stars near one another, as for example, those in the middle of the Great Bear's tail, differing very much from one another in clearness, notwithstanding our changing our position in our annual orbit round the sun, and that there would be a parallax were the star which is brighter nearer to us than the other, as is very probable it is, yet whatever part of the year you look upon them, they will not in the least have altered their distance. Those that have hitherto undertook to calculate their distance, have not been able perfectly to compass their design, by reason of the extreme niceness and almost impossibility of the observations requisite for their purpose. The only method that I see remaining, to come to any tolerable probability in so difficult a case, I shall here make use of. Seeing then that the stars, as I said before, are so many suns, if we do but suppose one of them equal to ours, it will follow that its distance from us is as much greater than that of the sun as its apparent diameter is less than the diameter of the sun. But the stars, even those of the first magnitude, though viewed through a telescope, are so very small, that they seem only like so many shining points, without any perceivable breadth. So that such observations can here do us no good. When I saw this would not succeed, I studied by what way I could lessen the diameter of the sun, as to make it not appear larger than the Dog, or any other of the chief stars. To this purpose I closed one end of my twelve-foot tube with a very thin plate, in the middle of which I made a hole not exceeding the twelfth part of a line, that is the hundred and forty-fourth part of an inch. That end I turned to the sun, placing my eye at the other, and I could see so much of the sun as was in diameter about the 182nd part of the whole. But still that little piece of him was brighter much than the Dog-star is in the clearest night. I saw that this would not do, but that I must lessen the diameter of the sun a great deal more. I made then such another hole in a plate, and against it I placed a little round glass that I had made use of in my microscope, of much about the same diameter with the former hole. Then looking again towards the sun (taking care that no light might come near my eye to hinder my observations) I found it appeared of much the same clearnesss with Sirius. But casting up my account, according to the rules of Dioptrics, I found his diameter now was but 1/152 part of that one hundred and eighty-second part of his whole diameter that I saw through the former hole. Multiplying 1/152 and 1/182 with one another, the product I found to be 1/27664. The sun therefore being contracted into such a compass, or being removed so far from us (for it is the same thing) as to make his diameter but the 27664th part of that we every day see, will send us just the same light as the Dog-star now doth. And his distance then from us will be to his present distance undoubtedly as 27664 is to 1; and his diameter little above four thirds, 4''' seeing then Sirius is supposed equal to the sun, it follows that his diameter is likewise 4''' and that his distance to the distance of the sun from us is as 27664 to 1. And what an incredible distance that is, will appear by the same way of reasoning that we used in measuring that of the sun. For if 25 years are required for a bullet out of a cannon, with its utmost swiftness, to travel from the sun to us; then by multiplying the number 27664 into 25 we shall find that such a bullet would spend almost seven hundred thousand years in its journey between us and the fixed stars. And yet when in a clear night we look upon them, we cannot think them above some few miles over our heads. What I have here enquired into is concerning the nearest of them. And what a prodigious number must there be besides of those which are placed in the vast spaces of heaven, as to be as remote from these as these are from the sun! For if with our bare eyes we can observe above a thousand, and with a telescope can discover ten or twenty times as many; what bounds of number can we set to those which are out of the reach even of these assistances, especially if we consider the infinite power of God! Really, when I have been reflecting thus with myself, methoughts all our arithmetic was nothing, and we are versed but in the very rudiments of numbers, in comparison of this great sum. For this requires an immense Treasury, not of twenty or thirty figures only, in our decuple progression, but of as many as there are grains of sand upon the shore. And yet who can say, that even this number exceeds that of the fixed stars? Some of the Ancients and Jordanus Brunus carried it farther, in declaring the number infinite: he would presuade us that he has proved it by many arguments, though in my opinion they are none of them conclusive. Not that I think the contrary can ever be made out. Indeed it seems to me certain that the universe is infinitely extended; but what God has been pleased to place beyond the region of the stars, is as much above our knowledge, as it is beyond our habitation.

Or what if beyond such a determinate space he has left an infinite vacuum; to show, how inconsiderable all that he has made is, to what his power could, had he pleased have produced? But I am falling, before I am aware, into that intricate dispute of infinity; therefore I shall wave this, and not, as soon as I am free of one, take upon me another difficult task.

THOMAS WRIGHT:

An Original Theory of the Universe

The Hypothesis, or Theory, fully explained and demonstrated, proving the sidereal Creation to be finite

Sir,

I know you are an enemy to all sorts of schemes where they are not absolutely necessary, and may possibly be avoided; and for that reason I have purposely omitted geometrical figures, and other representations in this work, which might have been inserted and in some places, especially here, I might have introduced diagrams, perhaps more explicit than words; but as you have frequently observed, they are only of use to the few learned, and contribute more to the taking away the little ideas and knowledge the more ignorant many may be endued with, by a prejudicial impression of imperfect images, rather than the adding any new light to their understanding, I have purposely avoided, as much as possible, both here and every where, all such complex diagrams as might be in danger of betraying any the least such conscious diffidence in you arising from the want of a proper *Precognita* in the sciences.

This imperfection, much to be lamented, as greatly to the disadvantage of all mathematical reasoning, I would willingly always prevent, in my readers, and to chuse in my friend; I shall therefore content myself with referring you to a few orbicular figures, concave and convex, as may best suggest to your fancy the simplest way, a just idea of the hypothesis I have framed, and naturally enough I hope, render my theory so intelligible, as to help you sufficiently to conceive the solution aimed at, of the important problem I have attempted.

As I have said before, we cannot long observe the beauteous parts of the visible creation, not only of this world on which we live, but also the myriads of bright bodies round us, with any attention, without being convinced, that a power supreme, and of a nature unknown to us, presides in, and governs it.

From *An Original Theory or New Hypothesis of the Universe* by Thomas Wright of Durham, 1750. A second edition of this work from which the following selection is taken (portions of "Letter VII") was printed in the United States by C. S. Rafinesque in 1837, pp. 103–12.

The course and frame of this vast bulk, display
A reason and fix'd law, which all obey.
                                    SHER. MANILIUS

And notwithstanding the many wonderful productions of nature in this our
known habitation, yet the Earth, when compared with other bodies of our own
system, seems far from being the most considerable in it; and it appears not
only very possible, but highly probable, from what has been said, and from
what we can farther demonstrate, that there is as great a multiplicity of worlds,
variously dispersed in different parts of the universe, as there are variegated
objects in this we live upon. Now, as we have no reason to suppose, that the
nature of our Sun is different from that of the rest of the Stars; and since we
can no way prove him superior even to the least of those surprising bodies,
how can we, with any show of reason, imagine him to be the general centre
of the whole, *i.e.* of the visible creation, and seated in the centre of the mundane space? This, in my humble opinion, is too weak even for conjecture,
their apparent distribution, and [See the Zodiacal Constellations, you'll find that
in some signs there are several Stars of the first, second, and third magnitude,
and in many others none of these at all.] irregular order argue so much against
it.

The Earth indeed has long possessed the chief seat of our system, and
peaceably reigned there, as in the centre of the universe for many ages past;
but it was human ignorance, and not divine wisdom, that placed it there;
some few indeed from the beginning have disputed its right to it, as judging
it no way worthy of such high eminence. Time at length has discovered the
truth to every body, and now it is justly displaced by the united consent of
all its inhabitants, and instead of being thought the most majestic of all nature's
lower works, now rather disgraces the creation, so much it is reduced in its
present state from what it had reason to expect in the former.

Now it is no longer the terrestrial globe in the Universe, but is proved to
be one of the least planets of the solar system, and surprisingly inferior to
some of its fellow worlds. The Sun, or rather the System, has almost as long
usurped the centre of infinity, with as little pretence to such preeminence; but
now, thanks to the sciences, the scene begins to open to us on all sides, and
truths scarce to have been dreamt of, before persons of observation had proved
them possible, invade our senses with a subject too deep for the human understanding, and where our very reason is lost in infinite wonders. How ought
this to humble every mind susseptible of reason!

In this place, I believe, you will pardon a digression; which, in answer to
part of your last letter, I judge will not be very impertinent, though perhaps
just here I cannot so well justify it.

Your late conversation with our friend Mr. * * *, I am persuaded, must have
been very entertaining; but I cannot help thinking his reflections upon the
wonders of nature and the wisdom of providence, though I must allow them
all to be very just and curious, instead of elevating the mind to the pitch he

would have it, rather as considered above, depress it below the proper, nay I might say necessary, standard of human ideas.

This, probably, you'll say is an odd turn, and may want some explanation, since every object in the chain of nature, must of force be granted, a subject worthy of our speculations, being altogether made, as in the maximum of wisdom: But what I mean is this, since nothing is more natural for beings in every state in search after their own advantages, and the enlargement of their ideas, to look upward, surely it may be presumed, that time may be mispent, if not lost in inspecting too narrowly things so little beneficial in states below us; as Mr. *Pope* says,

> Why has not man a microscopic eye?
> For this plain reason, man is not a fly.
> Say what the use, where finer opticks given,
> To inspect a mite, not comprehend the heaven.
>
> *Essay on Man*

Amusement alone can never be supposed to be the sole end of human life, where even true happiness is a thing we rather taste than enjoy. The mind we find capable of much more rational pleasure than can possibly fall within the reach of human power, either to promise or procure it; but then this very defect in our present state of existence affords us no less than a moral assurance, that some where in a future, we may, if we please, be entitled to the very *Plenum* of all enjoyments.

The peculiar business then of the human mind naturally precedes its amusements, as evidently ordained to soar above all the inferior beings of this world; and however our natures may, through indolence, or through ignorance, degenerate, that of the man can never be supposed to sink into the mole.

The properest way then surely for men to preserve their preeminence over the brute creation, is to make use of that reason and reflection, which so manifestly distinguishes their natural superiority. A right application of which, must of course then direct us to a forward, rather than a backward search in the vast visible chain of our existence, which clearly connects all beings and states as under the direction of one supreme agent.

This is all I would have understood by the foregoing position, which, in one word, implies no more than that the sublime philosophy ought in all reason to be preferred to the minute; but I hope you will not infer from this my seeming partiality for the celestial sciences, that I mean to insinuate, that the study of terrestrial physics is not a rational amusement.

Mr. * * * you say, seems to lament the taste of mankind in general much in the same degree as you do his I readly grant you; a man who can talk so well upon an ant, might make a more entertaining discourse upon the eagle; but I beg his pardon, and though we are all too ready, and most apt to condemn all such pleasures as vain or trifling, which we have no share in, or taste for ourselves; yet I don't think it follows, that those ingenious labours of his are useless. The pleasures arising from natural philosophy are all undoubtedly great ones, whether we consider nature in her highest, or in her lowest capacity; the beauties of the creation are every day varied to us below, as much they are every night above, and in both cases, through every object, the Creator shines so manifest, that we may justly consider him every where smiling full in the face of all his creatures, commanding as it were an awful reverence, and respect due, not only to his omnipotency, but also to his infinite goodness and endless indulgencies. This is the only return our gratitude can make for all those blessings he daily bestows upon us, and to this great Author of her laws, nature herself cries aloud through myriads of various objects, and after her own expressive and peculiar manner, seems to command us with an attractive grace, to observe her sovereign, and admire his wisdom. The majesty, power, and dominion of GOD is best displayed in the external direction of things, his wisdom and visible agency in the internal: hence, by proper objects, selected from both, attended with just reflections, we may certainly raise our ideas almost to the pitch of immortals; but how far the human imagination may possible go, or how much minds like ours may be improved, is a question not easily determined; but as natural knowledge evidently increases daily, astronomical enquiries are the most capable of opening our minds, and enlarging our conception, of consequence they must be most worthy our attention of all other studies. But of this I have said enough, and think it is now more than time to attempt the remaining part of my theory.

When we reflect upon the various aspects, and perpetual changes of the planets, both with regard to their [Not to mention their several conjunctions and apulces to fixed Stars, &c. see the state of the heavens in 1662. *December* the first, when all the known planets were in one sign of the zodiac, *viz. Sagittarius*] heliocentric and geocentric motion, we may readily imagine, that nothing but a like excentric position of the Stars could any way produce such an apparently promiscuous difference in such otherwise regular bodies. And that in like manner, as the planets would, if viewed from the Sun, there may be one place in the Universe to which their order and primary motions must appear most regular and most beautiful. Such a point, I may presume, is not unnatural to be supposed, although hitherto we have not been able to produce any absolute proof of it.

This is the great order of nature which I shall now endeavor to prove and thereby solve the Phænomena of the *Via Lactea;* and in order thereto, I want nothing to be granted but what may easily be allowed, namely that the *Milky Way* is formed of an infinite number of small Stars.

Let us imagine a vast infinite gulph, or medium, every way extended like a plane, and inclosed between two surfaces, nearly even on both sides, but of such a depth or thickness as to occupy a space equal to the double radius, or diameter of the visible creation, that is to take in one of the smallest Stars each way, from the middle station, perpendicular to the plane's direction, and, as near as possible, according to our idea of their true distance.

But to bring this image a little lower, and as near as possible level to every

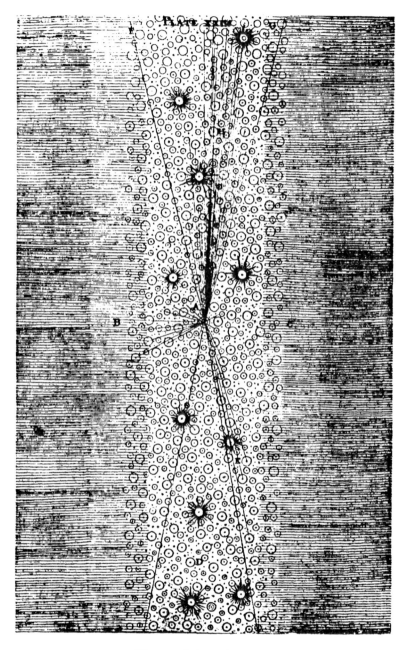

Diagram of the Milky Way

capacity, I mean such as cannot conceive this kind of continued Zodiac, let us suppose the whole frame of nature in the form of an artificial horizon of a globe, I do not mean to affirm that it really is so in fact, but only state the question thus to help your imagination to conceive more aptly what I would explain. *Plate* XXIII. will then represent a just section of it. Now in this space let us imagine all the Stars scattered promiscuously, but at such an adjusted distance from one another, as to fill up the whole medium with a kind of regular irregularity of objects. And next let us consider what the consequence would be to an eye situated near the centre point, or any where about the middle plane, as at the point A. Is it not, think you, very evident, that the Stars would there appear promiscuously dispersed on each side, and more and more inclining to disorder, as the observer would advance his station towards either surface, and nearer to B or C, but in a direction of the general plane towards H or D, by the continual approximation of the visual rays, crowding together as at H betwixt the limits D and G, they must infallibly terminate in the utmost confusion. If your optics fails you before you arrive at these external regions, only imagine how infinitely greater the number of stars would be in those remote parts, arising thus from their continual crowding behind one another, as all other objects do towards the horizon point of their perspective, which ends but with infinity: thus, all their rays at last so near uniting, must meeting in the eye appear, as almost in contact, and form a perfect zone of light; this I take to be the real case, and the true nature of our *Milky Way,* and all the irregularity we observe in it at the Earth, I judge to be entirely owing to our Sun's position in this great firmament, and may easily be solved by his excentricity, and the diversity of motion that may naturally be conceived amongst the stars themselves, which may here and there, in different parts of the Heavens, occasion a cloudy knot of stars as perhaps at E. . . .

How much a confirmation of this is to be wished, your own Curiosity may make you judge, and here I leave it for the opticians to determine. I shall content myself with observing that nature never leaves us without a sufficient guide to conduct us through all the necessary paths of knowledge; and it is far from absurd to suppose Providence may have every where throughout the whole universe, interspersed modules of every creation, as our divines tell us, man is the image of God himself.

Thus, sir, you have had my full opinion, without the least reserve, concerning the visible creation, considered as part of the finite universe; how far I have succeeded in my designed solution of the *Via Lactea,* upon which the theory of the whole is formed, is a thing that will hardly be known in the present century, as in all probability it may require some ages of observation to discover the truth of it.

It remains that I should now give you some idea of time and space; but this will afford matter sufficient for another letter.

I am now, &c.

IMMANUEL KANT:

Universal Natural History and Theory of the Heavens

Of the Systematic Constitution among the Fixed Stars

THE SCIENTIFIC THEORY of the Universal Constitution of the World has obtained no remarkable addition since the time of Huygens. At the present time nothing more is known than what was already known then, namely, that six planets with ten satellites, all performing the circle of their revolution almost in one plane, and the eternal comets which sweep out on all sides, constitute a system whose centre is the sun, towards which they all fall, around which they perform their movements, and by which they are all illuminated, heated, and vivified; finally, that the fixed stars are so many suns, centres of similar systems, in which everything may be arranged just as grandly and with as much order as in our system; and that the infinite space swarms with worlds, whose number and excellency have a relation to the immensity of their Creator.

The systematic arrangement which was found in the combination of the planets which move around their sun, seemed in the view of astronomers of that time to disappear in the multitude of the fixed stars; and it appeared as if the regulated relation which is found in the smaller solar system, did not rule among the members of the universe as a whole. The fixed stars exhibited no law by which their positions were bounded in relation to each other; and they were looked upon as filling all the heavens and the heaven of heavens without order and without intention. Since the curiosity of man set these limits to itself, he has done nothing further than from these facts to infer, and to admire, the greatness of Him who has revealed Himself in works so inconceivably great.

It was reserved for an Englishman, Mr. Wright of Durham, to make a

From *Universal Natural History and Theory of the Heavens* (1755), translated from the German by W. Hastie in *Kant's Cosmogony*, Glasgow, 1900, pp. 53–65 (from the "First Part"), 135–156 (from the seventh chapter, "Second Part").

happy step with a remark which does not seem to have been used by himself for any very important purpose, and the useful application of which he has not sufficiently observed. He regarded the Fixed Stars not as a mere swarm scattered without order and without design, but found a systematic constitution in the whole universe and a universal relation of these stars to the ground-plan of the regions of space which they occupy. We would attempt to improve the thought which he thus indicated, and to give to it that modification by which it may become fruitful in important consequences whose complete verification is reserved for future times.

Whoever turns his eye to the starry heavens on a clear night, will perceive that streak or band of light which on account of the multitude of stars that are accumulated there more than elsewhere, and by their getting perceptibly lost in the great distance, presents a uniform light which has been designated by the name *Milky Way*. It is astonishing that the observers of the heavens have not long since been moved by the character of this perceptibly distinctive zone in the heavens, to deduce from it special determinations regarding the position and distribution of the fixed stars. For it is seen to occupy the direction of a great circle, and to pass in uninterrupted connection round the whole heavens: two conditions which imply such a precise destination and present marks so perceptibly different from the indefiniteness of chance, that attentive astronomers ought to have been thereby led, as a matter of course, to seek carefully for the explanation of such a phenomenon.

As the stars are not placed on the apparent hollow sphere of the heavens, and as some are more distant than others from our point of view and are lost in the depths of the heavens, it follows from this, that at the distances at which they are situated away from us, one behind the other, they are not indifferently scattered on all sides, but must have a predominant relation to a certain plane which passes through our point of view and to which they are arranged so as to be found as near it as possible. This relation is such an undoubted phenomenon that even the other stars which are not included in the whitish streak, are yet seen to be more accumulated and closer the nearer their places are to the circle of the Milky Way; so that of the two thousand stars which are perceived by the naked eye, the greatest part of them are found in a not very broad zone whose centre is occupied by the Milky Way.

If we now imagine a plane drawn through the starry heavens and produced indefinitely, and suppose that all the fixed stars and systems have a general relation in their places to this plane so as to be found nearer to it than to other regions, then the eye which is situated in this plane when it looks out to the field of the stars, will perceive on the spherical concavity of the firmament the densest accumulation of stars in the direction of such a plane under the form of a zone illuminated by varied light. This streak of light will advance as a luminous band in the direction of a great circle, because the position of the spectator is in the plane itself. This zone will swarm with stars which, on account of the indistinguishable minuteness of their clear points that cannot be severally discerned and their apparent denseness, will present a uniformly whitish glimmer,—in a word, a Milky Way. The rest of the heavenly host whose relation to the plane described gradually diminishes, or which are situated nearer the position of the spectator, are more scattered, although they are seen to be massed relatively to this same plane. Finally, it follows from all this that our solar world, seeing that this system of the fixed stars is seen from it in the direction of a great circle, is situated in the same great plane and constitutes a system along with the other stars.

In order to penetrate better into the nature of the universal connection which rules in the universe, we will try to discover the cause that has made the positions of the fixed stars come to be in relation to a common plane.

The sun is not limited in the range of its attractive force to the narrow domain of the planetary system. According to all appearance the force of attraction extends *ad infinitum*. The comets, which pass very far beyond the orbit of Saturn, are compelled by the attraction of the sun to return again, and to move in orbits. Although, therefore, it is more in accordance with the nature of a force which seems to be incorporated in the essence of matter, to be un-limited, and it is also actually recognized to be so by those who accept Newton's principles, yet we would only have it granted that this attraction of the sun extends approximatively to the nearest fixed star; also that the fixed stars, as being so many suns, exercise an action around them in a similar range; and, consequently, that the whole host of them are striving to approach each other through their mutual attraction. Thus all the systems of the universe are found so constituted by their mutual approach, which is incessant and is hindered by nothing, that they will fall together sooner or later into one mass, unless this ruin is prevented by the action of the centrifugal forces, as in the globes of our planetary system. These forces, by deflecting the heavenly bodies from falling in a straight line, bring about, when combined with the forces of attraction, their perpetual revolutions, and thereby the structure of the crea-tion is secured from destruction and is adapted for an endless duration.

Thus all the suns of the firmament have movements of revolution, either round one universal centre or round many centres. But we may here apply the analogy which is observed in the revolutions of our Solar System, namely, that the same cause that has communicated to the planets the centrifugal force, in virtue of which they perform their revolutions, has also so directed their orbits that they are all related to one plane. And so also the cause, whatever it may be, which has given the power of revolving to the suns of the upper world, as so many wandering stars of a higher order of worlds, has likewise brought their orbits as much as possible into one plane, and has striven to limit the deviations from it.

Following out this idea, the System of the Fixed Stars may be viewed as represented in some measure by the System of the Planets, if the latter be regarded as indefinitely enlarged. If instead of the six planets with their ten companions, we suppose there to be as many thousands of them; if the twenty-eight or thirty comets which have been observed be multiplied a hundred or a thousand times, and if we suppose these same bodies to be self-luminous, then the eye of the spectator looking at them from the earth would see before it just the very appearance of what is presented to it by the fixed stars of the Milky Way. For these supposed planets, by their proximity to the common plane of their relation, being situated with our earth in the same plane, would present to us a zone densely illuminated by innumerable stars, and its direction would be that of a great circle of the celestial sphere. This streak of light would be seen everywhere thickly sown with stars, although, according to the hypothesis, they are wandering stars and consequently not fixed to one place; for by their displacement there would always be found stars enough on one side, although others had changed from that position.

The breadth of this luminous zone, which represents a sort of zodiac, will be determined by the different degrees of the deviation of the said wandering stars from the plane of their relation, and by the inclination of their orbits to this plane; and as most of them are near this plane, their number will appear more scattered according to the degree of their distance from it. But the comets which occupy all the regions of space without distinction, will cover the field of the heavens on both sides.

The form of the starry heavens is therefore due to no other cause than such a systematic constitution on the great scale as our planetary world has on the small—all the suns constituting a system whose universal relative plane is the Milky Way. Those suns which are least closely related to this plane, will be seen at the side of it; but on that account they are less accumulated, and are much more scattered and fewer in number. They are, so to speak, the comets among the suns.

This new theory attributes to the suns an advancing movement; yet everybody regards them as unmoved, and as having been fixed from the beginning to their places. The designation of them as 'Fixed' Stars, which they have received from that view of them, seems to be established and put beyond doubt by the observation of all the centuries. This difficulty raises an objection which would annihilate the theory here advanced, were it well founded. But in all probability this want of movement is merely apparent. It is either only excessively slow, arising from the great distance from the common centre around which the stars revolve, or it is due to mere imperceptibility, owing to their distance from the place of observation. Let us estimate the probability of this conception by calculating the motion which a fixed star near our sun would have, if we supposed our sun to be the centre of its orbit. If, following Huygens, its distance is assumed to be more than 21,000 times greater than the distance of the sun from the earth, then according to the established law of the periods of revolution—which are in the ratio of the square root of the cube of the distances from the centre—the time which such a star would require to revolve once round the sun would be more than a million and a half years; and this would only produce a displacement of its position by one degree in four thousand years. Now, as there are perhaps very few fixed stars so near the sun as Huygens has conjectured Sirius to be, as the distance of the rest of the heavenly host perhaps immensely exceeds that of Sirius, and would therefore require incomparably longer time for such a periodic revolution; and, moreover, as it is very probable that the movement of the suns of the starry heavens goes round a common centre, whose distance is incomparably great, so that the progression of the stars may therefore be exceedingly slow, it may be inferred with probability from this, that all the time that has passed since men began to make observations on the heavens is perhaps not yet sufficient to make perceptible the alteration which has been produced in their positions. We need not, however, give up the hope yet of discovering even this alteration, in the course of time. Subtle and careful observers, and a comparison of observations taken at a great distance from each other, will be required for it. These observations must be directed especially to the stars of the Milky Way,[1] which is the main plane of all motion. Mr. Bradley has observed almost imperceptible displacements of the stars. The ancients have noticed stars in certain places of the heavens, and we see new ones in others. Who knows but that these are just the former, which have only changed their place? The excellence of our instruments and the perfection of astronomy, give us well-founded hopes for the discovery of such peculiar and remarkable things.[2] The credibility of the fact itself, in accordance with the principles of nature and analogy which well support this hope, is such that they may stimulate the attention of the explorer of nature so as to bring about its realization.

The Milky Way is, so to speak, the zodiac of new stars which alternately show themselves and disappear, almost only in that region of the heavens. If this alteration of their visibility arises from their periodical removal and approach to us, then it appears from the systematic constitution of the stars here indicated, that such a phenomenon must be seen for the most part only in the region of the Milky Way. For as there are stars which revolve in very elongated orbits around other fixed stars like satellites around their planets, analogy with our planetary world, in which, only those heavenly bodies that are near the common plane of the movements have companions revolving round them, demands that those stars only which are in the Milky Way will also have suns revolving round them.

I come now to that part of my theory which gives it its greatest charm, by the sublime idea which it presents of the plan of the creation. The train of thought which has led me to it is short and natural; it consists of the following

---

1. Likewise to those clusters of stars, of which there are many found together in a small space, as, for example, the Pleiades, which perhaps make up a small system by themselves in the greater system.

2. De la Hire, in the *Mémoires* of the Paris Academy of the year 1693, remarks that from his own observations, as well as from comparison of them with those of Ricciolus, he has perceived a marked alteration in the positions of the stars of the Pleiades.

ideas. If a system of fixed stars which are related in their positions to a common plane, as we have delineated the Milky Way to be, be so far removed
from us that the individual stars of which it consists are no longer sensibly
distinguishable even by the telescope; if its distance has the same ratio to the
distance of the stars of the Milky Way as that of the latter has to the distance
of the sun; in short, if such a world of fixed stars is beheld at such an immense
distance from the eye of the spectator situated outside of it, then this world
will appear under a small angle as a patch of space whose figure will be circular
if its plane is presented directly to the eye, and elliptical if it is seen from the
side or obliquely. The feebleness of its light, its figure, and the apparent size
of its diameter will clearly distinguish such a phenomenon when it is presented,
from all the stars that are seen single.

We do not need to look long for this phenomenon among the observations
of the astronomers. It has been distinctly perceived by different observers. They
have been astonished at its strangeness; and it has given occasion for conjectures, sometimes to strange hypotheses, and at other times to probable conceptions which, however, were just as groundless as the former. It is the 'nebulous'
stars which we refer to, or rather a species of them, which M. de Maupertuis
thus describes: 'They are,' he says, 'small luminous patches, only a little more
brilliant than the dark background of the heavens; they are presented in all
quarters; they present the figure of ellipses more or less open; and their light
is much feebler than that of any other object we can perceive in the heavens.'[3]

The author of the *Astro-Theology* imagined that they were openings in the
firemament through which he believed he saw the Empyrean.[4] A philosopher
of more enlightened views, M. de Maupertuis, already referred to, in view of
their figure and perceptible diameter, holds them to be heavenly bodies of
astonishing magnitude which, on account of their great flattening, caused by
the rotatory impulse, present elliptical forms when seen obliquely.

Any one will be easily convinced that this latter explanation is likewise
untenable. As these nebulous stars must undoubtedly be removed at least as
far from us as the other fixed stars, it is not only their magnitude which would
be so astonishing—seeing that it would necessarily exceed that of the largest
stars many thousand times—but it would be strangest of all that, being self-
luminous bodies and suns, they should still with this extraordinary magnitude
show the dullest and feeblest light.

It is far more natural and conceivable to regard them as being not such
enormous single stars but systems of many stars, whose distance presents them
in such a narrow space that the light which is individually imperceptible from
each of them, reaches us, on account of their immense multitude, in a uniform
pale glimmer. Their analogy with the stellar system in which we find ourselves,
their shape, which is just what it ought to be according to our theory, the

3. *Discours sur la Figure des Astres.* Paris, 1742.
4. *Astro-Theology, or a Demonstration of the Being and Attributes of God from a Survey of
the Heavens,* by W. Derham. London, 1714.

feebleness of their light which demands a presupposed infinite distance: all this is in perfect harmony with the view that these elliptical figures are just universes and, so to speak, Milky Ways, like those whose constitution we have just unfolded. And if conjectures, with which analogy and observation perfectly agree in supporting each other, have the same value as formal proofs, then the certainty of these systems must be regarded as established.

The attention of the observers of the heavens, has thus motives enough for occupying itself with this subject. The fixed stars, as we know, are all related to a common plane and thereby form a co-ordinated whole, which is a World of worlds. We see that at immense distances there are more of such star-systems, and that the creation in all the infinite extent of its vastness is everywhere systematic and related in all its members.

It might further be conjectured that these higher universes are not without relation to one another, and that by this mutual relationship they constitute again a still more immense system. In fact, we see that the elliptical figures of these species of nebulous stars, as represented by M. de Maupertuis, have a very near relation to the plane of the Milky Way. Here a wide field is open for discovery, for which observation must give the key. The Nebulous Stars, properly so called, and those about which there is still dispute as to whether they should be so designated, must be examined and tested under the guidance of this theory. When the parts of nature are considered according to their design and a discovered plan, there emerge certain properties in it which are otherwise overlooked and which remain concealed when observation is scattered without guidance over all sorts of objects.

The theory which we have expounded opens up to us a view into the infinite field of creation, and furnishes an idea of the work of God which is in accordance with the infinity of the great Builder of the universe. If the grandeur of a planetary world in which the earth, as a grain of sand, is scarcely perceived, fills the understanding with wonder; with what astonishment are we transported when we behold the infinite multitude of worlds and systems which fill the extension of the Milky Way! But how is this astonishment increased, when we become aware of the fact that all these immense orders of star-worlds again form but one of a number whose termination we do not know, and which perhaps, like the former, is a system inconceivably vast—and yet again but one member in a new combination of numbers! We see the first members of a progressive relationship of worlds and systems; and the first part of this infinite progression enables us already to recognize what must be conjectured of the whole. There is here no end but an abyss of a real immensity, in presence of which all the capability of human conception sinks exhausted, although it is supported by the aid of the science of number. The Wisdom, the Goodness, the Power which have been revealed is infinite; and in the very same proportion are they fruitful and active. The plan of their revelation must therefore, like themselves, be infinite and without bounds.

Of the Creation in the Whole Extent of Its Infinitude in Space as Well as in Time

THE UNIVERSE, by its immeasurable greatness and the infinite variety and beauty that shine from it on all sides, fills us with silent wonder. If the presentation of all this perfection moves the imagination, the understanding is seized by another kind of rapture when, from another point of view, it considers how such magnificence and such greatness can flow from a single law, with an eternal and perfect order. The planetary world in which the sun, acting with its powerful attraction from the centre of all the orbits, makes the moving spheres of its system revolve in eternal circles, has been wholly formed, as we have seen, out of the originally diffused primitive stuff that constituted all the matter of the world. All the fixed stars which the eye discovers in the hollow depths of the heavens, and which seem to display a sort of prodigality, are suns and centres of similar systems. Analogy thus does not leave us to doubt that these systems have been formed and produced in the same way as the one in which we find ourselves, namely, out of the smallest particles of the elementary matter that filled empty space—that infinite receptacle of the Divine Presence.

If, then, all the worlds and systems acknowledge the same kind of origin, if attraction is unlimited and universal, while the repulsion of the elements is likewise everywhere active; if, in presence of the infinite, the great and small are small alike; have not all the universes received a relative constitution and systematic connection similar to what the heavenly bodies of our solar world have on the small scale—such as Saturn, Jupiter, and the Earth, which are particular systems by themselves, and yet are connected with each other as members of a still greater system? If in the immeasurable space in which all the suns of the Milky Way have formed themselves, we assume a point around which, through some cause or other, the first formation of nature out of chaos began, there the largest mass and a body of extraordinary attraction will have arisen which has thereby become capable of compelling all the systems in the process of being formed within an enormous sphere around it, to fall towards itself as their centre, and to build up a system around it on the great scale similar to that which the elementary matter that formed the planets has constructed in the small scale around the sun. Observation puts this conjecture almost beyond doubt. The host of the stars, by their relative positions towards a common plane, constitute a system just as much as do the planets of our Solar System around the sun. The Milky Way is the Zodiac of those higher worlds, which diverge as little as possible from its zone and whose strip of space is always illuminated by their light, just as the zodiac of the planets always glitters here and there although only in a few points with the splendour of these spheres. Every one of these suns, with its revolving planets, constitutes a particular system by itself; but this does not hinder them from being parts of a still greater system, just as Jupiter or Saturn, notwithstanding their being accompanied by satellites of their own, are embraced in the systematic constitution of a still greater system. With such an exact agreement in their constitution, can we not recognize the same cause and mode of production in them?

If, then, the fixed stars constitute a system whose extent is determined by the sphere of the attraction of that body which is situated in the centre, shall there not have arisen more Solar Systems and, so to speak, more Milky Ways, which have been produced in the boundless field of space? We have beheld with astonishment figures in the heavens which are nothing else than such systems of fixed stars confined to a common plane—Milky Ways, if I may so express myself, which, in their different positions to the eye, present elliptical forms with a glimmer that is weakened in proportion to their infinite distance. They are systems of, so to speak, an infinite number of times infinitely greater diameter than the diameter of the Solar System. But undoubtedly they have arisen in the same way, have been arranged and regulated by the same causes, and preserve themselves in their constitution by a mechanism similar to that which rules our own system.

If, again, these star-systems are viewed as members in the great chain of the totality of nature, then there is just as much reason as formerly to think of them as in mutual relation and in connections which, in virtue of the law of their primary formation that rules the whole of nature, constitute a new and greater system ruled by the attraction of a body of incomparably mightier attraction, and acting from the centre of their regulated positions. The attraction which is the cause of the systematic constitution among the fixed stars of the Milky Way acts also at the distance even of those worlds, so that it would draw them out of their positions and bury the world in an inevitably impending chaos, unless the regularly distributed forces of rotation formed a counterpoise or equilibrium with attraction, and mutually produced in combination that connection which is the foundation of the systematic constitution. Attraction is undoubtedly a property of matter extending as far as that coexistence which constitutes space, seeing that it combines substances by their mutual dependence; or, to speak more exactly, attraction is just that universal relation which unites the parts of nature in one space. It reaches, therefore, to the whole extent of space, even to all the distance of nature's infinitude. If light reaches us from these distant systems—light which is only an impressed motion—must not attraction, that original source of motion, which is prior to all motion, which needs no foreign cause and can be stopped by no obstacles, because it penetrates into the inmost recesses of matter without any impact even in the universal repose of nature; must not, I say, attraction have put those fixed star-systems, notwithstanding their immense distances, into motion when nature began to stir through the unformed dispersion of her material? And as we have seen on the small scale, is not this attraction the source of the systematic combination and the lasting persistence of the members of these systems, and that which secures them from falling to pieces?

But what is at last the end of these systematic arrangements? Where shall creation itself cease? It is evident that in order to think of it as in proportion to the power of the Infinite Being, it must have no limits at all. We come no nearer the infinitude of the creative power of God, if we enclose the space of its revelation within a sphere described with the radius of the Milky Way, than if we were to limit it to a ball an inch in diameter. All that is finite, whatever has limits and a definite relation to unity, is equally far removed from the infinite. Now, it would be absurd to represent the Deity as passing into action with an infinitely small part of His potency, and to think of His Infinite Power—the storehouse of a true immensity of natures and worlds—as inactive, and as shut up eternally in a state of not being exercised. Is it not much more reasonable, or, to say it better, is it not *necessary* to represent the system of creation as it must be in order to be a witness of that power which cannot be measured by any standard? For this reason the field of the revelation of the Divine attributes is as infinite as these attributes themselves.[5] Eternity is not sufficient to embrace the manifestations of the Supreme Being, if it is not combined with the infinitude of space. It is true that development, form, beauty, and perfection are relations of the elements and the substances that constitute the matter of the universe, and this is perceived in the arrangements which the wisdom of God adopts at all times. It is also most conformable to that wisdom that these relations and arrangements should be evolved out of their implanted universal laws by an unconstrained consecution. And hence it may be laid down, with good reason, that the arrangement and institution of the universe comes about gradually, as it arises out of the provision of the created matter of nature in the sequence of time. But the primitive matter itself, whose qualities and forces lie at the basis of all changes, is an immediate consequence of the Divine existence; and that same matter must therefore be at once so rich and so complete, that the development of its combinations in the flow of eternity may extend over a plane which includes in itself all that can be, which accepts no limit, and, in short, which is infinite.

If, therefore, the creation is infinite in space, or at least has really been so

---

5. The conception of an infinite extension of the world finds opponents among the metaphysicians, and has lately found one in M. Weitenkampf. If these gentlemen, on account of the supposed impossibility of a quantity without number and limits, cannot accommodate themselves to this idea, I would only ask in the meantime the question: Whether the future succession of Eternity will not embrace in it a true infinitude of multiplicities and changes, and whether this infinite series is not now already wholly present at once to the Divine intelligence? Now, if it was possible that God could make the conception of the infinitude which presents itself at once to His understanding, real in a series of facts following each other in succession, why should He not be able to exhibit the conception of another infinitude in a *combined connection* in space, and thereby make the extension of the world without limits? While any one is trying to answer this question, I will use the opportunity which may be furnished, to remove the supposed difficulty by an elucidation drawn from the nature of numbers, in so far as, on careful consideration, it may still be regarded as a question requiring explanation. And so I ask, whether that which a higher Power, accompanied with the highest wisdom, *has produced* in order to reveal itself, does not stand in the relation of a *differential* quantity to that which it *was able to produce*?

in its matter from the beginning, and is ready to become so in form or development, then the whole of space will be animated with worlds without number and without end. Will then that Systematic Connection, which we have already considered in particular in regard to all the parts of the world, extend to the whole and embrace the whole universe, the totality of nature in a single system, by the connecting power of attraction and centrifugal force? I say, Yes. If there existed only isolated systems which had no unified connection into a whole with one another, it might be imagined—were this chain of members assumed to be really infinite—that an exact equality in the attraction of their parts from all sides, could keep these systems secure from the destruction with which their inner mutual attraction threatens them. But this would evolve such an exact measured determination at the distance proportionate to the attraction, that even the slightest displacement would draw the ruin of the universe along with it; and after long periods, which, however, must finally come to an end, it would give it up to utter overthrow. A constitution of the world which did not maintain itself without a miracle, has not the character of that stability which is the mark of the choice of God. It is therefore much more in conformity with that choice to make the whole creation a single system which puts all the worlds and systems of worlds, that fill the whole of infinite space, into relation to a single centre. A scattered swarm of systems, however—for they might be separated from each other—would, by an unchecked tendency, hurry to disorder and destruction, unless a certain relative disposition were made by reference to a universal centre, the centre of the attraction of the universe, and unless means were taken for the maintenance of the whole of nature by systematic motions.

This universal centre of the attraction of the whole of nature, both in its crude and formed state, is the point at which is undoubtedly situated the mass of the most powerful attraction which embraces within the sphere of its attraction all the worlds and systems which time has produced, and which Eternity will produce; and it may with probability be assumed that around it nature made the beginning of its formation, and that the systems are accumulated most closely there, whereas further from that point, in the infinitude of space, they will disappear more and more with greater and greater degrees of dispersion. This law might be deduced from the analogy of our Solar System; and such a constitution may moreover serve this purpose, that at great distances not only the universal central body, but all the systems that revolve next round it, may combine their attraction and exercise it as if from one mass upon the systems which are at a still greater distance. This arrangement would then be subservient to embracing the whole of nature in all the infinitude of its extension in a single system.

Let us now proceed to trace out the construction of this Universal System of Nature from the mechanical laws of matter striving to form it. In the infinite space of the scattered elementary matter there must have been some one place where this primitive material had been most densely accumulated so as through the process of formation that was going on predominantly there, to have procured for the whole Universe a mass which might serve as its fulcrum. It indeed holds true that in an infinite space no point can properly have the privilege to be called the centre; but by means of a certain ratio, which is founded upon the essential degrees of the density of the primitive matter, according to which at its creation it is accumulated more densely in a certain place and increases in its dispersion with the distance from it, such a point may have the privilege of being called the centre; and it really becomes this through the formation of the central mass by the strongest attraction prevailing in it. To this point all the rest of the elementary matter engaged in particular formations is attracted; and, thereby, so far as the evolution of nature may extend it makes in the infinite sphere of the creation the whole universe into only one single system.

But it is important, and, if approved, is deserving of the greatest attention, that in consequence of the order of nature in this our system, the creation, or rather the development of nature, first begins at this centre and, constantly advancing, it gradually becomes extended into all the remoter regions, in order to fill up infinite space in the progress of eternity with worlds and systems. Let us dwell upon this idea for a moment with the silent satisfaction it brings. I find nothing which can raise the spirit of man to a nobler wonder, by opening to him a prospect into the infinite domain of omnipotence, than that part of my theory which concerns the successive realization of the creation. If it is admitted that the matter, which is the stuff for the formation of all the world, was not uniform in the whole infinite space to which God is present, but was spread out according to a certain law, perhaps proportioned to the density of the particles, and according to which the dispersion of the primitive matter increased from a certain point, as the place of densest accumulation, with the distance from this centre: then at the primary stirring of nature, formation will have begun nearest this centre; and in advancing succession of time the more distant regions of space will have gradually formed worlds and systems with a systematic constitution related to that centre. Every finite period, whose duration has a proportion to the greatness of the work to be accomplished, will always bring only a finite sphere to its development from this centre; while the remaining infinite part will still be in conflict with the confusion and chaos, and will be the further from the state of completed formation the farther its distance is away from the sphere of the already developed part of nature. In consequence of this, although from the place of our abode in the Universe, we look out upon a world wholly completed as it seems, and, so to speak, at an infinite host of worlds which are systematically combined; yet, strictly speaking, we find ourselves only in the neighbourhood of the centre of the whole of nature, where it has already evolved itself out of chaos and attained its proper perfection. If we could overstep a certain sphere we would there perceive chaos and the dispersion of the elements which, in the proportion in which they are found nearer this centre, lose in part their crude state and are nearer the perfection of their development, but in the degree in which they are removed from the centre, they are gradually lost in a complete dispersion. We would see how the infinite space, co-extensive with the Divine Presence, in which is to be found the provision for all possible natural formations, buried in a silent night, is full of matter which has to serve as material for the worlds that are to be produced in the future, and of impulses for bringing it into motion, which begin with a weak stirring of those movements with which the immensity of these desert spaces are yet to be animated. There had mayhap flown past a series of millions of years and centuries, before the sphere of the formed nature in which we find ourselves, attained to the perfection which is now embodied in it; and perhaps as long a period will pass before Nature will take another step as far in chaos. But the sphere of developed nature is incessantly engaged in extending itself. Creation is not the work of a moment. When it has once made a beginning with the production of an infinity of substances and matter, it continues in operation through the whole succession of eternity with ever increasing degrees of fruitfulness. Millions and whole myriads of millions of centuries will flow on, during which always new worlds and systems of worlds will be formed after each other in the distant regions away from the centre of nature, and will attain to perfection. Notwithstanding the Systematic Constitution embodied in their parts, they will obtain a universal relation to the centre which has become the first formative point and the centre of the creation, through the attractive power of its predominant mass. This infinity in the future succession of time, by which eternity is unexhausted, will entirely animate the whole range of space to which God is present, and will gradually put it into that regular order which is conformable to the excellence of His plan. And if we could embrace the whole of eternity with a bold grasp, so to speak, in one conception, we would also be able to see the whole of infinite space filled with systems of worlds and the creation all complete. But as, in fact, the remaining part of the succession of eternity is always infinite and that which has flowed is finite, the sphere of developed nature is always but an infinitely small part of that totality which has the seed of future worlds in itself, and which strives to evolve itself out of the crude state of chaos through longer or shorter periods. The creation is never finished or complete. It has indeed once begun, but it will never cease. It is always busy producing new scenes of nature, new objects, and new worlds. The work which it brings about has a relationship to the time which it expends upon it. It needs nothing less than an eternity to animate the whole boundless range of the infinite extension of space with worlds, without number and without end. We may say of it what the sublimest of the German poets writes of Eternity:

> Infinity! What measures thee?
> Before thee worlds as days, and men as moments flee!
> Mayhap the thousandth sun is rounding now;
> And thousands still remain behind!

Even as the clock its weight doth wind,
A sun by God's own power is driven;
And when its work is done, again in heaven
Another shines. But thou remain'st! To thee all numbers bow.

Von Haller

It is not a small pleasure to sweep in imagination beyond the boundary of the completed creation into the region of chaos, and to see the half crude nature in the neighbourhood of the sphere of the developed world losing itself gradually through all stages and shades of imperfection, throughout the whole range of unformed space. But it will be said: Is it not a reprehensible boldness to put forward an hypothesis, and to laud it as a subject for the entertainment of the understanding, which, perhaps, is only too arbitrary when it is asserted that nature is developed only in an infinitely small part, and that endless ranges of space are still involved in a conflict with chaos in order to bring forth through the succession of future ages whole hosts of worlds and systems, and to present them in all their proper order and beauty? I am not so devoted to the consequences of my theory that I should not be ready to acknowledge that the supposition of the successive expansion of the creation through the infinite regions of space which contain the matter for it, cannot entirely escape the reproach of its being undemonstrable. Nevertheless, I expect from those who are capable of estimating degrees of probability that such a chart of the infinite, comprehending, as it does, a subject which seems to be destined to be for ever concealed from the human understanding, will not on that account be at once regarded as a chimera, especially when recourse is had to analogy, which must always guide us in those cases in which the understanding cannot follow the thread of infallible demonstrations.

This analogy may, however, be supported by other tenable reasons; and the perspicuity of the reader, so far as I may flatter myself with his favourable consideration, will perhaps be able to increase them by adding other reasons more powerful still. For if it is considered that creation does not bring the character of stability with it, in so far as it does not oppose to the universal striving of attraction which acts through all its parts, as thorough a determination sufficient to resist the tendency of the former to destruction and disorder; if it has not distributed forces of impulsion which, in combination with the central tendency, establish a universal Systematic Constitution: then we are forced to a universal centre of the whole universe which holds all its parts together in a combined connection, and makes the totality of nature into only one system. If to this we add the conception of the formation of the world-bodies from the scattered elementary matter, as has been delineated above, yet do not limit it here to a separate system but extend it to the whole of nature: we shall be compelled to think of such a distribution of the primitive matter in the space of the original chaos, as bringing with it naturally a centre of the whole creation in order that the active mass, which embraces the whole of nature within its sphere, may be concentrated at this point, and that a thoroughgoing combination may be effected, by which all the worlds shall constitute only one single structure. But in the infinite space, another mode of distribution of the original material, which could posit a true centre and attracting point for the whole of nature, could hardly be conceived other than the one according to which it is arranged to all remote distances by a law of increasing dispersion from that point. But this law, moreover, involves a difference in the time which a system needs in the different regions of infinite space for the maturing of its development, so that this period is found the shorter the nearer the place of the formation of a world is found to the centre of creation, because the elements of matter are more densely accumulated there; and, on the other hand, longer time will be required the greater the distance is, because the particles are more dispersed there, and unite for formation later on.

If the whole hypothesis which I have sketched is examined in the extent of all I have said, as well as of what I shall still specially present, the boldness of its demands on the reader should at least not be regarded as being beyond excuse. The inevitable tendency which every world that has been brought to completion gradually shows towards its destruction, may even be reckoned among the reasons which may establish the fact that the universe will again be fruitful of worlds in other regions to compensate for the loss which it has suffered in any one place. The whole portion of nature which we know, although it is only an atom in comparison with what remains concealed above or below our horizon, establishes at least this fruitfulness of nature, which is unlimited, because it is nothing else than the exercise of the Divine omnipotence. Innumerable animals and plants are daily destroyed and disappear as the victims of time; but not the less does nature by her unexhausted power of reproduction, bring forth others in other places to fill up the void. Considerable portions of the earth which we inhabit are being buried again in the sea, from which a favourable period had drawn them forth; but at other places nature repairs the loss and brings forth other regions which were hidden in the depths of being in order to spread over them the new wealth of her fertility. In the same way worlds and systems perish and are swallowed up in the abyss of eternity; but at the same time creation is always busy constructing new formations in the heavens, and advantageously making up for the loss.

We need not be astonished at finding a certain transitoriness even in the greatest of the works of God. All that is finite, whatever has a beginning and origin, has the mark of its limited nature in itself; it must perish and have an end. The duration of a world has, by the excellence of its construction, a certain stability in itself which, according to our conceptions, approaches an endless duration. Perhaps thousands, mayhap millions, of centuries will not destroy it; but because the vanity which cleaves to finite natures works constantly for their destruction, eternity will contain in itself all the possible periods required to bring about at last by gradual decay the moment when the world shall perish. Newton, that great admirer of the attributes of God from the perfection

of His works, who combined with the deepest insight into the excellence of
nature the greatest reverence for the revelation of the Divine omnipotence,
saw himself compelled to predict the decay of nature by the natural tendency
which the mechanics of motion has to it. If a Systematic Constitution, by the
inherent consequence of its perishableness through great periods, brings even
the very smallest part which can be imagined nigh to the state of disorder,
there must be a moment in the infinite course of eternity at which this gradual
diminution will have exhausted all motion.

But we ought not to lament the perishing of a world as a real loss of
Nature. She proves her riches by a sort of prodigality which, while certain
parts pay their tribute to mortality, maintains itself unimpaired by numberless
new generations in the whole range of its perfection. What an innumerable
multitude of flowers and insects are destroyed by a single cold day! And how
little are they missed, although they are glorious products of the art of nature
and demonstrations of the Divine Omnipotence! In another place, however,
this loss is again compensated for to superabundance. Man who seems to be
the masterpiece of the creation, is himself not excepted from this law. Nature
proves that she is quite as rich and quite as inexhaustible in the production of
what is most excellent among the creatures, as of what is most trivial, and that
even their destruction is a necessary shading amid the multiplicity of her suns,
because their production costs her nothing. The injurious influences of infected
air, earthquakes, and inundations sweep whole peoples from the earth; but it
does not appear that nature has thereby suffered any damage. In the same way
whole worlds and systems quit the stage of the universe, after they have played
out their parts. The infinitude of the creation is great enough to make a world,
or a Milky Way of worlds, look in comparison with it, what a flower or an
insect does in comparison with the earth. But while nature thus adorns eternity
with changing scenes, God continues engaged in incessant creation in forming
the matter for the construction of still greater worlds.

> He sees with equal eye, as God of all,
> A hero perish, or a sparrow fall;
> Atoms or systems into ruin hurl'd,
> And now a bubble burst, and now a world.
>
> POPE

Let us then accustom our eye to these terrible catastrophes as being the
common ways of providence, and regard them even with a sort of complacency.
And in fact nothing is more befitting the riches of nature than such an attitude
towards her. For when a world-system in the long succession of its duration
exhausts all the manifold variation which its structure can embrace; when it
has at last become a superfluous member in the chain of beings; there is
nothing more becoming than that it should play the last part in the drama of
the closing changes of the universe, a part which belongs to every finite thing,
namely, that it should pay its tribute to mortality. Nature—as has been said—

already shows in the smallest part of her system that rule of procedure which eternal fate has prescribed to her on the whole. And, I say it again, the greatness of what has to perish, is not the least obstacle to it; for all that is great becomes small, nay, it becomes as it were a mere point, when it is compared with the Infinitude which creation has to exhibit in unlimited space throughout the succession of eternity.

It seems that this end which is to be the fate of the worlds, as of all natural things, is subject to a certain law whose consideration gives our theory a new feature to recommend it. According to that law the heavenly bodies that perish first, are those which are situated nearest the centre of the universe, even as production and formation did begin near this centre; and from that region deterioration and destruction gradually spread to further distances till they come to bury all the world that has finished its period, through a gradual decline of its movements, in a single chaos at last. On the other hand, Nature unceasingly occupies herself at the opposite boundary of the developed world, in forming worlds out of the raw material of the scattered elements; and thus, while she grows old on one side near the centre, she is young on the other, and is fruitful in new productions. According to this law the developed world is bounded in the middle between the ruins of the nature that has been destroyed and the chaos of the nature that is still unformed; and if we suppose, as is probable, that a world which has already attained to perfection may last a longer time than what it required to become formed, then, notwithstanding all the devastations which the perishableness of things incessantly brings about, the range of the universe will still generally increase.

But, finally, when admission is given to another idea which is just as probably in accordance with the constitution of the Divine works, the satisfaction which such a delineation of the changes of nature excites, is raised to the highest degree of complacency. Can we not believe that Nature, which was capable of developing herself out of chaos into a regular order and into an arranged system, is likewise capable of re-arranging herself again as easily out of the new chaos into which the diminution of her motions has plunged her, and to renew the former combination? Cannot the springs which put the stuff of the dispersed matter into motion and order, after the stopping of the machine has brought them to rest, be again put into action by extended forces; and may they not by the same general laws limit each other until they attain that harmony by which the original formation was brought about? It will not need long reflection to admit this, when it is considered that after the final exhaustion of the revolving movements in the universe has precipitated all the planets and comets together into the sun, its glowing heat must obtain an immense increase by the commingling of so many and so great masses; especially as the distant globes of the Solar System, in consequence of the theory already expounded, contain in themselves the lightest matter in all nature, and that which is most active on fire. This fire, thus put by new nourishment and the most volatile matter into the most violent conflagration, will undoubtedly not only resolve everything again into the smallest elements, but will also disperse and scatter these elements again in this way with a power of expansion proportional to the heat, and with a rapidity which is not weakened by any resistance in the intervening space; and they will thus be dissipated into the same wide regions of space which they had occupied before the first formation of nature. The result of this will be that, after the violence of the central fire has been subdued by an almost total dispersion of its mass, the forces of attraction and repulsion will again combine to repeat the old creations and the systematically connected movements, with not less regularity than before, and to present a new universe. If, then, a particular planetary system has fallen to pieces in this way, and has again restored itself by its essential forces, nay, when it has even repeated this play more than once, then at last the period will approach which will gather in the same way the great system of which the fixed stars are members into one chaos through the falling of their movements. Here it will still less be doubted that the reunion of such an infinite multitude of masses of fire as these burning suns are, together with the train of their planets, will disperse the matter of their masses when dissolved by the ensuing unspeakable heat into the old space of their sphere of formation, and will there furnish materials for new productions by the same mechanical laws, whereby the waste space will again be animated with worlds and systems. When we follow this Phoenix of nature, which burns itself only in order to revive again in restored youth from its ashes, through all the infinity of times and spaces; when it is seen how nature, even in the region where it decays and grows old, advances unexhausted through new scenes, and, at the other boundary of creation in the space of the unformed crude matter, moves on with steady steps, carrying on the plan of the Divine revelation, in order to fill eternity, as well as all the regions of space, with her wonders: then the spirit which meditates upon all this sinks into profound astonishment. But unsatisfied even yet with this immense object, whose transitoriness cannot adequately satisfy the soul, the mind wishes to obtain a closer knowledge of that Being whose Intelligence and Greatness is the source of that light which is diffused over the whole of nature, as it were, from one centre. With what reverence must not the soul regard even its own being, when it considers that it is destined to survive all these transformations! It may well say to itself what the philosophic poet says of Eternity:

> And when the World shall sink, and Nothing be once more,
>   When but its place remains, and all else is consumed;
>   And many another heaven, by other stars illumed,
> Shall vanish when its course is o'er:
>   Yet thou shalt be as far as ever from thy death,
>   And as to-day thou then shalt breathe eternal breath.
>   VON HALLER

Oh! happy will be the soul if, amid the tumult of the elements and the crash of nature, she is always elevated to a height from whence she can see the devastations which their own perishableness brings upon the things of the world as they thunder past beneath her feet. This happiness, which Reason of herself could not be bold enough even to aspire to, Revelation teaches us to hope for with full conviction. When the fetters which keep us bound to the vanity of the creatures, have fallen away at the moment which has been destined for the transformation of our being, then will the immortal spirit be liberated from dependence on finite things, and find in fellowship with the Infinite Being the enjoyment of its true felicity. All nature, which involves a universal harmonious relation to the self-satisfaction of the Deity, cannot but fill the rational creature with an everlasting satisfaction, when it finds itself united with this Primary Source of all perfection. Nature, seen from this centre, will show on all sides utter security, complete adaptation. The changeful scenes of the natural world will not be able to disturb the restful happiness of a spirit which has once been raised to such a height. And while it already tastes before-hand this blessed state with a sweet hopefulness, it may at the same time utter itself in those songs of praise with which all eternity shall yet resound.

> When Nature fails, and day and night
>   Divide Thy works no more,
> My ever grateful heart, O Lord,
>   Thy mercy shall adore.
>
> Through all Eternity to Thee
>   A joyful song I'll raise;
> For, oh! Eternity's too short
>   To utter all Thy praise.
>   ADDISON

JOHANN HEINRICH LAMBERT:

Cosmological Letters

The System of the Fixed Stars

LET us apply, by way of analogy, to space, comprehending the whole universe, what we know of space occupied by the solar world; and let us endeavour to rise, by just gradations, from system to system, till we at last arrive at the universal system.

The most simple system of which we can form an idea, is a planet, such as Jupiter, Saturn, or the Earth, with moons or satellites, which revolve round it. The Sun, with all its comets and planets, primary and secondary, forms a system of greater intricacy; and to each fixed star is attached a similar system. In short, the sum of all the fixed stars will constitute the universal system.

But have we not been too rapid in our arrangement? Have we not made too great a leap from the second to the third gradation? Nature's progress is more slow, her advances are more gradual, and there is every appearance, that, between the second and third step of the process, there is a vast gap to be filled up by intermediate divisions.

It is natural, and in the order of things, that number, space, and time, should increase proportionally, according as the system expands. The Earth has but one satellite, Jupiter has four, Saturn five, in the proportion of their distances from the Sun and of their masses, which render the borrowed light of these satellites more or less necessary. The Sun reigns over millions of globes. But in respect of a system of Suns, millions will be but a fraction. Let us conduct our enquiry then by this principle of analogy, and pursue the footsteps of nature.

First of all, let us consider the milky way separately from the rest of the heavens, from which it visibly separates itself, and make but one system of all the stars that are without it, including such as cover certain points of the milky way, but which, being seen by the naked eye, are at a very considerable distance from it. In this system our Sun has his place.

But do we not perceive, in the milky way itself, certain intervals, or sep-

From John Henry Lambert, *The System of the World*, translated ("digested") from the French by James Jacque, London, 1800, Part II, Chapters IV, VII, VIII, IX.

arations, which indicate a plurality of system? In various places we observe it as it were crushed and broken; and, in its higher altitudes, it may still contain numberless systems, which, hidden the one by the other, prevent our discovering the limits by which they are bounded.

All those systems of worlds resemble, though on a large scale, the solar system, in as much as in each, the stars of which it is composed, revolve round a common center, in the same manner as the planets and comets revolve round the Sun. It is even probable that several individual systems concur in forming more general systems, and so on. Such, for example, as are comprehended in the milky way, will make component parts of a more enlarged system; and this way will belong to other milky ways, with which it will constitute a whole.

If these last are invisible to us it is by reason of their immense distance. The fixed stars that are nearest to the Earth give us a certain degree of light, in as much as in a fine night, it so far subdues the darkness, as to enable us to see and distinguish the objects around us. This light, though weak, may render a still weaker light wholly imperceptible. Stars of the seventh and subsequent magnitudes escape our observation, nor do we see them even indistinctly, except when they are at once extremely numerous and crowded together; witness the *nebulæ*, seen by the telescope. Hence we may have an idea of the infinite number of stars that people the milky way, which is at such a prodigious distance from us; for it is to nothing but to the united rays of those stars that we are indebted for the impression they still make on the organ of vision.

It would not be at all astonishing then, if milky ways, situated still farther from us in the depths of the heavens, should make no impression on the eye whatever. But who knows whether the pale light that is observed in Orion, and through which Derham fancied he saw the empyreal heavens, is not one of those milky ways nearer to us than the rest? and, perhaps, by a diligent application of the telescope, we may discover elsewhere similar appearances. The variations said to be observed in this, ought not to discourage us. At our vast distance from the object, it is scarce possible that, seen across our atmosphere, and optical instruments, it should uniformly display the same aspect, and be equally well defined in the bottom of the eye.

Here then is the milky way parcelled out into various systems, each of which has its center of revolution, and the whole of this way taken together, still making but a very small part of a great system in which it is included, with an infinity of others of a similar description. In short, this great system has in like manner its center, round which revolve all the systems that compose it, as well as the particular center of each of those systems. Here we merely suggest an idea, which will be unfolded afterwards, when we come to consider more particularly the nature of those motions and centers.

We now proceed to make a remark which solicits our notice. All those systems of fixed stars, that in which we are included, as well as those in the milky way, are in the same plane, which, however, is extremely extensive. But we have seen in the constitution of the solar system, that in order to promote its population, it was of consequence, that the orbits of the celestial globes should not lie in the same plane, but be inclined as much as possible, to one another under different angles. Why then is this principle applicable as it is to the economy of great as well as small systems, not equally observed in both?

We might answer, that the firmament, as exhibited to our view, is but a small part of the whole; and that the milky way is, perhaps, to the universe, only what Jupiter or Saturn, with their satellites which are placed nearly in the same plane with their planet, are to the solar world. We might answer still farther, that as the systems of the milky way are not at rest, but in motion, we know not what aspect they may assume in time to come, and after the lapse of some myriads of ages. It would however be singular enough, should we happen to live precisely in the period when they occupy the same plane.

In the meantime let us attend to those breaks or chasms which seem to trace out even to our senses the limits of visible systems.

First of all we see the milky way evidently detached from the starry heavens, which as we conceive, make a single system only where we are pent up with our Sun and his whole train. In order to represent to ourselves this separation under a sensible image, let us fill up the void which is between the milky way and the rest of the heavens, by supposing it equally replenished with stars.

Let us conceive a vast illuminated area; place upon the ground some thousands of rows of lamps at equal distances; raise above them a second tier of the same extent, and everywhere of the same height; in short, above this, let there be a hundred similar tiers, one over the other; and let us place ourselves in the center of the illumination. If we turn our eyes vertically up and down, we shall see fifty lamps only, but the more we incline them towards the horizon, the more the number of lamps that come under our eye will be increased; and this increase will be nearly proportionate to the secants of the angles. As we thus survey the illumination, it will appear nowhere interrupted; we shall nowhere find in it any thing analogous to that intervening chasm or void so observable between the milky way and the rest of the heavens.

But let us now withdraw one great circle of intermediate lamps, allowing those to remain that are nearest, as well as those that are farthest from us. What will be the consequence? We shall see the former both larger and more luminous: the latter fainter, and apparently more close to one another. This is precisely the effect which the illuminated heavens produce on our senses. We see the stars of the milky way blended, and, as it were, touching each other. This proves that a vast void exists between them and the rest of the heavens; if this were not the case, the interruption we speak of would not appear so sudden and abrupt; it would be gradual, and we should see the stars decrease in number by little and little, even to the milky way. We must add, that this is the only satisfactory account that can be given of the apparent extreme proximity of the fixed stars in the milky way to one another.

But there are similar chasms in the milky way itself, though, by reason of its distance they are not so exactly defined. If we begin by considering such of its systems only as are most contiguous to the system of fixed stars, of which we make a part, and suppose them at nearly equal distances from one another, we can scarce conceive more than six of them in the same plane; the others are in higher altitudes and at greater distances. The apparent diameter of the most contiguous systems will scarcely be equal to the mean breadth of the milky way, and will be about ten degrees. Thus, if we take for our measure their true diameter, the intervals between those systems will correspond to six such diameters. Perhaps, however, the apparent diameter, instead of ten, will not exceed five or six degrees, and thence one might conclude, that the vacant intervening space between two systems, is ten or twelve times the breadth of those systems. In short, as the milky way, in spite of those interruptions, appears upon the whole to be continuedly one and the same, there must be above its visible axis, numberless systems which cover its vacuities. The milky way, however, has its limits, and, as we have seen above, it is possible to imagine an infinity of others of which this makes but a small part.

It is unnecessary to inform the reader, who has followed us with attention, of the use of those large voids between the systems of fixed stars; he readily perceives that they are for the purpose of admitting orbits which require a wider scope, in proportion as the system becomes more complicated and extensive.

In considering farther the oval form of the milky way, such as it appears to us, we find its distance from the arctic to be 35 deg. and 25 only from the antarctic pole. It cuts the equator nearly into two equal parts. Whence we may infer that our system of fixed stars is not only somewhat out of the plane of the milky way, but likewise nearer its circumference than its center. It is probable enough, that the part of the milky way to which we approach the nearest, is that which cuts the colure of capricorn, in the spot where its breadth is double, where it seems as if it were divided, where the small axis appears to pass through and opposite to the constellation of Orion, from which besides it was right to remove somewhat our Sun, in order to give the series of stars ranged the one behind the other, a greater length.

With respect to the position of the milky way, relatively to others of the same description that is not easily determined, because we know nothing positively of any other but this. Hitherto we have only shewn that the pale light seen in Orion may be something of the same kind, and that more of those lucid appearances may be discovered by diligent observers.

In fine, we may also proceed eastward in our system of fixed stars. We shall at least perceive, that our Sun is not in the extremity of the system of which he makes a part. This system being separated from the rest by an immense inverval, were we placed on its confines, we should see the stars of the first magnitude in one hemisphere of the Sun only. But we see them in part everywhere, and consequently the Sun is nearer the center; though we must not carry him too near it neither, for reasons that shall afterwards be assigned. But at what distance, in what precise situation he is placed, we know not.

Centres

. . . Nothing is either great or small in immensity: and, since on the wing of light we can traverse the vast regions of the Heavens, masses and volumes ought no longer to excite our astonishment. Beginning with the satellites, even suns are but bodies of the first magnitude, the centres of the fixed stars of the fourth, those of groups of systems or milky ways of the fifth, and so of the rest.

The sum of the milky ways taken together have, in like manner, their common centre of revolution; but how far soever we may thus extend the scale we must necessarily stop at last; and where? At the centre of centres, at the centre of creation, which I should be inclined to term the capital of the universe, in as much as thence originates motion of every kind, and there stands the great wheel in which all the rest have their indentations. From thence, in one word, the laws are issued which govern and uphold the universe, or, rather, there they resolve themselves into one law of all others the most simple.

But who would be competent to measure the space and time which all the globes, all the worlds, all the worlds of worlds employ in revolving round that immense body, the Throne of Nature, and the Footstool of the Divinity? What painter, what poet, what imagination is sufficiently exalted to describe the beauty, the magnificence, the grandeur of this source of all that is beautiful, great, magnificent, and from which order and harmony flow in eternal streams through the whole bounds of the universe?

But, as those centres are bodies of such prodigious magnitude, would it be impossible to discover where they reside? Would it be impossible even to see them? Might we not, at least, discover that of the system to which our Sun belongs? Let us not despair: time, observation, the sagacity of astronomers, and diligent research may carry us much farther than we are apt to imagine.

The change of place in the fixed stars being fully ascertained, the parallax of the solar orbit would be the best mean of discovering the centre of their revolution. Here the case is the same as would be that of the inhabitants of the Moon if they could not see the Sun; for the Earth is to him in respect of his motion round the centre of fixed stars, what the Moon is to the Earth in respect of her motion round the Sun.

If we suppose, therefore, the inhabitants of the Moon are as able astronomers as ourselves, they will have two methods of discovering the Sun's place.

The first will consist in determining the ellipse which the Moon would describe round the Earth if she were governed by the Earth's attraction alone, and then in finding the places of the greatest deviation from this ellipse produced by the attraction of the Sun. In respect of velocity, for example, the deviation would be most sensible either after conjunction or opposition, and these places being given would shew that of the Sun.

By the other method they would deduce from a certain number of observations the true place of the planets; an operation which they would perform with the assistance of Kepler's laws, nearly in the same manner as we find the places of comets, though with somewhat more difficulty. The revolution of the Moon round the Earth, and the distance of the one from the other would give them a parallax which they might employ with much more advantage than we can the annual parallax of the Earth's orbit in relation to the fixed stars. But this last would also render them the same services which we hope to derive from the parallax of the Sun's orbit.

It is by similar methods, then, that we ought to proceed in order to get acquainted with the system of fixed stars to which we belong, and to discover its centre. This may be found practicable in process of time; and we will then know our distance from that point. We have no reason to imagine that our Sun is placed at the centre; but neither does it appear that he is farther removed from it than by a few stars: so that his motion ought to be more rapid than that of stars situated at a more remote distance. All this will some day contribute to determine the place of the central mass.

But, if we suppose this body is of an enormous bulk, and illuminated besides by one or more fixed stars, it should certainly seem that we ought to perceive it in whole or in part with the help of the telescope. As we are at no prodigious distance from it, its apparent diameter even to us may be very considerable; and, how weak soever its reflected light, it cannot be enfeebled in its passage to us in such a degree as to be rendered imperceptible. Enlightened by one or more Suns, this body ought to present phases analogous to those of the Moon. Perhaps it may have spots like the other globes; a circumstance which ought also to create a variety in its aspects. The diameter of the Sun is about double the diameter of the Moon's orbit; that of our central body then may, without exaggeration, be supposed to be greater than the diameter of the orbit of Saturn. Such a body, were it illuminated only in an equal degree with that planet, ought not to be invisible.

But are we sure that we perceive nothing like it? Has not that pale light in Orion which we were at first inclined to take for a milky way, a greater resemblance to the enlightened side of a central body? Derham did not view it as a luminous body, but rather as a sort of opening through which we discover something illuminated, probably, by the reflection of the empireal heavens. The appearance corresponds infinitely better to an illuminated body than to a collection of stars shining with their native lustre. Certain variations have been observed in the visible form of this light; a circumstance which accords better with the spots and aspects of a central body than with a milky way; where such variations could not be perceived in so small an interval of time, unless we were disposed to explain them from a greater or less degree of transparency in our atmosphere at different periods.

In fine, this body would be precisely in the region of the heavens, where it should seem we ought to seek for the centre of our system. The Sun, by which it is illuminated, may appear to us but a small star; while the central body itself, which is incomparably larger, may be visible to us under a sensible diameter. The stars that are observed under the pale light of Orion, are certainly arranged in series one behind another, at different distances, otherwise they would be too much crowded. Finally it would be requisite that we observe the variations of this light but a few years, in order to determine whether from their regularity we are not warranted to infer either phases or a rotatory motion.

If we could see the centre of our system of fixed stars, it would not be impossible that we might discover also other centres; for their magnitude increases in proportion to the largeness of the system, and the largeness of the systems in proportion to their complexity. I say we might see those centres, unless the weakness of reflection, the vast length of way the light has to travel, or the superior brilliancy of the fixed stars should rob us of that spectacle. Who knows but the eye, assisted by the telescope, may at length penetrate all the way to the centres of the milky way, and why not even to the centre of the universe?

Besides, a principle of analogy should seem to require it. The central body ought to extend its influence even to the extremities of its system; and, consequently ought to appear under a sensible diameter, or at least be visible to the telescope. It is thus that the satellites see their planets, and the planets their Sun: the most distant planet still sees him under an angle of more than three minutes: the comet of 1759 sees him from the summit of its aphelion under an angle of a minute: a comet, whose aphelion is at sixty times a greater distance, sees him from the same point under the angle of a second; and this, in a good telescope, would exceed two minutes. It may be questioned whether any comet retires to so great a distance from the Sun; for a comet of this description taken at the lowest computation would require thirty-five thousand years to accomplish its revolution.

Again, the attractive force of a body decreases as the square of the sine of its apparent semi-diameter; and this semi-diameter cannot be invisible in any place to which its attractive force and its sphere of activity extend.

Our Earth belongs, by a chain of gradations, to several systems, and at last to the system of the universe: all the centres of those systems as well as the universal centre, exerts their influence over her. The whole of those centres then, ought, in respect of the Earth, to occupy a sensible space in the heavens; at least we have a right to suppose that we ought to see them with the telescope. Nothing but the inconveniences arising from the transmission of light as detailed above, could intercept our view of them. For as to their magnitude, it is such as it would be necessary to render them visible.

Here then we have all the systems of the universe reduced to order, and enchased in one another. But what is our position amidst those systems?

Where are we? As to this point we can speak indefinitely, negatively, and by approximation only.

The Earth is not at the centre of the solar system. The Sun is not at the centre of his system of fixed stars; a centre which is either in the region of Orion or Sirius. This system is neither at the centre nor in the plane of the milky way, though it seems to project over it a little; the portion of this way which it approaches the nearest, is that which passes by the colure of Capricorn, where its breadth is double. But where is the milky way itself in relation to other milky ways? Here ends all our science with the utmost stretch of our eyes and instruments.

Ellipses Changed into Cycloids—Universal Motion

Hitherto we have proceeded on the supposition that the heavenly bodies revolve in ellipses. The new point of view to which our theory leads us, will produce an entire change; and we shall see that we have reviewed a suite of hypotheses which overturn each other, in proportion as we advance in our enquiry.

The Moon, it is said, describes an ellipse round the Earth. This would be true were the Earth at rest; but as she moves round the Sun, and obliges the Moon to participate in her motion, the orbit of this last cannot be an ellipse, but a cycloid. The ellipse of the Earth vanishes for the same reason, the moment the Sun ceases to be immoveable, and is found to describe an orbit round a new centre. Then the ellipse of the Earth becomes a cycloid of the first degree, that of the Moon of the second, and the velocity of their motion increases in the same proportion.

But this order continues no longer than the new centre, or body of the fourth degree, reckoning from the Moon, is supposed immoveable. As soon as we give motion to it, the ellipse of the sun vanishes in its turn; he then describes a cycloid of the first degree, the Earth of the second, the Moon of the third.

We easily perceive, that what is here said of the Moon, the Earth, and the Sun, applies equally to all the satellites, planets, comets, and fixed stars, without exception. There is not a heavenly body which does not partake of these motions, more or less complex; while each shares them in a degree suited to its particular circumstances.

We observe likewise, that as we pass on from centre to centre, these motions become more and more complicated; and their combinations only terminate at the universal centre, which alone is in a state of real and absolute rest. If, beginning by the Moon, we suppose that the body which occupies that centre is in the thousandth; the cycloid of the Earth will be in the nine hundred and ninety-eighth degree. There, and there alone, will be the true orbit of the Earth, while the velocity with which she describes it, will be her true velocity.

But who is in condition to determine it, to describe the nature of her cycloid, to trace the perplexed path of our planet, and the strange bounds or skips she makes in the regions of the Heavens?

Nothing, however, is more evident. The Earth as well as all the other globes revolve, properly speaking, round the universal centre alone. With respect to the Sun, she only attends him in the same route, and as his fellow traveller, avails herself of his company, by partaking in his light and heat. She undoubtedly makes many circuitous, and, as they may appear to us, useless trips, but which, as the law of gravitation supplies no other means of keeping two or more bodies together, are nevertheless necessary. She gravitates towards all the centres on which she depends; with the Moon towards the Sun, with the Sun towards a body of the fourth degree, with this last towards a body of the fifth, or towards the centre of the milky way, and so on of the rest. Thus, from system to system, our cycloid takes new inflections, which increase in magnitude as we advance in our career.

The satellites exhibit in small, every thing that happens in great: as they unquestionably move in cycloids, we easily perceive that the elliptic is not the only species of motion that obtains in the world. Our theory, however, requires motion which becomes more complicated in proportion as bodies are distant from the universal centre. By this means the ellipses are transferred far from us, and confined to those bodies alone which depend immediately on the centre of centres: these communicate motion to bodies in their immediate dependence, and force them to move in cycloids of the first degree. From thence motion passing on to the utmost limits of the universe, becomes progressively more complex. It is in this manner that the wheels of the great machine mutually clinch and support each other; and it affords a fresh proof of what we have frequently had occasion to observe, that things become simple in proportion as we approach the sum of the whole.

As it is by no means probable that we shall ever come to the knowledge of the true cycloid of the Earth, we may continue, in our calculations concerning the planets and comets, to employ the ellipse which is sufficient for all our present purposes. We may in like manner content ourselves with a cycloid of the first degree, so long as we only take an interest in the system of fixed stars in which we are placed. Twenty centuries hence, perhaps, when we may be in condition to give a just arrangement to the systems of the milky way, we shall adopt a cycloid of the second degree. From one epoch to another we shall go on approximating the truth, and enlarging our measurement in the same proportion. Our present measurement, as applied in the Copernican hypothesis, is the radius of the Earth's orbit for space, and the Sun's annual revolution for time. We are in the way of gaining another step; and presently our measurement will be the radius of the Sun's orbit, and his periodical time. The third epoch will give measurements on a still more extended scale.

There is this additional convenience, that complicated theories may easily be resolved into such as are more simple. This we actually experience in respect of the Moon's orbit, which we never consider as a cycloid but when we wish to account for anomalies occasioned by the action of the Sun: in all other cases we regard it as an ellipse, and endeavour, with all the address in our power to make it cover all her irregularities; a mode of proceeding which facilitates the calculation of her motion, the ellipse being a curve of a more simple and uniform description than the cycloid. We will apply the same method to the comets and planets, as soon as the periodical motion of the Sun shall have enabled us to observe the anomalies of their motion.

In astronomy the scenery is continually shifting, and the modes of language vary in proportion as this inexhaustible science makes progress in improvement, and supplies us with new theories. Ptolemy spake the language of the people: to Copernicus we are indebted for the language of astronomy; which Tycho Brahe in some measure confounded: Kepler and Newton rectified his faults, and gave to astronomical language a superior degree of elegance and perfection. The discoveries of the present and future times will introduce in this respect farther changes. All these different modes of language will, nevertheless, continue to be always intelligible; and may always be preserved in a certain degree, and within certain limitations.

The astronomers of the Moon may have suspected before us the periodical motion of the Sun as well as the universal motion; because that of the Moon depends on the Earth, and that of the Earth on the Sun. After having found that the Moon has a motion round the Earth, they would have occasion to remark that the Earth herself is not at rest, but has, in like manner, a motion round the Sun. They would then have discovered two periodical motions which would lead them to conjecture the third, and by and by the others. If the Moon herself had a satellite, this satellite having three periodical motions as a model, would, with still greater ease, come at the knowledge of the rest.

We may compare the universal motion to that of the waves of the sea, which succeed and impel one another. We see them rise and fall alternately, leaving a cavity between them, which vanishes the moment the uneven surface recovers its level. These elevations and depressions exhibit in their section a representation of the cycloids, which the heavenly bodies describe in the regions of space. The small waves shew us, in a single elevation and depression, a cycloid of the first degree. But, in proportion as the wave swells, it becomes composed of small ones, each of which retains its own curvature, and which, nevertheless, follows the inflections of the large wave rising and falling along with it. Nature seems to be fond of this sort of undulations, since she contrives to introduce them into almost all her motions. It is in this manner a ship ploughs the plains of the ocean; the lever or wave which makes her rise and descend must be of a greater size in proportion as the vessel is of a larger volume: the shallop attached to her rises and falls repeatedly in the time the ship requires to do the same thing but once; but the sum of elevations and depressions of the shallop is equivalent to those of the ship. Thus it is that the celestial globes undulate in the ocean of space. They all make the tour of the centre; but the larger the mass of which a body is composed, the greater the resistance it opposes, and the more its undulations are simple and uniform. Their motion, therefore, is in its nature much more quiet and regular than that of the agitated waves of the sea; they are carried on, so to speak, by a constant and equable breeze, and they have no fear of a storm.

The cycloid of the Moon in the system of Copernicus, has but a very small inflection: one of its oscillations, which comprehends an elevation and a descent, amounts to thirty degrees of the Earth's orbit in length, and, consequently, about one-fourth of the whole; whilst the elevation and descent, taken together and considered in themselves, scarcely extend to the 365th part of this diameter.[1] Thus the length of the cycloid exceeds its height nearly four score and ten times.

The orbit of the Earth, if we suppose it become a cycloid of the first degree by the Sun's motion, will be one whose length exceeds its height a great deal more than in the proportion of 365 to one. In fact, we are entirely ignorant whether, since the days of Hypparchus, that is to say, two thousand years, the Sun has traversed one degree of his orbit, nor are we less in the dark as to the dimensions of those degrees.[2] The length of these cycloids depends on the rapidity of the Sun's course; if we suppose it two or three times greater than that of the Earth according to the Copernican system, the length of the cycloids will then be very considerably extended, whilst their curvature becomes greatly diminished.

This, however, does not appear to be universally necessary. The satellites of the two higher planets, in the system of Copernicus, describe a particular species of cycloid which cannot be compared to the undulations of the waves. As they move with a greater velocity than their planets themselves in their ellipses, when they arrive between their planets and the Sun, their motion becomes retrograde, and their cycloids cross each other.

What extraordinary curves must not the heavenly bodies describe in their true motion? The mind loses itself when it endeavours to follow them. Let us try to form a conception of the orbit of one of those comets which travel in hyperbolas from world to world, I mean its true orbit in relation to all intermediate centres as well as the universal centre. All our ideas get into confusion, and we are obliged to desist from the attempt.

The question still returns: where are we? In what district of the universe does our Earth proceed in her course? Through how many inflections does she steer in order to arrive at a succeeding one of a greater magnitude? Within what distance is she permitted to approach the centre of centres? What is the farthest limit to which she ventures to recede from it? In what point of each inflection would she be found at this moment? And what is the true velocity

---

1. Indeed, it may be computed at no more than a 386th part; since the distance of the Moon from the Earth is to the distance of the Earth from the Sun, at the utmost, as 854 to 330,000.

2. But we know, that in this space of time, the Earth has made 2000 oscillations, Saturn 70, and the other planets in the same proportion. A comet, whose period should be of some ages, would not have oscillated so often.

of her present motion? Is she actually stationary or retrograde, or spinning in
the highest pitch of her velocity, which is compounded of all those velocities
that the bodies of each degree to which she belongs, would have in their
ellipses, if the body of the immediately superior order remained immoveable!
Would not this velocity be almost infinite? Or would it be tempered and
controlled by the different directions of moving bodies, some of which are
before, and some rather below. Were that the case, nothing could be more
fantastical than the motion of the Earth and the other heavenly bodies; no
rocket in an exhibition of fireworks presents motions whimsical enough to give
us the smallest idea of it.

As to all those points we are, and will long remain ignorant. But this we
clearly perceive, that all things are made by weight and measure, that the
motions, the most capricious in appearance, are regulated by eternal laws; that
centres, systems, and motions, are in a chain of subordination one to another,
so as to make all act in concert for the preservation of each individual and
the harmony of the whole.

General Conclusions—Recapitulation

In the progress of this work we have interspersed some reflections which
we will here collect in one general point of view.

There is a difference which deserves to be remarked between the general
aspect of the Earth and that of the universe. The former presents us with such
apparent disorder as we cannot unravel, but by considering it in connection
with the whole, where it falls easily under rule; whereas the latter carries on
the face of it apparent order and regularity, and that of the simplest kind, but
which becomes more and more complicated in proportion as we proceed in
its investigation.

The rising and setting of the Sun, the firmament turning round us with its
numberless stars in the space of twenty-four hours; can anything be more
simple or uniform? But in the progress of more minute and accurate observation, this uniformity disappears: we see the Moon going backwards; we see
the planets swimming against a current which overpowers them, whilst,
besides a motion common to all, they obey motions which are peculiar to
themselves. This led to the system of Copernicus, who introduced a certain
arrangement into the world; an arrangement, however, which is already, as
you observe, become much more complex. By and by the Sun and the fixed
stars begin to revolve in orbits; whole systems of fixed stars, and systems of
systems get into motion; the order of the universe becomes progressively more
and more complicated, till at last we come to it in the greatest possible degree
of complexity, where we have just surveyed it.

Is it not truly wonderful, that, in the constitution of the universe, time and
space should every where be so happily combined, that notwithstanding the

infinity of wheels and springs which mutually depend on each other, and which are all necessary to the play of the machine, the visible order of nature should, nevertheless, every where preserve the same air of simplicity and uniformity. But does not the purpose of this arrangement readily present itself? It can be no other than that the visible firmament should be in all its aspects, at all times, and for the inhabitants of all the spheres, a clock accommodated to all places, to all divisions of time, and to the exigences of the inhabitants of each individual globe. Such, in respect of us, are the diurnal revolution of the heavens, the diurnal and annual motions of the Sun; elsewhere the same clock which measures our hours, our days, our years, measures under different aspects, but with equal uniformity, centuries, thousands, myriads, millions of ages, the whole in due proportion to the situation, motion, and distance of the different globes. It even seems highly probable, that this amazing degree of intricate combination was indispensibly necessary in the first and primitive arrangements, in order to produce an appearance every where so simple, and from which we derive such important advantages.

On the other hand, if we could penetrate across superficial appearances, and ascend to the reality, we should find motion in the greatest degree of simplicity as we set off from the universal centre, and we should see it become more complicated in proportion as we removed from the central point. The bodies over which this centre exerts its immediate influence, revolve with a sort of majesty in their ellipses, and have a motion suited to the dignity of their order. Then come the cycloids, the epicycloids, then those of the succeeding gradation. Here the astronomer will make an admirable discovery, he will have under his eye all the links of that vast chain depending on one another by virtue of a series, all the terms of which are formed from the preceding term by an invariable law.

This series, in fact, is the most simple and perfect of what they call recurrents, 1, 1, 2, 3, 5, 8, 13, &c. by deducting each term from that by which it is followed, the same series is reproduced. This leads us back to the supposition of an immoveable body which we have considered as moving in an ellipse; by which means the cycloids that depend on it become more simple by one degree, the series remaining the same. We might carry this operation so far as to set at rest any body whatever of a given order, or even the length of reducing its cycloid to an ellipse. This is exactly what we do, when, according to the Copernican system, we consider the Sun immoveable, and the orbits of the comets and planets as ellipses.

The system of Copernicus is, in fact, only a theory; but we have seen that astronomy can come at the truth only by carefully bringing under a review every possible hypothesis. This science, however, has made astonishing progress; by converting, at different periods, apparent arrangements into such as are more conformable to what actually exists, we have left behind various theories founded in appearances, and penetrated, if not fully and demonstrably, at least in the way of fair conjecture, even to the real and genuine order of things

Indeed, we may say, thanks to the science of astronomy, we know the heavens greatly better than we do the Earth, where we are still very far from being in condition to unravel the apparently chaotic disorder which reigns in the physical as well as the moral world. We shall have sooner arranged a system of fixed stars, and ascertained their motion, than reduced the changes of the weather and the variations of the barometer to any fixed and determined rule. These last phenomena depend on too many minute causes, and too great a variety of particular circumstances, to be brought under a general principle. There is not an inequality on the surface of the globe, nor a mountain, nor a valley, nor a spring of water, &c. which may not be concerned in the production of these effects.

Upon the whole, we may draw this conclusion, that the heavens are made to endure, and the things of the earth to pass away. Nature changes in small, and is maintained and preserved in great. The vast clock of the firmament can only develope its springs in the course of numberless epoques which succeed each other, and each in the epoque to which it is assigned. Hitherto we scarcely discern the needle which points the minutes and seconds.

Let us recapitulate and have done.—The law of gravitation extends universally over all matter. The fixed stars obeying central forces move in orbits. The milky way comprehends several systems of fixed stars; those that appear out of the tract of the milky way form but one system which is our own. The sun being of the number of fixed stars, revolves round a centre like the rest. Each system has its centre, and several systems taken together have a common centre. Assemblages of their assemblages have likewise theirs. In fine, there is a universal centre for the whole world round which all things revolve. Those centres are not void, but occupied by opaque bodies. Those bodies may borrow their light from one or more Suns, and hence become visible with phases. Perhaps the pale light seen in Orion is our centre. The real orbits of comets, planets, and suns, are not ellipses, but cycloids of different degrees. The orbits of those bodies which are immediately subject to the action of the universal centre can alone be ellipses.

WILLIAM HERSCHEL:

On the Construction of the Heavens

THE SUBJECT of the Construction of the Heavens, on which I have so lately ventured to deliver my thoughts to this Society, is of so extensive and important a nature, that we cannot exert too much attention in our endeavours to throw all possible light upon it; I shall, therefore, now attempt to pursue the delineations of which a faint outline was begun in my former paper.

By continuing to observe the heavens with my last constructed, and since that time much improved instrument, I am now enabled to bring more confirmation to several parts that were before but weakly supported, and also to offer a few still further extended hints, such as they present themselves to my present view. But first let me mention that, if we would hope to make any progress in an investigation of this delicate nature, we ought to avoid two opposite extremes, of which I can hardly say which is the most dangerous. If we indulge a fanciful imagination and build worlds of our own, we must not wonder at our going wide from the path of truth and nature; but these will vanish like the Cartesian vortices, that soon gave way when better theories were offered. On the other hand, if we add observation to observation, without attempting to draw not only certain conclusions, but also conjectural views from them, we offend against the very end for which only observations ought to be made. I will endeavour to keep a proper medium; but if I should deviate from that, I could wish not to fall into the latter error.

That the milky way is a most extensive stratum of stars of various sizes admits no longer of the least doubt; and that our sun is actually one of the heavenly bodies belonging to it is as evident. I have now viewed and gaged this shining zone in almost every direction, and find it composed of stars whose number, by the account of these gages, constantly increases and decreases in proportion to its apparent brightness to the naked eye. But in order to develop the ideas of the universe, that have been suggested by my late observations, it will be best to take the subject from a point of view at a considerable distance both of space and of time.

From "On the Construction of the Heavens" (1785) in *Collected Papers*, edited by J. L. E. Dreyer, London, 1912, I, 223 ff.

Theoretical View.—Let us then suppose numberless stars of various sizes, scattered over an indefinite portion of space in such a manner as to be almost equally distributed throughout the whole. The laws of attraction, which no doubt extend to the remotest regions of the fixed stars, will operate in such a manner as most probably to produce the following remarkable effects.

Formation of Nebulæ.—Form I. In the first place, since we have supposed the stars to be of various sizes, it will frequently happen that a star, being considerably larger than its neighbouring ones, will attract them more than they will be attracted by others that are immediately around them; by which means they will be, in time, as it were, condensed about a center; or, in other words, form themselves into a cluster of stars of almost a globular figure, more or less regularly so, according to the size and original distance of the surrounding stars. The perturbations of these mutual attractions must undoubtedly be very intricate, as we may easily comprehend by considering what SIR ISAAC NEWTON says in the first book of his "Principia," in the 38th and following problems; but in order to apply this great author's reasoning of bodies moving in ellipses to such as are here, for a while, supposed to have no other motion than what their mutual gravity has imparted to them, we must suppose the conjugate axes of these ellipses indefinitely diminished, whereby the ellipses will become straight lines.

Form II. The next case, which will also happen almost as frequently as the former, is where a few stars, though not superior in size to the rest, may chance to be rather nearer each other than the surrounding ones; for here also will be formed a prevailing attraction; in the combined centre of gravity of them all, which will occasion the neighbouring stars to draw together; not indeed so as to form a regular or globular figure, but, however, in such a manner as to be condensed towards the common center of gravity of the whole irregular cluster. And this construction admits of the utmost variety of shapes, according to the number and situation of the stars which first gave rise to the condensation of the rest.

Form III. From the composition and repeated conjunction of both the foregoing forms, a third may be derived, when many large stars, or combined small ones, are situated in long extended, regular, or crooked rows, hooks, or branches; for they will also draw the surrounding ones, so as to produce figures of condensed stars coarsely similar to the former which gave rise to these condensations.

Form IV. We may, likewise, admit of still more extensive combinations; when, at the same time that a cluster of stars is forming in one part of space, there may be another collecting in a different, but perhaps not far distant quarter, which may occasion a mutual approach towards their common center of gravity.

Form V. In the last place, as a natural consequence of the former cases, there will be formed great cavities or vacancies by the retreat of the stars towards the various centers which attract them: so that upon the whole there is evidently a field of the greatest variety for the mutual and combined attractions of the heavenly bodies to exert themselves in. I shall, therefore, without extending myself farther upon this subject, proceed to a few considerations, that will naturally occur to every one who may view this subject in the light I have here done.

Objections Considered.—At first sight then it will seem as if a system, such as it has been displayed in the foregoing paragraphs, would evidently tend to a general destruction, by the shock of one star's falling upon another. It would here be a sufficient answer to say, that if observation should prove this really to be the system of the universe, there is no doubt but that the great Author of it has amply provided for the preservation of the whole, though it should not appear to us in what manner this is effected. But I shall moreover point out several circumstances that do manifestly tend to a general preservational; as, in the first place, the indefinite extent of the sidereal heavens, which must produce a balance that will effectually secure all the great parts of the whole from approaching to each other. There remains then only to see how the particular stars belonging to separate clusters will be preserved from rushing on to their centers of attraction. And here I must observe, that though I have before, by way of rendering the case more simple, considered the stars as being originally at rest, I intended not to exclude projectile forces; and the admission of them will prove such a barrier against the seeming destructive power of attraction as to secure from it all the stars belonging to a cluster, if not for ever, at least for millions of ages. Besides, we ought perhaps to look upon such clusters, and the destruction of now and then a star, in some thousands of ages, as perhaps the very means by which the whole is preserved and renewed. These clusters may be the *Laboratories* of the universe, if I may so express myself, wherein the most saluatary remedies for the decay of the whole are prepared.

Optical Appearances.—From this theoretical view of the heavens, which has been taken, as we observed, from a point not less distant in time than in space, we will now retreat to our own retired station, in one of the planets attending a star in its great combination with numberless others; and in order to investigate what will be the appearances from this contracted situation, let us begin with the naked eye. The stars of the first magnitude being in all probability the nearest, will furnish us with a step to begin our scale; setting off, therefore, with the distance of Sirius or Arcturus, for instance, as unity, we will at present suppose, that those of the second magnitude are at double, and those of the third at treble the distance, and so forth. It is not necessary critically to examine what quantity of light or magnitude of a star entitles it to be estimated of such or such a proportional distance, as the common coarse estimation will answer our present purpose as well; taking it then for granted, that a star of the seventh magnitude is about seven times as far as one of the first, it follows, that an observer, who is inclosed in a globular cluster of stars, and not far from the center, will never be able, with the naked eye, to see to the end of it: for, since, according to the above estimations, he can only extend his view to about seven times the distance of Sirius, it cannot be expected that his eyes should reach the borders of a cluster which has perhaps not less than fifty stars in depth every where around him. The whole universe, therefore, to him will be comprised in a set of constellations, richly ornamented with scattered stars of all sizes. Or if the united brightness of a neighbouring cluster of stars should, in a remarkable clear night, reach his sight, it will put on the appearance of a small, faint, whitish, nebulous cloud, not to be perceived without the greatest attention. To pass by other situations, let him be placed in a much extended stratum, or branching cluster of millions of stars, such as may fall under the III^d form of nebulæ considered in a foregoing paragraph. Here also the heavens will not only be richly scattered over with brilliant constellations, but a shining zone or milky way will be perceived to surround the whole sphere of the heavens, owing to the combined light of those stars which are too small, that is, too remote to be seen. Our observer's sight will be so confined, that he will imagine this single collection of stars, of which he does not even perceive the thousandth part, to be the whole contents of the heavens. Allowing him now the use of a common telescope, be begins to suspect that all the milkiness of the bright path which surrounds the sphere may be owing to stars. He perceives a few clusters of them in various parts of the heavens, and finds also

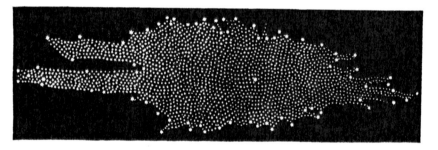

The Stellar System According to Herschel's Disc-Theory

that there are a kind of nebulous patches; but still his views are not extended so far as to reach to the end of the stratum in which he is situated, so that he looks upon these patches as belonging to that system which to him seems to comprehend every celestial object. He now increases his power of vision, and, applying himself to a close observation, finds that the milky way is indeed no other than a collection of very small stars. He perceives that those objects which had been called nebulæ are evidently nothing but clusters of stars. He finds their number increase upon him, and when he resolves one nebula into stars he discovers ten new ones which he cannot resolve. He then forms the idea of immense strata of fixed stars, of clusters of stars and of nebulæ; till, going on with such interesting observations, he now perceives that all these appearances must naturally arise from the confined situation in which we are placed. *Confined* it may justly be called, though in no less a space than what before appeared to be the whole region of the fixed stars; but which now has assumed the shape of a crookedly branching nebula; not, indeed, one of the least, but perhaps very far from being the most considerable of those number- less clusters that enter into the construction of the heavens.

Modern Theories
of the Universe

Introduction

THE INCREASE IN ATTENTION to cosmological questions since the second decade of the present century is due primarily to the promise held out for a successful prosecution of studies in this area due to the use of new instruments, both optical and conceptual, that made their appearance during this period. It is true that the use of new instruments, however radical their innovations, are generally but improvements in one form or another upon older models. And this we find to be the case with the more powerful telescopes that the astronomer uses to gather his data. It is similarly true of the various mathematical tools that the physicist uses to construct his theoretical models. Yet the difference from the old may be so great, that they warrant an altogether fresh outlook. When for example, Hubble began his studies with the use of the 100-inch telescope, or Einstein incorporated into his relativity theory the basic ideas of non-Euclidean geometry and the tensor calculus, these more refined instruments brought such a wealth of information or insight, that they helped to raise the subject of cosmology to a new level altogether. Questions that had been endlessly debated in earlier periods, either now found definite answers or were seen in an altogether different light because of the rich field of possibilities opened up.

We must recognize, at the same time, a certain parallel in the problems confronting cosmology today and those which engaged the attention of cosmologists at the beginnings of modern astronomy following upon the Copernican revolution. In that earlier period, as we have seen, the assimilation of the earth as a member of the planetary family of the sun, and the sun as a member of the system of stars, raised the fundamental question as to the extent and possible structure of the universe, where the latter is taken as made up of the stars as its basic astronomical units. Today the scale has shifted from stars to galaxies. We know now what was only surmised in the eighteenth century by Wright, Kant, Herschel and others. Our Galaxy is a vast but finite system of stars that is located among an enormously extensive population of galactic systems of different types, shapes and structures. The universe, as far as present knowledge indicates, is made up of such galaxies as its basic units. The problem with which cosmology is now primarily concerned is then: How extensive is the universe of galaxies? What spatial structure does it have? Is the "space" or "geometry" to be used in describing it of a finite or an infinite sort? Is the universe, when regarded from a temporal point of view, one that had a finite origin in the past or has it always been in existence? If the latter, has it always

271

possessed roughly the same structure as it does now? These alternatives and the several refinements and combinations that result from their analyses yield a variety of theories actively discussed in the present phase of our subject.

The general theory of relativity, which Einstein proposed in 1915, undertook to widen and generalize the ideas contained in the special theory of relativity of 1905. It also provided through its new interpretation of gravitational phenomena, a schematic base for encompassing the whole range of mechanical phenomena dealt with in classical physics. Its superiority over the older physics consisted in its greater systematic simplicity and the capacity it had both for making predictions and for offering explanations over a wider range and with a greater precision than had been possible formerly. The first major success of the general theory of relativity was obtained in dealing with the dynamics of the solar system. It was here that the famous three "crucial" tests of the theory were obtained. It was Einstein himself, in a paper published in 1917, who went on to show that the general theory of relativity had another useful line of theoretical application. This time its basic equations were designed to assist in understanding the structure of the physical universe as a whole, as distinguished from the relatively restricted problems encountered in the dynamics of the solar system. In this first cosmological application of general relativity (briefly and simply expounded by Einstein in his discussion of "Considerations on the Universe as a Whole"), the way was pointed out for making use of non-Euclidean geometry in framing a world-view. The universe was pictured as being at once both finite and unbounded. Since then, not only Einstein, but others as well have worked out numerous other possibilities within the framework of "relativistic cosmology" on the basis of the leads furnished by this path-breaking investigation.

What Hubble established by 1924 in his epoch-making observational studies of "the realm of the nebulae" was the general answer to the much-debated question as to the status of the nebulae. The two possibilities, that they were independent stellar systems lying beyond the confines of our own Galaxy, or on the other hand, were proper parts of our own Galaxy, was settled in favor of their existence as independent systems. The preliminary "reconnaissance" of the observable region of the nebulae, as recounted by Hubble, established two main empirical results. One was that the extra-galactic nebulae (or, as some writers prefer to call them, simply, "the galaxies") are distributed throughout space in a uniform and homogeneous manner. The other was that the spectra of these galaxies display a shift toward the red—the greater the shift the more distant they are. This latter feature is commonly interpreted as signifying that the galaxies are receding from us and from one another in such a fashion that the more distant the galaxy, the faster is its velocity of recession. It is this fact which lends support to the belief that the universe is "expanding." A further selection by Hubble, given below, sketches the observational program to be undertaken by the 200-inch Palomar telescope. It indicates both the progress made since the initial exploratory studies and some of the typical problems still facing the astronomer.

From the time of the construction of the "static" Einstein model ("static" because it did not envisage the possibility of the recession of the galaxies or the expansion of the system as a whole) and the exploratory observational studies by Hubble and others of the properties of the galaxies, the field of scientific cosmology has developed rapidly. In De Sitter's discussion of "Relativity and Modern Theories of the Universe," the reader will find an expert summary of the development of relativistic cosmologies up to the period of the early nineteen-thirties. It is written by one who himself played an important role in that development. Relativistic cosmologies find their common base in the "field-equations" of the general theory of relativity. Since the field-equations of that theory, being its most fundamental part, are deliberately stated in a most general form, they leave unspecified the particular values to be given to certain key variables. The specification of these to suit, for example, the cosmologic field of their application, is made in accordance with reasons that seem compelling to a given author or at a given stage of research. The series of resulting equations, being in effect tools of calculation and mathematical devices of representation, constitute what we mean by "models of the universe as a whole." It was in this fashion that the original "Einstein model" and the one proposed by De Sitter shortly afterward (known as the "De Sitter model") were constructed. These "Einstein" and "De Sitter" models, however, were essentially "static" models. They were soon to be replaced by a group of models which took as their central feature the "expansion" of space. These are designated as "the expanding-universe models of general relativity." Eddington and his pupil Lemaître shared a prominent role in the development and popularization of these ideas. Each of them offers below his own highly instructive account of this side of the subject.

In the mid-thirties, the English cosmologist E. A. Milne undertook to challenge the almost exclusive reliance hitherto put upon general relativity as a base from which to construct theoretical models of the universe. In a system of thought which he identified by the name "Kinematic Relativity," Milne offered a most ingenious conceptual framework in terms of which to construct a model of the expanding universe. It is an approach in which the notion of time assumes a central importance as contrasted with the more "geometrizing" emphases of relativity theory. His approach illustrates a mode of thought that would seem, however, for all its suggestiveness, not to have won any considerable following. One of the reasons, perhaps, for the reluctance of scientists in general to accept Milne's point of view, is its highly rationalistic character. It places its primary emphasis on deuctions made from allegedly "self-evident" premises rather than on the leads and checks furnished by observational experience.

The more conventional, orthodox, and antirationalistic viewpoint of scientists is illustrated in the discussion by H. P. Robertson of "Geometry as a Branch of Physics." Robertson writes primarily from the vantage point of "orthodox" relativistic cosmology, to which he himself made important contributions. He stresses the necessary role of observational experience in scientific cosmology. Such a role, he argues, arises particularly with respect to the choice of a "metric" or geometry to be used in constructing an adequate model of the universe as a whole.

Since the end of the Second World War the subject of cosmology has continued to receive important contributions of both an observational and theoretical sort. The use of more powerful instruments such as the 200-inch Hale telescope, the 48-inch Schmidt telescope, and others, has increased in a significant way the quantity and quality of observational data. Among the most important results obtained in this direction is the recent recalibration of the distance-scale for galaxies, effected by Baade at Palomar Observatory. This has given grounds for a considerable revision upward in the calculated range to which our instruments may be taken as probing in space. By this change in the scale to be used for the calculation of cosmic distances, some of the difficulties faced by earlier models of the "expanding universe" no longer hold. These included particularly the problem of how to accept the so-called "age of the universe," which came out on some views to be *less* than the calculated ages of some of the constituents of the universe, such as the earth. Another development which promises to hold important consequences for the whole subject of cosmology is the relatively recent use of radio astronomy as a means of gathering data, in addition to the conventional reliance upon optical telescopes.

Meanwhile on a conceptual level the current field of interest is divided among those who favor a "continuous creation" theory of the expanding universe as against those who uphold an "evolutionary" type of cosmology. In the arguments offered by Bondi, Sciama and Hoyle, who are among the chief protagonists of the "continuous creation" theory, we find the universe assigned an infinite time-scale. In its gross features, the universe is thought of as being ever the same, no matter at what point in time it is considered or from whatever station in space it is viewed. Matter, it is claimed, is created continuously in elementary form. It is out of such randomly produced particles that eventually agglomerations of matter of the size of galaxies and clusters of galaxies come to be formed. While the continuing expansion of the universe removes some galaxies from our range of observability, other newly-formed systems come to take their place and thus help to keep the universe in a "steady-state."

In opposition to this view are those who favor an "evolutionary" world picture, like Lemaître or Gamow. They would account for the variety of observational facts, including the relative abundances of the various elements as found throughout space, in terms of a universe that "originated" at some finite epoch in the past. The event which marked such an origin is pictured as both cataclysmic in its proportions and unique in the thermonuclear and other physical conditions which it possessed. It is from such a beginning, conceived as a "Primeval Atom" by Lemaître and as an extremely compressed state of matter called "ylem" by Gamow, that the present constitution of the universe is taken as having evolved.

ALBERT EINSTEIN:

Considerations on the Universe as a Whole

Cosmological Difficulties of Newton's Theory

IF WE PONDER over the question as to how the universe, considered as a whole, is to be regarded, the first answer that suggests itself to us is surely this: As regards space (and time) the universe is infinite. There are stars everywhere, so that the density of matter, although very variable in detail, is nevertheless on the average everywhere the same. In other words: However far we might travel through space, we should find everywhere an attenuated swarm of fixed stars of approximately the same kind and density.

This view is not in harmony with the theory of Newton. The latter theory rather requires that the universe should have a kind of centre in which the density of the stars is a maximum, and that as we proceed outwards from this centre the group-density of the stars should diminish, until finally, at great distances, it is succeeded by an infinite region of emptiness. The stellar universe ought to be a finite island in the infinite ocean of space.[1]

This conception is in itself not very satisfactory. It is still less satisfactory because it leads to the result that the light emitted by the stars and also individual stars of the stellar system are perpetually passing out into infinite space, never to return, and without ever again coming into interaction with other objects of nature. Such a finite material universe would be destined to become gradually but systematically improverished.

From Albert Einstein, *Relativity: The Special and General Theory*, translated by R. W. Lawson, 1920, Part III, Chapters XXX, XXXI, XXXII. Reprinted by kind permission of the publisher, Peter Smith, New York.

1. *Proof.*—According to the theory of Newton, the number of "lines of force" which come from infinity and terminate in a mass m is proportional to the mass m. If, on the average, the mass-density ρ_0 is constant throughout the universe, then a sphere of volume V will enclose the average mass $\rho_0 V$. Thus the number of lines of force passing through the surface F of the sphere into its interior is proportional to $\rho_0 V$. For unit area of the surface of the sphere the number of lines of force which enters the sphere is thus proportional to $\rho_0 \frac{V}{F}$ or to $\rho_0 R$. Hence the intensity of the field at the surface would ultimately become infinite with increasing radius R of the sphere, which is impossible.

In order to escape this dilemma, Seeliger suggested a modification of Newton's law, in which he assumes that for great distances the force of attraction between two masses diminishes more rapidly than would result from the inverse square law. In this way it is possible for the mean density of matter to be constant everywhere, even to infinity, without infinitely large gravitational fields being produced. We thus free ourselves from the distasteful conception that the material universe ought to possess something of the nature of a centre. Of course we purchase our emancipation from the fundamental difficulties mentioned, at the cost of a modification and complication of Newton's law which has neither empirical nor theoretical foundation. We can imagine innumerable laws which would serve the same purpose, without our being able to state a reason why one of them is to be preferred to the others; for any one of these laws would be founded just as little on more general theoretical principles as is the law of Newton.

The Possibility of a "Finite" and Yet "Unbounded" Universe

But speculations on the structure of the universe also move in quite another direction. The development of non-Euclidean geometry led to the recognition of the fact, that we can cast doubt on the *infiniteness* of our space without coming into conflict with the laws of thought or with experience (Riemann, Helmholtz). These questions have already been treated in detail and with unsurpassable lucidity by Helmholtz and Poincaré, whereas I can only touch on them briefly here.

In the first place, we imagine an existence in two-dimensional space. Flat beings with flat implements, and in particular flat rigid measuring-rods, are free to move in a *plane*. For them nothing exists outside of this plane: that which they observe to happen to themselves and to their flat "things" is the all-inclusive reality of their plane . . . In contrast to ours, the universe of these beings is two-dimensional; but, like ours, it extends to infinity. In their universe there is room for an infinite number of identical squares made up of rods, *i.e.* its volume (surface) is infinite. If these beings say their universe is "plane," there is sense in the statement, because they mean that they can perform the constructions of plane Euclidean geometry with their rods. In this connection the individual rods always represent the same distance, independently of their position.

Let us consider now a second two-dimensional existence, but this time on a spherical surface instead of on a plane. The flat beings with their measuring-rods and other objects fit exactly on this surface and they are unable to leave it. Their whole universe of observation extends exclusively over the surface of the sphere. Are these beings able to regard the geometry of their universe as being plane geometry and their rods withal as the realisation of "distance"? They cannot do this. For if they attempt to realise a straight line, they will obtain a curve, which we "three-dimensional beings" designate as a great circle, *i.e.* a self-contained line of definite finite length, which can be measured up by means of a measuring-rod. Similarly, this universe has a finite area, that can be compared with the area of a square constructed with rods. The great charm resulting from this consideration lies in the recognition of the fact that *the universe of these beings is finite and yet has no limits.*

But the spherical-surface beings do not need to go on a world-tour in order to perceive that they are not living in a Euclidean universe. They can convince themselves of this on every part of their "world," provided they do not use too small a piece of it. Starting from a point, they draw "straight lines" (arcs of circles as judged in three-dimensional space) of equal length in all directions. They will call the line joining the free ends of these lines a "circle." For a plane surface, the ratio of the circumference of a circle to its diameter, both lengths being measured with the same rod, is, according to Euclidean geometry of the plane, equal to a constant value π, which is independent of the diameter of the circle. On their spherical surface our flat beings would find for this ratio the value

$$\pi \frac{\sin\left(\dfrac{r}{R}\right)}{\left(\dfrac{r}{R}\right)},$$

i.e. a smaller value than π, the difference being the more considerable, the greater is the radius of the circle in comparison with the radius R of the "world-sphere." By means of this relation the spherical beings can determine the radius of their universe ("world"), even when only a relatively small part of their world-sphere is available for their measurements. But if this part is very small indeed, they will no longer be able to demonstrate that they are on a spherical "world" and not on a Euclidean plane, for a small part of a spherical surface differs only slightly from a piece of a plane of the same size.

Thus if the spherical-surface beings are living on a planet of which the solar system occupies only a negligibly small part of the spherical universe, they have no means of determining whether they are living in a finite or in an infinite universe, because the "piece of universe" to which they have access is in both cases practically plane, or Euclidean. It follows directly from this discussion, that for our sphere-beings the circumference of a circle first increases with the radius until the "circumference of the universe" is reached, and that it thence-forward gradually decreases to zero for still further increasing values of the radius. During this process the area of the circle continues to increase more and more, until finally it becomes equal to the total area of the whole "world-sphere."

Perhaps the reader will wonder why we have placed our "beings" on a sphere rather than on another closed surface. But this choice has its justification in the fact that, of all closed surfaces, the sphere is unique in possessing the property that all points on it are equivalent. I admit that the ratio of the circumference c of a circle to its radius r depends on r, but for a given value of r

it is the same for all points of the "world-sphere"; in other words, the "world-sphere" is a "surface of constant curvature."

To this two-dimensional sphere-universe there is a three-dimensional analogy, namely, the three-dimensional spherical space which was discovered by Riemann. Its points are likewise all equivalent. It possesses a finite volume, which is determined by its "radius" $(2\pi^2 R^3)$. Is it possible to imagine a spherical space? To imagine a space means nothing else than that we imagine an epitome of our "space" experience, *i.e.* of experience that we can have in the movement of "rigid" bodies. In this sense we *can* imagine a spherical space.

Suppose we draw lines or stretch strings in all directions from a point, and mark off from each of these the distance r with a measuring-rod. All the free end-points of these lengths lie on a spherical surface. We can specially measure up the area (F) of this surface by means of a square made up of measuring-rods. If the universe is Euclidean, then $F=4\pi r^2$; if it is spherical, then F is always less than $4\pi r^2$. With increasing values of r, F increases from zero up to a maximum value which is determined by the "world-radius," but for still further increasing values of r, the area gradually diminishes to zero. At first, the straight lines which radiate from the starting point diverge farther and farther from one another, but later they approach each other, and finally they run together again at a "counter-point" to the starting point. Under such conditions they have traversed the whole spherical space. It is easily seen that the three-dimensional spherical space is quite analogous to the two-dimensional spherical surface. It is finite (*i.e.* of finite volume), and has no bounds.

It may be mentioned that there is yet another kind of curved space: "elliptical space." It can be regarded as a curved space in which the two "counter-points" are identical (indistinguishable from each other). An elliptical universe can thus be considered to some extent as a curved universe possessing central symmetry.

It follows from what has been said, that closed spaces without limits are conceivable. From amongst these, the spherical space (and the elliptical) excels in its simplicity, since all points on it are equivalent. As a result of this discussion, a most interesting question arises for astronomers and physicists, and that is whether the universe in which we live is infinite, or whether it is finite in the manner of the spherical universe. Our experience is far from being sufficient to enable us to answer this question. But the general theory of relativity permits of our answering it with a moderate degree of certainty, and in this connection the difficulty mentioned above[2] finds its solution.

The Structure of Space According to the General Theory of Relativity

According to the general theory of relativity, the geometrical properties of space are not independent but they are determined by matter. Thus we can

_____
2. p. 275 f.

draw conclusions about the geometrical structure of the universe only if we base our considerations on the state of the matter as being something that is known. We know from experience that, for a suitably chosen co-ordinate system, the velocities of the stars are small as compared with the velocity of transmission of light. We can thus as a rough approximation arrive at a conclusion as to the nature of the universe as a whole, if we treat the matter as being at rest.

We already know from our previous discussion that the behaviour of measuring-rods and clocks is influenced by gravitational fields, i.e. by the distribution of matter. This in itself is sufficient to exclude the possibility of the exact validity of Euclidean geometry in our universe. But it is conceivable that our universe differs only slightly from a Euclidean one, and this notion seems all the more probable, since calculations show that the metrics of surrounding space is influenced only to an exceedingly small extent by masses even of the magnitude of our sun. We might imagine that, as regards geometry, our universe behaves analogously to a surface which is irregularly curved in its individual parts, but which nowhere departs appreciably from a plane: something like the rippled surface of a lake. Such a universe might fittingly be called a quasi-Euclidean universe. As regards its space it would be infinite. But calculation shows that in a quasi-Euclidean universe the average density of matter would necessarily be nil. Thus such a universe could not be inhabited by matter everywhere; it would present to us that unsatisfactory picture which we portrayed.[3]

If we are to have in the universe an average density of matter which differs from zero, however small may be that difference, then the universe cannot be quasi-Euclidean. On the contrary, the results of calculation indicate that if matter be distributed uniformly, the universe would necessarily be spherical (or elliptical). Since in reality the detailed distribution of matter is not uniform, the real universe will deviate in individual parts from the spherical, i.e. the universe will be quasi-spherical. But it will be necessarily finite. In fact, the theory supplies us with a simple connection[4] between the space-expanse of the universe and the average density of matter in it.

3. p. 275 f.
4. For the "radius" R of the universe we obtain the equation

$$R^2 = \frac{2}{\kappa\rho}$$

The use of the C.G.S. system in this equation gives $\frac{2}{\kappa} = 1 \cdot 08.10^{27}$; ρ is the average density of the matter.

EDWIN P. HUBBLE:

The Exploration of Space

THE EXPLORATION of space[1] has penetrated only recently into the realm of the nebulæ. The advance into regions hitherto unknown has been made during the last dozen years with the aid of great telescopes. The observable region of the universe is now defined and a preliminary reconnaissance has been completed. The chapters which follow are reports on various phases of the reconnaissance.

The earth we inhabit is a member of the solar system—a minor satellite of the sun. The sun is a star among the many millions which form the stellar system. The stellar system is a swarm of stars isolated in space. It drifts through the universe as a swarm of bees drifts through the summer air. From our position somewhere within the system, we look out through the swarm of stars, past the borders, into the universe beyond.

The universe is empty, for the most part, but here and there, separated by immense intervals, we find other stellar systems, comparable with our own. They are so remote that, except in the nearest systems, we do not see the individual stars of which they are composed. These huge stellar systems appear as dim patches of light. Long ago they were named "nebulæ" or "clouds"— mysterious bodies whose nature was a favorite subject for speculation.

But now, thanks to great telescopes, we know something of their nature, something of their real size and brightness, and their mere appearance indicates the general order of their distances. They are scattered through space as far as telescopes can penetrate. We see a few that appear large and bright. These are the nearer nebulæ. Then we find them smaller and fainter, in constantly increasing numbers, and we know that we are reaching out into space, farther and ever farther, until, with the faintest nebulæ that can be detected with the greatest telescope, we arrive at the frontiers of the known universe.

From Edwin Hubble, *The Realm of the Nebulae*, Yale University Press, 1936, Chapter I. Reprinted with the kind permission of Yale University Press. From "Explorations in Space: The Cosmological Program for the Palomar Telescopes," *Proceedings of the American Philosophical Society*, Volume 95, 1951.

1. This summary of nebular research may be compared with a "progress report" of results obtained up to the end of 1928, published under the same title in *Harper's Magazine* for May, 1929. Some of the material in the earlier report is included in the present summary, with the permission of Harper & Brothers.

This last horizon defines the observable region of space. It is a vast sphere, perhaps a thousand million light-years in diameter. Throughout the sphere are scattered a hundred million nebulæ—stellar systems—in various stages of their evolutionary history. The nebulæ are distributed singly, in groups, and occasionally in great clusters, but when large volumes of space are compared, the tendency to cluster averages out. To the very limits of the telescope, the large-scale distribution of nebulæ is approximately uniform.

One other general characteristic of the observable region has been found. Light which reaches us from the nebulæ is reddened in proportion to the distance it has traveled. This phenomenon is known as the velocity-distance relation, for it is often interpreted, in theory, as evidence that the nebulæ are all rushing away from our stellar system, with velocities that increase directly with distances.

Receding Horizons

This sketch roughly indicates the current conception of the realm of the nebulæ. It is the culmination of a line of research that began long ago. The history of astronomy is a history of receding horizons. Knowledge has spread in successive waves, each wave representing the exploitation of some new clew to the interpretation of observational data.

The exploration of space presents three such phases. At first the explorations were confined to the realm of the planets, then they spread through the realm of the stars, and finally they penetrated into the realm of the nebulæ.

The successive phases were separated by long intervals of time. Although the distance of the moon was well known to the Greeks, the order of the distance of the sun and the scale of planetary distances was not established until the latter part of the seventeenth century. Distances of stars were first determined almost exactly a century ago, and distances of nebulæ, in our own generation. The distances were the essential data. Until they were found, no progress was possible.

The early explorations halted at the edge of the solar system, facing a great void that stretched away to the nearer stars. The stars were unknown quantities. They might be little bodies, relatively near, or they might be gigantic bodies, vastly remote. Only when the gap was bridged, only when the distances of a small, sample collection of stars had been actually measured, was the nature determined of the inhabitants of the realm beyond the solar system. Then the explorations, operating from an established base among the now familiar stars, swept rapidly through the whole of the stellar system.

Again there was a halt, in the face of an even greater void, but again, when instruments and technique had sufficiently developed, the gap was bridged by the determination of the distances of a few of the nearer nebulæ. Once more, with the nature of the inhabitants known, the explorations swept even more rapidly through the realm of the nebulæ and halted only at the limits of the greatest telescope.

The Theory of Island Universes

This is the story of the explorations. They were made with measuring rods, and they enlarged the body of factual knowledge. They were always preceded by speculations. Speculations once ranged through the entire field, but they have been pushed steadily back by the explorations until now they lay undisputed claim only to the territory beyond the telescopes, to the dark unexplored regions of the universe at large.

The speculations took many forms and most of them have long since been forgotten. The few that survived the test of the measuring rod were based on the principle of the uniformity of nature—the assumption that any large sample of the universe is much like any other. The principle was applied to stars long before distances were determined. Since the stars were too far away for the measuring instruments, they must necessarily be very bright. The brightest object known was the sun. Therefore, the stars were assumed to be like the sun, and distances could be estimated from their apparent faintness. In this way, the conception of a stellar system, isolated in space, was formulated as early as 1750. The author was Thomas Wright (1711–86) an English instrument maker and private tutor.[2]

But Wright's speculations went beyond the Milky Way. A single stellar system, isolated in the universe, did not satisfy his philosophical mind. He imagined other, similar systems and, as visible evidence of their existence, referred to the mysterious clouds called "nebulæ."

Five years later, Immanuel Kant (1724–1804) developed Wright's conception in a form that endured, essentially unchanged, for the following century and a half. Some of Kant's remarks[3] concerning the theory furnish an excellent example of reasonable speculation based on the principle of uniformity.[4]

The theory, which came to be called the theory of island universes,[5] found a permanent place in the body of philosophical speculation. The astronomers themselves took little part in the discussions: they studied the nebulæ. Toward the end of the nineteenth century, however, the accumulation of observational data brought into prominence the problem of the status of the nebulæ and, with it, the theory of island universes as a possible solution.

2. *An Original Theory or New Hypothesis of the Universe* (London, 1750).
3. *Allgemeine Naturgeschichte und Theorie des Himmels*, published first in 1755. The passages are found in the First Part.
4. Cf. above for selections from Kant's *Universal Natural History and Theory of the Heavens.*
5. The realm of the stars was once known as the "universe of stars" and the term persisted after the isolation of the stellar system was recognized. The multiplication of stellar systems led to the term "Weltinseln"—Island Universes—used in von Humboldt's *Kosmos* (Vol. III [1850]), presumably for the first time. In the familiar English translation by Otté (1855), the word is translated literally as "world islands" (Vol. III, 149, 150). The transition to "island universes" is an obvious step, but the writer has not ascertained the first use of the term.

The Nature of the Nebulæ

A. THE FORMULATION OF THE PROBLEM

A few nebulæ had been known to the naked-eye observers and, with the development of telescopes, the numbers grew, slowly at first, then more and more rapidly. At the time Sir William Herschel (1738–1822), the first outstanding leader in nebular research, began his surveys, the most extensive published lists were those by Messier, the last of which (1784) contained 103 of the most conspicuous nebulæ and clusters. These objects are still known by the Messier numbers—for example, the great spiral in Andromeda is M31. Sir William Herschel catalogued 2,500 objects, and his son, Sir John (1792–1871), transporting the telescopes to the southern hemisphere (near Capetown in South Africa) added many more.[6] Positions of about 20,000 nebulæ are now available, and perhaps ten times that number have been identified on photographic plates. The mere size of catalogues has long since ceased to be important. Now the desirable data are the numbers of nebulæ brighter than successive limits of apparent faintness, in sample areas widely distributed over the sky.

Galileo, with his first telescopes, resolved a typical "cloud"—Præsepe—into a cluster of stars. With larger telescopes and continued study, many of the more conspicuous nebulæ met the same fate. Sir William Herschel concluded that all nebulæ could be resolved into star-clusters, if only sufficient telescopic power were available. In his later days, however, he revised his position and admitted the existence, in certain cases, of a luminous "fluid" which was inherently unresolvable. Ingenious attempts were made to explain away these exceptional cases until Sir William Huggins (1824–1910), equipped with a spectrograph, fully demonstrated in 1864 that some of the nebulæ were masses of luminous gas.

Huggins' results clearly indicated that nebulæ were not all members of a single, homogeneous group and that some kind of classification would be necessary before they could be reduced to order. The nebulæ actually resolved into stars—the star-clusters—were weeded out of the lists to form a separate department of research. They were recognized as component parts of the galactic system, and thus had no bearing on the theory of island universes.

Among unresolved nebulæ, two entirely different types were eventually differentiated. One type consisted of the relatively few nebulæ definitely known to be unresolvable—clouds of dust and gas mingled among, and intimately associated with, the stars in the galactic system. They were usually found within

---

6. Sir John Herschel's general catalogue, representing the first systematic survey of the entire sky to a fairly uniform limit of apparent faintness, was published in 1864, and contained about 4,630 nebulæ and clusters observed by his father and himself, together with about 450 discovered by others. The catalogue was replaced by Dreyer's *New General Catalogue* in 1890.

the belt of the Milky Way and were obviously, like the star-clusters, members of the galactic system. For this reason, they have since been called "galactic" nebulæ. They are further subdivided into two groups, "planetary" nebulæ and "diffuse" nebulæ, frequently shortened to "planetaries" and "nebulosities."

The other type consisted of the great numbers of small, symmetrical objects found everywhere in the sky except in the Milky Way. A spiral structure was found in most, although not in all, of the conspicuous objects. They had many features in common and appeared to form a single family. They were given various names but, to anticipate, they are now known as "extragalactic" nebulæ[7] and will be called simply "nebulæ."

The status of the nebulæ, as the group is now defined, was undetermined because the distances were wholly unknown. They were definitely beyond the limits of direct measurement, and the scanty, indirect evidence bearing on the problem could be interpreted in various ways. The nebulæ might be relatively nearby objects and hence members of the stellar system, or they might be very remote and hence inhabitants of outer space. At this point, the development of nebular research came into immediate contact with the philosophical theory of island universes. The theory represented, in principle, one of the alternative solutions of the problem of nebular distances. The question of distances was frequently put in the form: Are nebulæ island universes?

B. THE SOLUTION OF THE PROBLEM

The situation developed during the years between 1885 and 1914; from the appearance of the bright nova in the spiral M31, which stimulated a new interest in the question of distances, to the publication of Slipher's first extensive list of radial velocities of nebulæ, which furnished data of a new kind and encouraged serious attempts to find a solution of the problem.

The solution came ten years later, largely with the help of a great telescope, the 100-inch reflector, that had been completed in the interim. Several of the most conspicuous nebulæ were found to be far beyond the limits of the galactic system—they were independent, stellar systems in extragalactic space. Further investigations demonstrated that the other, fainter nebulæ were similar systems at greater distances, and the theory of island universes was confirmed.

The 100-inch reflector partially resolved a few of the nearest, neighboring nebulæ into swarms of stars. Among these stars various types were recognized which were well known among the brighter stars in the galactic system. The intrinsic luminosities (candle powers) were known, accurately in some cases, approximately in others. Therefore, the apparent faintness of the stars in the nebulæ indicated the distances of the nebulæ.

The most reliable results were furnished by Cepheid variables, but other types of stars furnished estimates of orders of distance, which were consistent with the Cepheids. Even the brightest stars, whose intrinsic luminosities appear

---

7. The term "external galaxies," revived by Shapley, is also widely used, as is a third term "anagalactic" nebulæ, introduced by Lundmark.

to be nearly constant in certain types of nebulæ, have been used as statistical criteria to estimate mean distances for groups of systems.

The Inhabitants of Space

The nebulæ whose distances were known from the stars involved, furnished a sample collection from which new criteria, derived from the nebulæ and not from their contents, were formulated. It is now known that the nebulæ are all of the same order of intrinsic luminosity. Some are brighter than others, but at least half of them are within the narrow range from one half to twice the mean value, which is 85 million times the luminosity of the sun. Thus, for statistical purposes, the apparent faintness of the nebulæ indicates their distances.

With the nature of the nebulæ known and the scale of nebular distances established, the investigations proceeded along two lines. In the first place the general features of the individual nebulæ were studied; in the second, the characteristics of the observable region as a whole were investigated.

The detailed classification of nebular forms has led to an ordered sequence ranging from globular nebulæ, through flattening, ellipsoidal figures, to a series of unwinding spirals. The fundamental pattern of rotational symmetry changes smoothly through the sequence in a manner that suggests increasing speed of rotation. Many features are found which vary systematically along the sequence, and the early impression that the nebulæ were members of a single family appears to be confirmed. The luminosities remain fairly constant through the sequence (mean value, 8.5×10^7 suns, as previously mentioned), but the diameters[8] steadily increase from about 1,800 light-years for the globular nebulæ to about 10,000 light-years for the most open spirals. The masses are uncertain, the estimates ranging from 2×10^9 to 2×10^{11} times the mass of the sun.

The Realm of the Nebulæ

A. THE DISTRIBUTION OF NEBULAE

Investigations of the observable region as a whole have led to two results of major importance. One is the homogeneity of the region—the uniformity of the large-scale distribution of nebulæ. The other is the velocity-distance relation.

The small-scale distribution of nebulæ is very irregular. Nebulæ are found singly, in pairs, in groups of various sizes, and in clusters. The galactic system is the chief component of a triple nebula in which the Magellanic Clouds are the other members. The triple system, together with a few additional nebulæ,

---

8. The numerical values refer to the main bodies, which, as will be explained later, represent the more conspicuous portions of the nebulæ.

forms a typical, small group that is isolated in the general field of nebulæ. The members of this local group furnished the first distances, and the Cepheid criterion of distance is still confined to the group.

When large regions of the sky, or large volumes of space, are compared, the irregularities average out and the large-scale distribution is sensibly uniform. The distribution over the sky is derived by comparing the numbers of nebulæ brighter than a specified limit of apparent faintness, in sample areas scattered at regular intervals.

The true distribution is confused by local obscuration. No nebulæ are seen within the Milky Way, and very few along the borders. Moreover, the apparent distribution thins out, slightly but systematically, from the poles to the borders of the Milky Way. The explanation is found in the great clouds of dust and gas which are scattered throughout the stellar system, largely in the galactic plane. These clouds hide the more distant stars and nebulæ. Moreover, the sun is embedded in a tenuous medium which behaves like a uniform layer extending more or less indefinitely along the galactic plane. Light from nebulæ near the galactic poles is reduced about one fourth by the obscuring layer, but in the lower latitudes, where the light-paths through the medium are longer, the absorption is correspondingly greater. It is only when these various effects of galactic obscuration are evaluated and removed, that the nebular distribution over the sky is revealed as uniform, or isotropic (the same in all directions).

The distribution in depth is found by comparing the numbers of nebulæ brighter than successive limits of apparent faintness, that is to say, the numbers within successive limits of distance. The comparison is effectively between numbers of nebulæ and the volumes of space which they occupy. Since the numbers increase directly with the volumes (certainly as far as the surveys have been carried, probably as far as telescopes will reach), the distribution of the nebulæ must be uniform. In this problem, also, certain corrections must be applied to the apparent distribution in order to derive the true distribution. These corrections are indicated by the velocity-distance relation, and their observed values contribute to the interpretation of that strange phenomenon.

Thus the observable region is not only isotropic but homogeneous as well— it is much the same everywhere and in all directions. The nebulæ are scattered at average intervals of the order of two million light-years or perhaps two hundred times the mean diameters. The pattern might by represented by tennis balls fifty feet apart.

The order of the mean density of matter in space can also be roughly estimated if the (unknown) material between the nebulæ is ignored. If the nebular material were spread evenly through the observable region, the smoothed-out density would be of the general order of 10^{-29} or 10^{-28} grams per cubic centimeter—about one grain of sand per volume of space equal to the size of the earth.

The size of the observable region is a matter of definition. The dwarf nebulæ can be detected only to moderate distances, while giants can be recorded far out in space. There is no way of distinguishing the two classes, and thus the limits of the telescope are most conveniently defined by average nebulæ. The faintest nebulæ that have been identified with the 100-inch reflector are at an average distance of the order of 500 million light-years, and to this limit about 100 million nebulæ would be observable except for the effects of galactic obscuration. Near the galactic pole, where the obscuration is least, the longest exposures record as many nebulæ as stars.

B. THE VELOCITY-DISTANCE RELATION[9]

The foregoing sketch of the observable region has been based almost entirely upon results derived from direct photographs. The region is homogeneous and the general order of the mean density is known. The next—and last—property to be discussed, the velocity-distance relation, emerged from the study of spectrograms.

When a ray of light passes through a glass prism (or other suitable device) the various colors of which the light is composed are spread out in an ordered sequence called a spectrum. The rainbow is, of course, a familiar example. The sequence never varies. The spectrum may be long or short, depending on the apparatus employed, but the order of the colors remains unchanged. Position in the spectrum is measured roughly by colors, and more precisely by wave-lengths, for each color represents light of a particular wave-length. From the short waves of the violet, they steadily lengthen to the long waves of the red.

The spectrum of a light source shows the particular colors or wave-lengths which are radiated, together with their relative abundance (or intensity), and thus gives information concerning the nature and the physical condition of the light source. An incandescent solid radiates all colors, and the spectrum is *continuous* from violet to red (and beyond in either direction). An incandescent gas radiates only a few isolated colors and the pattern, called an *emission* spectrum, is characteristic for any particular gas.

A third type, called an *absorption* spectrum and of special interest for astronomical research, is produced when an incandescent solid (or equivalent source), giving a continuous spectrum, is surrounded by a cooler gas. The gas absorbs from the continuous spectrum just those colors which the gas would radiate if it were itself incandescent. The result is a spectrum with a continuous background interrupted by dark spaces called absorption lines. The pattern of dark absorption lines indicates the particular gas or gases that are responsible for the absorption.

The sun and the stars give absorption spectra and many of the known elements have been identified in their atmospheres. Hydrogen, iron, and

9. A more extensive, nontechnical discussion of the velocity-distance relation by the writer will be found in *Red-Shifts in the Spectra of Nebulæ*, being the Halley Lecture delivered at Oxford University in 1934. Some of the material in the lecture is included in the present summary, with the permission of the Clarendon Press.

calcium produce very strong lines in the solar spectrum, the most conspicuous being a pair of calcium lines in the violet, known as H and K.

The nebulæ in general show absorption spectra similar to the solar spectrum, as would be expected for systems of stars among which the solar type predominated. The spectra are necessarily short—the light is too faint to be spread over long spectra—but the H and K lines of calcium are readily identified and, in addition, the G-band of iron and a few hydrogen lines can generally be distinguished.

Nebular spectra are peculiar in that the lines are not in the usual positions found in nearby light sources. They are displaced toward the red of their normal position, as indicated by suitable comparison spectra. The displacements, called red-shifts, increase, on the average, with the apparent faintness of the nebula that is observed. Since apparent faintness measures distance, it follows that red-shifts increase with distance. Detailed investigation shows that the relation is linear.

Small microscopic shifts, either to the red or to the violet, have long been known in the spectra of astronomical bodies other than nebulæ. These displacements are confidently interpreted as the results of motion in the line of sight—radial velocities of recession (red-shifts) or of approach (violet-shifts). The same interpretation is frequently applied to the red-shifts in nebular spectra and has led to the term "velocity-distance" relation for the observed relation between red-shifts and apparent faintness. On this assumption, the nebulæ are supposed to be rushing away from our region of space, with velocities that increase directly with distance.

Although no other plausible explanation of red-shifts has been found, the interpretation as velocity-shifts may be considered as a theory still to be tested by actual observations. Critical tests can probably be made with existing instruments. Rapidly receding light sources should appear fainter than stationary sources at the same distances, and near the limits of telescopes the "apparent" velocities are so great that the effects should be appreciable.

The Observable Region as a Sample of the Universe

A completely satisfactory interpretation of red-shifts is a question of great importance, for the velocity-distance relation is a property of the observable region as a whole. The only other property that is known is the uniform distribution of nebulæ. Now the observable region is our sample of the universe. If the sample is fair, its observed characteristics will determine the physical nature of the universe as a whole.

And the sample may be fair. As long as explorations were confined to the stellar system, the possibility did not exist. The system was known to be isolated. Beyond lay a region, unknown, but necessarily different from the star-strewn space within the system. We now observe that region—a vast sphere, through which comparable stellar systems are uniformly distributed.

There is no evidence of a thinning-out, no trace of a physical boundary. There is not the slightest suggestion of a supersystem of nebulæ isolated in a larger world. Thus, for purposes of speculation, we may apply the principle of uniformity, and suppose that any other equal portion of the universe, selected at random, is much the same as the observable region. We may assume that the realm of the nebulæ is the universe and that the observable region is a fair sample.

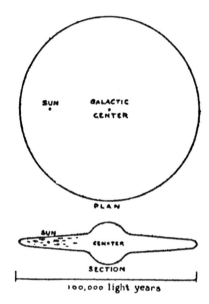

Structure of our galaxy limits the view of the 200-inch, as it does that of any other earth-bound telescope. Solar system's view of outer space is obscured by dark clouds toward edge and center of galaxy.

The conclusion, in a sense, summarizes the results of empirical investigations and offers a promising point of departure for the realm of speculation. That realm, dominated by cosmological theory, will not be entered in the present summary. The discussions will be largely restricted to the empirical data—to reports of the actual explorations—and their immediate interpretations.

Yet observation and theory are woven together, and it is futile to attempt their complete separation. Observations always involve theory. Pure theory may be found in mathematics but seldom in science. Mathematics, it has been said, deals with possible worlds—logically consistent systems. Science attempts to discover the actual world we inhabit. So in cosmology, theory presents an infinite array of possible universes, and observation is eliminating them, class by class, until now the different types among which our particular universe must be included have become increasingly comprehensible.

The reconnaissance of the observable region has contributed very materially to this process of elimination. It has described a large sample of the universe, and the sample may be fair. To this extent the study of the structure of the universe may be said to have entered the field of empirical investigation.

Cosmological Theory

. . . Much of modern cosmology is based upon the assumption that the universe will look very much the same from whatever position it is inspected. In other words, there is no center and no boundary; if the universe appears to be expanding around us, it will appear to be expanding around all other observers, no matter where they are placed. Attempts have been made to find what types of universes are permitted by this so-called Cosmological Principle, together with the principle of general relativity and the general laws of nature. The observed homogeneity of the observable region encouraged and simplified the task. It was found that such universes would be unstable, and in general would be either expanding or contracting. Theory did not predict either the direction or the rate of the change, and at this point the theorists turned to the observed law of red-shifts. The law was interpreted as evidence that the universe is expanding, and expanding rapidly. Thus emerged the various models of homogeneous, expanding universes of general relativity.

It was clear that the theories could be tested critically and weeded out, if we possessed accurate, reliable data on the distribution of matter in space (which indicates the curvature of space), and on the precise form of the law of red-shifts (which indicates the nature of the expansion). Furthermore, such data might permit us to determine directly whether or not red-shifts are velocity-shifts, and thus to test the fundamental assumptions of the cosmologists.

These were the problems to which answers were sought with the 100-inch, and for which the 100-inch proved inadequate. Answers were found, and they may be significant, but they involved uncertainties near the limit of the telescope which could not be evaluated at the time. Therefore the answers are regarded as suggestive rather than definitive. Nevertheless, the results have encouraged reexamination of the theories on more general assumptions. Non-homogeneous universes are under consideration, and a bold hypothesis of the continuous creation of matter has been introduced to save some of the suggested phenomena.

The Program for Palomar

The cosmological program for the Palomar telescopes has been formulated to get new answers to the observational problems, free from systematic errors and with the accidental errors sufficiently small to be unimportant. Specifically, the problems are first, to find the mean density of matter in space, and the rate of increase of red-shifts, in our immediate vicinity—say within 50 million light-years of our own system—and second, to determine whether or not there are any appreciable systematic changes with distance or direction in either of the two data. Hopes for success in the venture are found in the new order of accuracy attained with post-war techniques, and the far greater range over which the techniques can be applied with the 200-inch.

Any systematic departures from homogeneity in the distribution of nebulæ, or from linearity in the law of red-shifts, would be of first importance in distinguishing between theories. Absence of such departures would probably mean that the observable region is too small to serve as a fair sample of the universe, and that cosmological theory must still be regarded as speculative.

The Cosmic Distance-Scale

Thus the special problem for the 200-inch has been described as a search for second-order effects of great distances. Such effects are predicted by all of the current theories but the pioneer work suggests that if they are found they will be small, at any rate until vast distances are reached. Under the circumstances they may easily be masked by, or confused with, small errors in the distances. Consequently, the first step in the new program is an accurate redetermination of the cosmic distance-scale.

The current scale is a good first approximation. However, several sources of possible uncertainties are known to exist, and with resources now available they can be removed. For instance, the scale of apparent magnitudes (which measures apparent faintness) was established by photography, and the method is inherently difficult when large magnitude intervals must be bridged. The magnitude-scale is now being reconstructed with the aid of extremely sensitive photoelectric cells of new types developed during the war. We can trust the new magnitude-scale down to stars or nebulæ several million times fainter than the faintest stars seen with the naked eye. With these very faint standards well determined, the scale can be readily extended by photography over the narrow interval to the extreme limit of the 200-inch. Thus one source of uncertainty will be removed from the distance-scale.

Another example is furnished by Cepheid variables. The pioneer explorations of the observable region were based on these giant, easily recognized stars whose luminosities had been derived (by Hertzsprung and later by Shapley) from all the scanty data then available. When Cepheids were first found in nearby nebulæ these luminosities were used to derive distances and to calibrate other distance-indicators. But now we find that something is wrong somewhere. The luminosities assigned to the Cepheids led us to expect that globular clusters in M 31 would be readily resolved with the 200-inch, and that their brightest stars, as well as comparable stars in the main body of the nebula, could be studied individually with ease. It was found, however, that the stars in question were fainter than expected by a considerable fraction of a magnitude, and that they could be recorded and studied only with difficulty. The discrepancy is important because M 31 has been explored on the basis of Cepheids while our own system has been explored, in a sense, on the basis of globular clusters. There are other inconsistencies but there is no need to discuss them. They can all be removed or explained satisfactorily by a re-determination of the distance-scale using new methods that are free from any criticisms.

The Distance of Messier 31

The new program calls for a concentrated attack on the distance of a single nebula, selected by a compromise between nearness and abundance of distance-indicators included among its stars. The selected nebula is inevitably M 31; there is no competition. This giant spiral, less than a million light-years away, contains all kinds of stars in large numbers. If its distance can be established with confidence it will furnish luminosities of many types of giant stars, and all types of supergiants. Among them the most important distance-indicators are Cepheids (with luminosities of the order of a few hundred to a few thousand suns, depending on their periods) and normal novae (which reach maxima of the order of 100,000 suns). Cepheids can be studied with the 200-inch out to 3 or 4 million light-years, and novae to perhaps 10 million light-years. Several dozen Cepheids are known in the spiral, and novae outbursts occur at the rate of 20 to 30 each year.[10]

Our own stellar system is very similar to M 31. Both contain the two different types of stellar populations first distinguished by Baade.[11] The spiral

---

10. Normal novae should not be confused with supernovae. The latter outbursts are the most gigantic explosions known in the universe. They sometimes reach maxima of the order of 100,000,000 suns or as bright as an average stellar system. They occur at the rate of about one per stellar system per three or four centuries. None has been seen in our own system since telescopes came into use but a few can be identified in the pretelescopic annals, and in two of these cases, in A.D. 1054 and 1572, the novae were visible in broad daylight. In 1885, an outburst occurred in M 31 and reached almost to naked eye visibility.

Some thirty or more supernovae have been found on photographs of nebulae, at distances ranging out to about 50 million light-years. Although they could be detected out to far greater limits, they are too rare, and vary too much among themselves to serve as good distance-indicators, at least for individual nebulae.

11. The recognition of two distinct types of stellar populations, announced and fully documented by Baade during the last war, is the most important contribution of the past decade and more to our knowledge of the inhabitants of the universe. It brings order out of apparent confusion in the interpretation of nebular contents, guides the formulation of research programs and suggests inviting lines of speculation on evolution.

Type I populations are found in late type spirals and, in the purest form, in irregular nebulae such as the Large Magellanic Cloud. Because of the dominance of blue super-giants in the populations the integrated light of the systems is bluer than average, and some of the individual stars can be recorded out to great distances. Type II populations, in their pure forms, are found in globular clusters, elliptical nebulae, early spirals, and certain kinds of irregular and dwarf nebulae. Because of the absence of blue supergiants the integrated colors of these nebulae are deep yellow or red, and resolution is restricted to systems less than a million light-years away.

Mixtures of the two types are found in intermediate spirals, such as M 31, where the nuclear regions are pure type II, and as mentioned above, the outer arms, with their type I populations are embedded in a faint substratum of type II stars.

arms, with their blue supergiants and other distinctive features, are embedded in a substratum of Type II stars in which supergiants are exceedingly rare or absent. Globular clusters with their pure Type II populations are scattered through both nebulae in considerable numbers. The very brightest stars of this population can just be reached in M 31 with the 200-inch, both in the globular clusters and in the main body of the nebula itself.

The similarity of the two stellar systems has suggested a new attack on the problem of the distance of M 31, using globular clusters in our own system as an intermediate step. These clusters are compact masses of many thousands of stars ranging from an abrupt upper limit of luminosity downward in rapidly increasing numbers to the limit of telescopes. A few of the clusters are within 40,000 light-years of the earth, and in them stars like the sun or even fainter can be studied in detail with the largest telescopes.

Because the sun is immersed in the Type II substratum of the galactic system, the stars in our immediate vicinity, say within 50 light-years of the earth, are (almost all of them) the kind of stars found in globular clusters. These neighbors are the stars whose distances are known accurately from direct triangulation, and consequently whose luminosities are best determined. The number is so small that only a few giants are included; most neighbors are dwarfs comparable with or fainter than the sun. Among these latter stars a curiously precise relation has been found between luminosity and color. It is a part of the general Hertzsprung-Russell diagram but a part in which the detailed pattern is especially well determined. The same pattern should exist in globular clusters, and the bright end of the pattern has already been detected. When the diagram for a cluster has been extended to stars as faint as the sun, it can be accurately superposed on the diagram for our neighboring stars. The distance of the cluster is then derived from the apparent faintness of stars of known luminosities.

These distances determine accurately the luminosities of the brightest stars in the clusters, and among them are distance-indicators that can be recognized with the 200-inch, in M 31. Thus the distance of M 31, with its Cepheids and novae, is being derived in terms of the dwarf stars in the vicinity of the sun.

The detailed formulation of this phase of the program, as well as the general supervision of the necessary investigations, is largely the work of Baade. It represents a natural extension of his long and fruitful study of M 31 as a stellar system. Accurate luminosities and dimensions will increase greatly the significance of comparisons with the galactic system. The spiral M 31 we see in its entirety but from a great distance. Our own system we know intimately within a limited region but we find it difficult to "see the forest because of the trees." From the study of the two systems together, especially because of their similarity, there is emerging a complete picture of a stellar system that can be accepted with some confidence.

Shapley, in his classical study of globular clusters, used Cepheids to measure distances of the clusters; today, as we invade the next decimal place, we are reversing the process, and using globular clusters to measure the Cepheids.

Cepheid Variables as Distance-Indicators

Cepheids are now being used to derive distances of a few spirals, each the largest of a pair or group, and all beyond the reach of the 100-inch. This program is well advanced, and accurate distances of a dozen normal nebulae, together with a scattering of dwarf nebulae, will be available shortly, in units of the distance of M 31. These results will furnish tests of the assumption that the stellar populations in M 31 are in no way peculiar but may be used as representative of stellar systems in general.

Novae as Distance-Indicators

In one of these nebulae, M 81 in Ursa Major, novae outbursts are being recorded at the rate of more than one a month. The accumulating data furnish an independent value of the relative luminosity of Cepheids and novae for comparison with that found in M 31. Moreover, M 81 is a giant intermediate type spiral, similar to M 31 and our own system, and with the same mixture of the two stellar populations. Since novae are frequent in all three spirals (in contrast to the Large Magellanic Cloud and later spirals such as M 33 and M 101), we associate the high frequency with this particular type of nebulae. In technical terms, high frequency of normal novae is associated with stellar populations of Type II, but whether of Type II stars in general, or those in intermediate spirals in particular, remains to be determined.

A nova outburst has been described in these words—"a star becomes unstable and blows its cover off." We do not know why the explosions occur but we watch a star suddenly flare out to a maximum luminosity of the order of 100,000 suns, and within hours start to fade, ever more slowly, until months or years later it may reach its pristine level. The average rate of fading is known, so that although the maximum itself may not be observed, it can be estimated if the nova is found within a week or two after maximum.[12] Thus novae are good distance-indicators. The average apparent faintness at maxima

---

12. For the informed students, it may be emphasized that the precise maxima, the flash phenomena, are not very useful as distance-indicators. They are rarely observed, and the rates of fading vary widely and rapidly during the first few days. However, the records of many novae observed in M 31 indicate that the dispersion around mean luminosities at, say, 10 or 15 days after maximum is small enough to permit the derivation of the relative distances of nebulae, and the reliability of the results increases steadily with the accumulation of data. The procedure in the study of M 31 consisted in using only those novae whose maxima were observed or which were first seen following an unobserved interval of less than 30 or 40 days, and assuming that maxima occurred, on the average, at the midpoint of the unobserved intervals.

The dispersion from 71 novae, including both the real scatter in luminosities at 14 days after maxima and effects of errors in times of maxima, was about 0.5 mag., hence the probable error of the mean result was about 0.05 mag. (Mt. Wilson Contr., No. 376; Astrophys. Jour. 64:103, 1929). A reexamination of the problem using more novae, a reduced interval, and improved magnitudes is under way, and should still further reduce the scatter around mean values derived in the earlier study. A critical test of the method is furnished by the relative distances of M 31 and M 81 as derived from the novae, because in these nebulae the results can be checked by Cepheids.

measures the relative distances of nebulae in which the novae appear, and with the passage of time and accumulation of data the precision of the measures can be steadily improved.

Distances of the Virgo Cluster and Ursa Major Cloud of Nebulae from Novae

Because novae at maxima are the brightest stars in nebulae, they can be observed out to great distances—at least to ten million light-years with the 200-inch. Now within this limit are the Virgo Cluster and the Large Ursa Major Cloud, each organization containing several hundred nebulae of all types. Distances are currently estimated as of the order of 8 and 6 million light-years respectively. Cepheids cannot be reached at these distances, but novae should be observable in the giant, intermediate spirals belonging to both groups, and perhaps in earlier type nebulae as well.

The "Average" Nebula as a Statistical Distance-Indicator

A search for novae was initiated last winter, using a selected list of likely members in each group, but the weather was persistently bad, and progress during this first season is not impressive. Success, however, is only a matter of time and with it we shall have reliable distances and hence luminosities, of perhaps a thousand nebulae—a sample collection for the study of nebulae themselves. It will then be possible to redetermine the average luminosity of nebulae together with the dispersion about the average. These quantities permit statistical estimates of distances from apparent faintness of the nebulae themselves, out to the extreme limit of the telescope. The distribution of nebulae in space is then derived from counts to successive limits of apparent faintness which furnish the numbers of nebulae in spheres or shells of successively greater radii.

Brightest Stars in Nebulae

Two other distance-indicators can be calibrated from nebulae in the Virgo Cluster and Ursa Major Cloud; they are brightest stars in nebulae and brightest nebulae in clusters. There is an upper limit to the luminosities which permanent (that is, stable) stars can ever attain; this limit is approached in all giant, late type spirals for which information is available. The limit is known to be of the order of 60,000 suns. The many late spirals in the cluster and the cloud will furnish an accurate mean value of the upper limit for nebulae of a given total luminosity together with the dispersion about the mean, and also the variation of the limits with the total luminosities of the nebulae. Brightest stars can then be used to indicate the distances of all late spirals in which even a few stars can be seen. Because of the dispersion about the mean value, individual distances will be somewhat uncertain, but results for groups of nebulae containing several late spirals should be quite reliable. The criterion can be used with the 200-inch out to distances of the order of nearly ten million light-years.

Brightest Nebulae in Clusters

The brightest nebulae in clusters are analogous to the brightest stars in stellar systems. There seems to be an upper limit to the luminosities of nebulae, and the limit is approached in all the great clusters of nebulae for which information is available. The limit is well represented by M 31—of the order of 2,500 million suns.[13] A dozen nebulae seem to be comparable with M 31 but not a single one is known to be brighter.

The great clusters are so similar that the apparent faintness of their 1st, 3rd, 5th, 10th, or any other serial nebulae might be used to indicate their relative distances. Currently, the 5th brightest, of the order of 700 million suns, is used because the 1st or 3rd might be confused with foreground nebulae, seen by projection on the cluster, and fainter nebulae are more difficult to measure.

The criterion of brightest nebulae is of fundamental importance because it permits us to assign very great distances to individual objects with confidence. It is for this reason that clusters are used for investigations concerning effects of great distances. For instance, the law of red-shifts was formulated almost entirely from clusters. Once formulated however, it serves as an excellent criterion of distance for all nebulae whose spectra can be recorded, and it has the virtue that the percentage uncertainty actually decreases as the distance increases.

Effects of Red-Shifts on the Distance-Scale

The last step in the problem of the distance-scale concerns effects of red-shifts on apparent faintness. The combination of atmosphere, telescope, and photographic plate(or other receiver) acts as a window restricting the light which reaches the observer, to a limited range of colors. The results are readily understood if we imagine the light of a nebula spread out as a spectrum, and examined through a fixed aperture. Because intensities vary widely through the spectrum, the measures of the amount of light received will depend upon the location of the window in the array of colors.

Now as nebulae are observed at successively greater distances, the spectra

---

13. M 31, of course, is not in a cluster of nebulae. It is, however, a member of the "Local Group," a loose group of some 16 nebulae, including the galactic system, which is more or less isolated in the general field of nebulae. Until recently, our knowledge of Cepheids and normal novae was restricted to the Local Group. One of the notable achievements of the 200-inch has been to break away from the group and to invade the general field.

march past the fixed window according to the law of red-shifts. Consequently, the measures of apparent faintness made through the window refer to different regions of the spectra, and corrections must be applied in order to reduce the measures to a standard region. Only then can distance be inferred accurately from apparent faintness.

Corrections are derived from the intensity distributions along the spectra of nebulae. They are difficult to determine, and are still imperfectly known. A general investigation of the problem is scheduled for the near future, but meanwhile it is possible to apply corrections with some confidence under certain special conditions. Fortunately, the region of nebular spectra from blue-green to red is the best known, and it is here that the differences in intensities (the corrections for red-shifts) are the least. Much of the uncertainty can be avoided by measuring apparent faintness of nebulae in the red through a red filter. For distant nebulae, these measures represent the intensities normally found in the orange, yellow or green (depending on the distance) but displaced by the red-shifts to the particular region observed. Differences of intensity in the various colors are small, and are fairly well known, therefore errors in corrections for the differences are very small indeed. The method is practical because red-sensitive photographic emulsions have been highly developed during recent years, and with the help of photoelectric cells over much of the range, the measures can be pushed out to the very frontiers of the observable region. As a compromise, until accurate red magnitudes are available, yellow (or photovisual) magnitudes are being used.

Assumptions Underlying the Distance-Scale

Estimations of distance from apparent faintness involve two assumptions that have not yet been mentioned. The first assumption is that internebular space is sensibly transparent, that nebulae are not noticeably dimmed by dust or gas scattered between the systems. There is a great deal of dust and gas within our own stellar system (and in the other nebulae) and the resulting obscuration is very troublesome. But in space between the nebulae no clear evidence has yet been found of widespread general obscuration. Until such evidence emerges, the explorations will proceed on the assumption of transparency.

The second assumption concerns the rate of nebular evolution. It is a speculative problem and among the speculators a controversial one. When we look out into space, we look back into time, far back into the history of the universe. We must compare neighboring nebulae as they were only a few million years ago with remote nebulae as they were several hundred million years ago. Since we have no more recent information concerning the distant nebulae, we proceed on the assumption that evolutionary changes may be ignored, that total luminosities have not altered materially, during the time the light has travelled to reach us.

Geologists have reasons to suppose that the luminosity of the sun has not changed materially during the last 1,000 million years. Stellar systems may perhaps be equally stable, or they may not. Current theories even now suggest that giant stars, unlike the sun, may change rapidly. The best we can do is to start with the simple assumption of stability and keep alert for trouble in the form of systematic variation with distance which can be interpreted as evolutional changes.

The subject has a necessary place in cosmological studies, and a special program has been drafted for its investigation. Evidently the work must start with the comparative study of clusters of nebulae at successive distances because only in these clusters can large individual distances be assigned to many objects with confidence. Even now, a systematic reddening of early-type nebulae in clusters, observed by Stebbins and Whitford out to about 250 million light-years, is widely discussed as possible evidence of evolutional changes. The suggestion is exciting but it is only one of the possible explanations of the phenomena, and more information will be required before the possibilities can be weeded out.[14] Meanwhile it is recognized that the question of the rate of evolution of stellar systems seems to be the most serious of the undetermined factors that might confuse the distance-scale.

The program for the distance-scale may be summarized as (1) the use of globular clusters to establish the distance of M 31 in order to calibrate Cepheids and novae; (2) the use of novae to measure distances of the Virgo Cluster and the Ursa Major Cloud in order to calibrate the nebulae themselves, and (3) the measurement of the intensity distribution over the spectra of nebulae in order to furnish accurate corrections for effects of red-shifts on apparent faintness. The various lines of advance are being followed simultaneously, and progress to date seems to assure the success of the program, subject only to the unknown factor of a possibly rapid rate of evolutional changes in nebulae during the transit time of the light. With this restriction we are confident that the program will provide the proper tools for explorations of the whole of the vast observable region of space.

The Use of the Distance-Scale

THE LAW OF RED-SHIFTS

In conclusion I shall mention briefly some of the plans and prospects involved in those explorations. The first definite result to be expected is the

---

14. Some of the necessary information should be furnished by one project in the program of evolutional studies, formulated by Minkowski. Although supernovae are not accurate distance-indicators for individual nebulae, they are reliable in statistical investigations. By keeping watch (one or two plates per month) over several remote clusters of nebulae, it should be possible to assemble data on a sufficient number of supernovae to furnish, within 5 to 10 years, a reliable mean distance for the clusters as a group. A comparison of mean luminosities of these nebulae with luminosities of similar neighboring nebulae will indicate whether or not there are systematic changes depending on time which can be attributed to evolution.

accurate formulation of the law of red-shifts out to about twice the distance previously attained. Clusters selected from those found on 48-inch schmidt plates furnish excellent observing lists out to perhaps 350 million light-years. Beyond the reach of the 48-inch we must depend upon accidental finds with the 100-inch or 200-inch, and the time required for discovery is unpredictable. Humason has already recorded red-shifts corresponding to velocities ranging up to more than one-fifth the velocity of light (50 per cent greater than the limit reached with the 100-inch), and he believes that readable spectra can be obtained out to distances of the order of 500 million light-years, when suitable clusters are found. With accurate distances available, the data should indicate at once whether the law of red-shifts departs from a strictly linear relation, and, if so, whether the rate of expansion of the universe has been speeding up or slowing down during the immediate past. These data will furnish estimates of the "age of the universe"—the elapsed time since the expansion began. Furthermore, there are fairly good prospects that the same data will furnish a critical test of whether or not the red-shifts really are velocity-shifts, whether or not the universe is expanding at all, at the rate suggested by the red-shifts.

THE DISTRIBUTION OF NEBULAE

The second problem, concerning the homogeneity of the observable region, is much more laborious. We wish to know the distribution of nebulae in space, in other words, the average number of nebulae per unit volume, and to search these data for systematic variations, or trends depending either upon direction or upon distance. Furthermore, we wish to know the masses of nebulae in order to express the data as the density distribution of matter in space (in grams per cc).

The distribution of nebulae over the sky is not included in the Palomar program. Such an investigation is being carried out at Lick Observatory by counting the nebulae on survey plates with the fine 20-inch camera which in time will cover the entire sky observable from that latitude. Dr. Shane has described the immense project at a recent meeting of this Society. The final report will present the distribution of more than a million nebulae to about the 18th magnitude or fainter, in the form of a contour map with lines showing equal numbers per unit area. The data will test the current assumption of large-scale uniformity over the sky (isotropy, or "no favored direction") and will describe the small-scale distribution, or tendency towards clustering, in quantative terms. The results will represent a major, fundamental contribution to cosmology.

Distribution in depth, on the other hand, is a part of the Palomar program. It will be derived from counts of nebulae between successive limits of apparent faintness or, in other words, the numbers of nebulae in successive shells of space. Beyond the reach of the 20-inch at Lick, the numbers of nebulae per unit area increase so rapidly that complete counts over the available sky are out of the question. Instead, a number of sample regions spaced over the sky, each covering 100 square degrees, or about 3 or 4 times the area of the Big Dipper, will be studied on 48-inch schmidt plates. Each region will furnish between 40 and 50 thousand nebulae, to about mag. 19.5, and the counts will be made both on blue and on red photographs, to several different limits of apparent faintness. Then in small sample areas scattered over the counted regions the 200-inch will probe out to its extreme limits, again in both blue and red. By this sampling technique we hope to establish the distribution in depth, in the two colors, out to the very boundaries of the observable region, out perhaps to distances of the order of 1,000 million light-years.

The data, when assembled and analyzed, should indicate whether or not the average number of nebulae per unit volume of space varies systematically with distance. If such a trend is found, the interpretation will depend upon the nature of the variation. For instance, a fading out of numbers might mean obscuration by diffuse matter in space, or that the galactic system is located within a super-system of nebulae; either a fading out or increasing numbers might indicate and measure the curvature of space that is involved in most of the cosmological theories.

The isolated areas covered by the sampling surveys do not indicate the small-scale distribution of nebulae (the clustering) but instead furnish the average, smoothed-out values of numbers of nebulae per unit volume of space, known as the large-scale distribution. These data alone measure the homogeneity of the observable region, but their significance will be enormously increased when they can be expressed numerically as the density of matter in space. The transformation, however, involves the masses of nebulae, and these quantities are now rather controversial; the various methods of investigation lead to values differing by factors between 10 and 100. All methods, of course, depend upon velocity, and velocities at great distances are derived only from velocity-shifts in spectra. Velocities measure the gravitational fields in which they occur, and consequently the masses of the elements in those fields.

The program includes determination of masses from spectrographic rotations in single nebulae, differential velocities in close pairs, and velocity dispersions in clusters of nebulae, and less immediately, from masses of globular clusters, stepped up to early type nebulae in proportion to relative luminosities. Objections have been raised to each method, but as the work progresses the validity of the objections can be tested by the agreement or nonagreement of the results for the different procedures.

At present the greatest confidence is placed on spectrographic rotations, and the most reliable single mass is that assigned to M 31—of the order of 100,000 million suns. M 31, as previously mentioned, is the largest and most luminous nebula known, and presumably the most massive. Since nebulae are stellar systems, the orders of the masses of similar, fainter spirals may be derived from their relative luminosities. A nebula 1/100 as luminous as M 31 may contain 1/100 as many stars, and consequently may be about 1/100 as massive. The procedure may be improved by determining reliable masses for a few examples of each of the standard nebular types and extending the results within each type by relative luminosities. Justification is found in accumulating evidence suggesting a mass-luminosity coefficient which varies systematically through the sequence of classification of nebulae.

Time does not permit further discussion of the problems of masses and the time scale (which bear directly upon the curvature and the expansion of space) nor can we even mention the evolution of stars, stellar systems and the universe itself. For these ultimate problems as well as the more immediate problems of the laws of red-shifts and of nebular distribution, the necessary preliminary is reliable information concerning distances, luminosities, and masses; and first of all, concerning distances. This consideration has determined the emphasis of the discussion.

We know the preliminary objectives can be, and are being, attained. When they are won and consolidated, we shall turn to the great problems of the universe with new confidence. Observational results can be stated positively, with limits of uncertainties evaluated accurately. Then theory after theory can be eliminated. The long array of possible worlds can be reduced to a few that are compatible with the existing body of knowledge. And possibly, just possibly, we may be able to identify, in the shortened array, the specific type that must include the universe we inhabit.

WILLEM De SITTER:

Relativity and Modern Theories
of the Universe

THE THEORY of relativity may be considered as the logical completion of Newton's theory of gravitation, the direct continuation of the line of thought which dominates the development of the science of mechanics, from Archimedes—who may be considered as the first relativist—through Galileo to Newton. Newton's theory had celebrated its greatest triumphs in the eighteenth and nineteenth centuries; one after another all the irregularities in the motions of the planets and the moon had been explained by the mutual gravitational action of these bodies. In the beginning of the nineteenth century Laplace's monumental work completed the application of the theory on the motions of the planets.

The final triumph came in 1846 by the discovery of Neptune, verifying the prediction by Adams and Leverrier, based on the theory of gravitation.

Gradually Newton's law of gravitation had become a model on which physical laws were framed, and all physical phenomena were reduced to laws which were formulated as attractions or repulsions inversely proportional to some power of the distance, such as, e.g., Laplace's theory of capillarity, which was even published as a chapter in his "Mécanique céleste." Gradually, however, during the second half of the nineteenth century, the uncomfortable feeling of dislike of the action at a distance, which had been so strong in Huygens and other contemporaries of Newton, but had subsided during the eighteenth century, began to emerge again, and gained strength rapidly.

This was favoured by the purely mathematical transformation (which can be compared in a sense with that from the Ptolemaic to the Copernican system), replacing Newton's finite equations by the differential equations, the potential becoming the primary concept, instead of the force, which is only the gradient of the potential. These ideas, of course, arose first in the theory of electricity and magnetism—or perhaps one should say in the brain of Faraday. In electromagnetism also the law of the inverse square had been

From W. De Sitter, *Kosmos,* Harvard University Press, 1932, Chapter VI. Reprinted by the kind permission of the publishers, Harvard University Press.

supreme, but, as a consequence of the work of Faraday and Maxwell, it was superseded by the field. And the same change took place in the theory of gravitation. By and by the material particles, electrically charged bodies, and magnets—which are the things that we actually observe—come to be looked upon only as "singularities" in the field. So far this transformation from the force to the potential, from the action at a distance to the field, is only a purely mathematical operation. Whether we talk of a "particle of matter" or of a "singularity in the gravitational field" is only a question of a name. But this giving of names is not so innocent as it looks. It has opened the gate for the entrance of hypotheses. Very soon the field is materialised, and is called aether. From the mathematical point of view, of course, "aether" is still just another word for "field," or, perhaps better, for "space"—the absolute space of Newton—in which there may or may not be a field. From the point of view of physical theory (and it is especially in the theory of electromagnetism that this evolution took place), however, the "aether of space," as it used to be called about forty years ago, is not simply space, it is something substantial, it is the carrier of the field, and mechanical models, consisting of racks and pinions and cogwheels, are devised to explain how it does the carrying. These mechanical models have, of course, been given up long ago: they were too crude. But hypotheses have kept cropping up on all sides: electrons, atomic nuclei, protons, wave-packets, etc. At first the imagining of mechanical models went on. Fifteen, or even ten, years ago, although an atom was no longer, as the name implies, just a piece of matter that could not be cut into smaller pieces, atoms, electrons, and protons were still thought of as mechanical structures, models of the atom were imagined, having the mechanical properties of ordinary matter. The inconsistency of first explaining matter by atoms and then explaining atoms by matter was only slowly realised, and it is only comparatively recently that we have come to see that there is nothing paradoxical in the fact that an atom or an electron, which are not matter, may have properties different from those of matter, and must be allowed to do things that a material particle could not do.

However, whilst in all other domains of physics hypotheses have been found successful in accounting for the observed facts, and replacing the formal laws, the case of gravitation stands apart. Gravitation has been insusceptible to this general infection. By using this word I do not mean to suggest that the luxuriant growth of hypotheses in physics is a contagious disease,—it is not a disease, but a natural development,—but it is certainly contagious. Gravitation, however, seems to be immune to it. In the course of history a great number of hypotheses have been proposed in order to "explain" gravitation, but not one of these has ever had the least chance, they have all been failures. Why is that? How does it come about that we have been able to find satisfactory hypotheses to explain electricity and magnetism, light and heat, in short all other physical phenomena, but have been unsuccessful in the case of gravitation? The explanation must be sought in the peculiar position that gravitation occupies amongst the laws of nature. In the case of other physical phenomena there is something to get hold of, there are circumstances on which the action depends. Gravitation is entirely independent of everything that influences other natural phenomena. It is not subject to absorption or refraction, no velocity of propagation has been observed. You can do whatever you please with a body, you can electrify or magnetise it, you can heat it, melt or evaporate it, decompose it chemically, its behaviour with respect to gravitation is not affected. Gravitation acts on all bodies in the same way, everywhere and always we find it in the same rigorous and simple form, which frustrates all our attempts to penetrate into its internal mechanism. Gravitation is, in its generality and rigour, entirely similar to inertia, which has never been considered to require a particular hypothesis for its explanation, as any ordinary special physical law or phenomenon. Inertia has from the beginning been admitted as one of the fundamental facts of nature, which have to be accepted without explanation, like the axioms of geometry.

But gravitation is not only similar to inertia in its generality, it is also measured by the same number, called the mass. The inertial mass is what Newton calls the "quantity of matter": it is a measure for the resistance offered by a body to a force trying to alter its state of motion. It might be called the "passive mass." The gravitational mass, on the other hand, is a measure of the force exerted by the body in attracting other bodies. We might call it the "active" mass. The equality of active and passive, or gravitational and inertial, mass was in Newton's system a most remarkable accidental co-incidence, something like a miracle. Newton himself decidedly felt it as such, and made experiments to verify it, by swinging a pendulum with a hollow bob which could be filled with different materials. The force acting on the pendulum is proportional to its gravitational mass, the inertia to its inertial mass: the period of its swing thus depends on the ratio between these two masses. The fact that the period is always the same therefore proves that the gravitational and inertial masses are equal. Gradually, during the eighteenth century, physicists and philosophers had become so accustomed to Newton's law of gravitation, and to the equality of gravitational and inertial mass, that the miraculousness of it was forgotten and only an acute mind like Bessel's perceived the necessity of repeating those experiments. By the experiments of Bessel about 1830 and of Eötvös in 1909 the equality of gravitational and inertial mass has become one of the best ascertained empirical facts in physics.

In Einstein's general theory of relativity the identity of these two coefficients, the gravitational and the inertial mass, is no longer a miracle, but a necessity, because gravitation and inertia are identical.

There is another side to the theory of relativity. We have pointed out in the beginning how the development of science is in the direction to make it less subjective, to separate more and more in the observed facts that which belongs to the reality behind the phenomena, the absolute, from the subjective element, which is introduced by the observer, the relative. Einstein's theory is a great step in that direction. We can say that the theory of relativity is intended to remove entirely the relative and exhibit the pure absolute.

The physical world has three space dimensions and one time dimension; the position of a material particle at a certain time t is defined by three space coördinates, x, y, z. In Newton's system of mechanics this is unhesitatingly accepted as a property of the outside world: there is an absolute space and an absolute time. In Einstein's theory time and space are interwoven, and the way in which they are interwoven depends on the observer. Instead of three plus one we have four dimensions.

Is the fact that we observe the outside world as a four-dimensional continuum a property of this outside world, or is it a consequence of the particular nature of our consciousness, does it belong to the absolute or to the relative? I do not think the answer to that question can yet be given. For the present we may accept it as an empirically ascertained fact.

The sequence of different positions of the same particle at different times forms a one-dimensional continuum in the four-dimensional space-time, which is called the *world-line* of the particle. All that physical experiments or observations can teach us refers to intersections of world-lines of different material particles, light-pulsations, etc., and how the course of the world-line is between these points of intersection is entirely irrelevant and outside the domain of physics. The system of intersecting world-lines can thus be twisted about at will, so long as no points of intersection are destroyed or created, and their order is not changed. It follows that the equations expressing the physical laws must be invariant for arbitrary transformations.

This is the mathematical formulation of the theory of relativity. The metric properties of the four-dimensional continuum are described, as is shown in treatises on differential geometry, by a certain number (ten, in fact) of quantities denoted by $g_{\alpha\beta}$, and commonly called "potentials." The physical status of matter and energy, on the other hand, is described by ten other quantities, denoted by $T_{\alpha\beta}$, the set of which is called the "material tensor." This special tensor has been selected because it has the property which is mathematically expressed by saying that its divergence vanishes, which means that it represents something permanent. The fundamental fact of mechanics is the law of inertia, which can be expressed in its most simple form by saying that it requires the fundamental laws of nature to be differential equations of the second order. Thus the problem was to find a differential equation of the second order giving a relation between the metric tensor $g_{\alpha\beta}$ and the material tensor $T_{\alpha\beta}$. This is a purely mathematical problem, which can be solved without any reference to the physical meaning of the symbols. The simplest possible equation (or rather set of ten equations, because there are ten g's) of that kind that can be found was adopted by Einstein as the fundamental equation of his theory. It defines the space-time continuum, or the "field." The world-lines of material particles and light quanta are the geodesics in the four-dimensional continuum defined by the solutions $g_{\alpha\beta}$ of these field-equations. The equations of the geodesic thus are equivalent to the equations of motion of mechanics. When we come to solve the field-equations and substitute the solutions in the equations of motion, we find that in the first approximation, i.e. for small material velocities (small as compared with the velocity of light), these equations of motion are the same as those resulting from Newton's theory of gravitation. The distinction between gravitation and inertia has disappeared; the gravitational action between two bodies follows from the same equations, and *is* the same thing, as the inertia of one body. A body, when not subjected to an extraneous force (i.e. a force other than gravitation), describes a geodesic in the continuum, just as it described a geodesic, or straight line, in the absolute space of Newton under the influence of inertia alone.

The field-equations and the equations of the geodesic together contain the whole science of mechanics, including gravitation.

In the first approximation, as has been said just now, the new theory gives the same results as Newton's theory of gravitation. The enormous wealth of experimental verification of Newton's law, which has been accumulated during about two and a half centuries, is therefore at the same time an equally strong verification of the new theory. In the second approximation there are small differences, which have been confirmed by observations, so far as they are large enough for such a confirmation to be possible. Thus especially the anomalous motion of the perihelion of Mercury, which had baffled all attempts at explanation for over half a century is now entirely accounted for. Further the theory of relativity has predicted some new phenomena, such as the deflection of rays of light that pass near the sun, which has actually been observed on several occasions during eclipses; and the redshift of spectral lines originating in a strong gravitational field, which is also confirmed by observations, e.g. in the spectrum of the sun, and also in the spectrum of the companion of Sirius, which, being a so-called white dwarf, i.e. a small star with very high density and consequently a strong gravitational field, gives a considerable redshift. We cannot stop to explain these phenomena in detail. It must suffice just to mention them.

Two points should be specially emphasised in connection with the general theory of relativity.

First that it is a purely *physical* theory, invented to explain empirical physical facts, especially the identity of gravitational and inertial mass, and to coördinate and harmonise different chapters of physical theory, and simplify the enunciation of the fundamental laws. There is nothing metaphysical about its origin. It has, of course, largely attracted the attention of philosophers, and has, on the whole, had a very wholesome influence on metaphysical theories. But that is not what it set out to do, that is only a by-product.

Second that it is a pure generalisation, or abstraction, like Newton's system of mechanics and law of gravitation. It contains *no hypothesis,* as contrasted with other modern physical theories, electron theory, quantum theory, etc., which are full of hypotheses. It is, as has already been said, to be considered as the logical sequence and completion of Newton's Principia.

A special feature of the development of physics in the nineteenth century has been the arising of general principles beside the special laws, such as the principles of conservation of mass and of energy, the principle of least action, and the like. These differ from the special laws, not only by being more general, but they aspire, so to say, to a higher status than the laws. Their claim is that they express fundamental facts of nature, general rules, to which all special laws have to conform. And they accordingly exclude a priori all attempts at "explanation" by hypotheses or mechanical models. It is characteristic of the theory of relativity that it enables us to include all these principles of conservation in one single equation.

We have a direct knowledge only of that part of the universe of which we can make observations. I have already called this "our neighbourhood." Even within the confines of this province our knowledge decreases very rapidly as we get away from our own particular position in space and time. It is only within the solar system that our empirical knowledge of the quantities determining the state of the universe, the potentials $g_{\alpha\beta}$, extends to the second order of smallness (and that only for g_{44}, and not for the others), the first order corresponding to about one unit in the eighth decimal place. How the $g_{\alpha\beta}$ outside our neighbourhood are, we do not know, and how they are at infinity, of either space or time, we shall never know, otherwise it would not be infinity. That is what Archimedes meant when he said that the universe could not be infinite. The universe that we know cannot be infinite, because we ourselves are finite. Infinity is not a physical, but a mathematical concept, introduced to make our equations more symmetrical and elegant. From the physical point of view everything that is outside our neighbourhood is pure extrapolation, and we are entirely free to make this extrapolation as we please, to suit our philosophical or aesthetical predilections—or prejudices. It is true that some of these prejudices are so deeply rooted that we can hardly avoid believing them to be above any possible suspicion of doubt, but this conviction is not founded on any physical basis. One of these convictions, on which extrapolation is naturally based, is that the particular part of the universe in which we happen to be is in no way exceptional or privileged, in other words that the universe, when considered on a large enough scale, is isotropic and homogeneous. It should, however, be remembered that there have been epochs in the evolution of mankind when this was by no means thought self-evident, and the contrary conviction was rather generally held.

During the last years the limits of our "neighbourhood" have been enormously extended by the observations of extragalactic nebulae, made chiefly at the Mount Wilson Observatory. These wonderful observations have enabled us to make fairly reliable estimates of the distances of these objects and to say something about their distribution in space. It appears that they are distributed approximately evenly over "our neighbourhood." They also are all of roughly the same size, so that we can make an estimate of the density of matter in space. Further the observations have disclosed the remarkable fact that in their spectra there is a displacement of the lines towards the red corresponding to a receding velocity increasing with the distance, and, so far as the determinations of the distances are reliable, proportional with it. If the velocity is proportional to the distance, then not only the distance of any nebula from us is increasing, but *all* mutual distances between any two of them are increasing at the same rate. Our own galactic system is only one of a great many, and observations made from any of the others would show exactly the same thing: all systems are receding, not from any particular centre, but *from each other:* the whole system of galactic systems is *expanding.*

It is perhaps somewhat difficult to imagine the expansion of three-dimensional space. A two-dimensional analogy may help to make it clear. Let the universe have only two dimensions, and let it be the surface of an indiarubber ball. It is only the *surface* that is the universe, not the ball itself. Observations can only be made, distances can only be measured, along the surface, and evidently no point of the surface is different from any other point. Let there be specks of dust fixed to the surface to represent the different galactic systems. If the ball is inflated, the universe expands, and these specks of dust will recede from each other, their mutual distances, measured along the surface, will increase in the same rate as the radius of the ball. An observer in any one of the specks will see all the others receding from himself, but it does not follow that he is the centre of the universe. The universe (which is the surface of the ball, not the ball itself) has no centre.

It is, of course, not essential that we have chosen for our illustration a rubber *ball.* We might just as well have taken any other surface, it is not even necessary that it should be a closed surface. Even a plane sheet of rubber might do just as well, if only the stretching to which it must be subjected to illustrate the expanding universe is the same in all directions.

These then are the two observational facts about our neighbourhood, which have to be accounted for by the theory: there is a finite density of matter, and there is expansion, i.e. the mutual distances are increasing, and therefore the density is decreasing. Of course we can only be certain of these facts so far as our observations reach, i.e. for our "neighbourhood," but, in agreement with our principle of extrapolation, we extend these statements to the whole of the universe.

We have thus to find a universe—i.e. a set of potentials $g_{\alpha\beta}$ satisfying the field-equations of the general theory of relativity—that has both a finite density of matter and an expansion. And, since we only consider the universe on a very large scale, and make abstraction of all details and local irregularities, our universe must be homogeneous and isotropic. It follows at once from this condition of homogeneity and isotropy that the three-dimensional space of it must be what mathematicians call a space of constant curvature. Even so mathematics offers us a free choice between different kinds of space. The curvature may be positive, negative, or zero. It is not possible to picture, or imagine, the different kinds of three-dimensional space. We think that we have a mental picture of euclidian, or flat, space, i.e. space of which the curvature is everywhere zero, but I am not sure that this is not a self-deception, caused by the fact that the geometry of this special space has been taught in the schools for the last two thousand or more years. It is certain that for physical phenomena on the scale which our sense organs are able to perceive, i.e. neither too small nor too large, the euclidian space is a very close approximation to the true physical space, but for the electron, and for the universe, the approximation breaks down. To help us to understand three-dimensional spaces, two-dimensional analogies may be very useful (though also sometimes misleading). We can imagine different kinds of two-dimensional space, since we are able to place ourselves outside them. A two-dimensional space of zero curvature is a plane, say a sheet of paper. The two-dimensional space of positive curvature is a convex surface, such as the shell of an egg. It is bent away from the plane towards the same side in all directions. The curvature of the egg, however, is not constant: it is strongest at the small end. The surface of constant positive curvature is the sphere, say our indiarubber ball of a moment ago. The two-dimensional space of negative curvature is a surface that is convex in some directions and concave in others, such as the surface of a saddle or the middle part of an hour glass. Of these two-dimensional surfaces we can form a mental picture because we can view them from outside, living, as we do, in three-dimensional space. But for a being, who would be unable to leave the surface on which he was living, that would be impossible. He could only decide of which kind his surface was by studying the properties of geometrical figures drawn on it. For the geo-metrical figures have different properties on the different surfaces. On the sheet of paper the sum of the three angles of a triangle is equal to two right angles, on the egg, or the sphere, it is larger, on the saddle it is smaller. On the flat paper—and on the saddle-shaped surface—we can proceed indefinitely in the same direction, on the egg or the sphere, if we continue to move in the same direction we ultimately come back to our starting point. The spaces of zero and negative curvature are infinite, that of positive curvature is finite. Thus the inhabitant of the two-dimensional surface could determine its curvature if he were able to study very large triangles or very long straight lines. If the curva-ture were so minute that the sum of the angles of the largest triangle that he could measure would still differ from two right angles by an amount too small to be appreciable with the means at his disposal, then he would be unable to determine the curvature, unless he had some means of communicating with somebody living in the third dimension. Now our case with reference to three-dimensional space is exactly similar. We have no intuitive knowledge of the kind of space we live in. So we must find out which kind it is by studying the triangles and other geometrical figures in it. As we are concerned with *physical* space, the triangles that we must investigate are those formed by the tracks of material particles and rays of light, and naturally, in order to be able to distinguish different kinds of space, we must study very large triangles and rays of light coming from very great distances. Thus the decision must necessarily depend on astronomical observations.

Even the most refined astronomical observations, however, fail to show any trace of curvature. The triangles that we can measure are not large enough, and, I fear, never will be large enough, to detect the curvature. Fortunately, however, we are, in a way, able to communicate with the fourth dimension. The theory of relativity has given us an insight into the structure of the real universe: it does not consist of a three-dimensional space *and* a one-dimensional time, existing independently of each other, as in Newton's system of mechanics, but is a *four*-dimensional structure. The study of the way in which the three space-dimensions are interwoven with the time-dimension affords a kind of outside point of view of the three-dimensional space, and it is conceivable that from this outside point of view we might be able to perceive the curvature of the three-dimensional world.

At one time it was thought that this was so, and that we could actually prove that the curvature must be positive. However, recent mathematical investigation has shown that this was a mistake. We shall never be able to say anything about the curvature without introducing certain hypotheses. These rather vague statements will become clearer when we penetrate more deeply into the nature of the cross-connection between space and time in the four-dimensional universe. It is, of course, difficult to explain these without the use of mathematical formulae, but, by following the historical line of development, I shall endeavour to lead up to an understanding of the present position without using too technical language.

Let us begin by considering the finite density of matter in the universe. The average density is very small. Matter is actually distributed very unevenly, it is conglomerated into stars and galactic systems. The average density is the density that we should get if all these great systems could be evaporated into atoms of hydrogen, or protons, and these distributed evenly over the whole of space. There would then probably not be more than three or four protons in every cubic foot. That is a very small density indeed: it is about a million million times less than that of the most perfect vacuum that we can produce in our physical laboratories. The universe thus consists mostly of emptiness, and it appears natural to consider a universe without any matter at all, an empty universe, as a good approximation to begin with for our grand scale model. The galactic systems are details which can be put in afterwards. But we may also take as our first approximation a universe containing the same amount of matter as the actual one, but equally distributed, i.e. having a finite average density of three or four protons per cubic foot. The local deviations from the average, caused by the conglomeration of matter into stars and stellar systems, are then disregarded in the grand scale model, and are only taken into account when we come to study details.

Now fifteen years ago, in the beginning of 1917, two solutions of the field-equations for a homogeneous isotropic universe had been found, which I shall provisionally call the solutions "A" and "B." It should be mentioned that at that time only *static* solutions were looked for. It was thought that the universe must be a stable structure, which would retain its large scale properties unchanged for all time, or at least change them so slowly that the change could be disregarded. In one of these solutions (B) the average density was zero, it was empty; the other one (A) had a finite density. Both, of course, were, as was well appreciated, only approximations to the actual universe. In B, to get the real universe, we should have to put in a few galactic systems, in A we should have to condense the evenly distributed matter into galactic systems. The universe A is really and essentially static, there can be no systematic motions in it. It has an average density, but no expansion. It is therefore called the *static universe*. B, on the other hand, is not really static, it expands, and it could only parade in the garb of a static universe because there is nothing in it to show the expansion. B is therefore called the *empty universe*. Thus we had two approximations: the static universe with matter and without expansion, and the empty one without matter and with expansion. The actual universe, as we have seen, has both matter and expansion, and can, therefore, be neither A nor B. In 1917 this dilemma had not yet become urgent, and was hardly realised. The actual value of the density was still entirely unknown, and the expansion had not yet been discovered.

Now in both the solutions A and B the curvature is positive, in both three-dimensional space is finite: the universe has a definite size, we can speak of its radius, and, in the case A, of its total mass. In the case A, the static universe, there is a definite relation between the curvature and the density, in fact the density is proportional to the curvature, the factor of proportionality being a pure number ($1/4\pi$, if appropriate units are used). Thus, if we wish to have a finite density in a static universe, we must have a finite positive curvature.

At this point we must say a few words about the famous *lambda*. The field-equations, in their most general form, contain a term multiplied by a constant, which is denoted by the Greek letter λ (lambda), and which is sometimes called the "cosmical constant." This is a name without any meaning, which was only conferred upon it because it was thought appropriate that it should have a name, and because it appeared to have something to do with the constitution of the universe; but it must not be inferred that, since we have given it a name, we know what it means. We have, in fact, not the slightest inkling of what its real significance is. It is put in the equations in order to give them the greatest possible degree of mathematical generality, but, so far as its mathematical function is concerned, it is entirely undetermined: it may be positive or negative, it might also be zero. Purely mathematical symbols have no meaning by themselves; it is the privilege of pure mathematicians, to quote Bertrand Russell, not to know what they are talking about. They—the symbols—only get a meaning by the interpretation that is put on the equations when they are applied to the solution of physical problems. It is the physicist, and not the mathematician, who must know what he is talking about. At first, in Einstein's paper of November 1915, in which the theory reached its final form, the term with λ was simply omitted, in other words λ was supposed to have the special value *zero*. That was the simplest way of avoiding the responsibility of attaching a label to it, and, of course, an entirely legitimate way: it is very bad physics to introduce more arbitrary constants than are needed for the representation of the phenomena. But fifteen months later, in February 1917, it was found that a static solution with a positive curvature—the solution A—was not possible without the λ. In fact the curvature is proportional to λ (in solution A, λ is equal to the curvature; in B, when treated as a static solution, it is three times the curvature). Thus, at the time when we had only the two static solutions A and B, and thought that these were the only possible ones, here was a plausible physical interpretation of the meaning of λ: it was the curvature of the world, and the square root of its reciprocal, the radius of curvature, could be conceived as providing a natural unit of length. It gave the electron something to measure itself by, so that it might know how large it ought to be, as Sir Arthur Eddington has expressed it. Recently Eddington has replaced this interpretation by another one of the same nature, involving the gravitational mass and the electric charge of the electron. But recent developments have made it very difficult to maintain these interpretations, as we shall presently see.

We must now take up the thread of the narrative where we left it a little while ago. We were in the position of having two possible solutions: the static universe with matter but without expansion, and the empty universe with expansion but without matter.

Now the observed rate of expansion is large: the universe doubles its size in about fifteen hundred million years, which is a short time, astronomically speaking. In the "static" universe expansion is impossible, the "empty" universe does expand. Therefore we may be tempted to consider the empty universe as the most likely approximation; and we can proceed to compute the radius of curvature of the universe, supposing it to be of the empty type, from the observed rate of expansion. It comes out as about two thousand million light-years.

The universe, however, is not empty, but contains matter. The point is how much matter. Is the density anywhere near that corresponding to the static universe, or is it so small that we can consider the empty universe as a good approximation? We have seen that the universe is some million million times as empty as our most perfect vacuum. But this is not the correct way to measure the emptiness of the universe. We must use as a standard of comparison, not our terrestrial experience, but the theoretical density of the static universe. It is easy to compute the density of a static universe of a radius of two thousand million lightyears, and it comes out only very little larger than the observed density. The actual universe is thus very far from empty, it is, on the contrary, nearly full.

We thus come to the conclusion, which was already foreshadowed above, that the actual universe is neither the static nor the empty one. It differs so much from both of these that neither can be used as an appropriate grand scale model. We must thus look for other solutions of the general field-equations. On account of the expansion our solution must necessarily be a non-static one, and it must have a finite density. There is only one possible static solution possessing a finite density, viz. our old friend A, but of non-static solutions with finite density there exists a great variety. I will now depart from the strictly historical narrative and enumerate these different possible solutions, not in the order in which they have been discovered, but in the sequence of a natural classification.

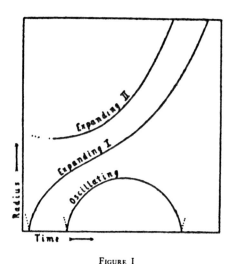

FIGURE I

The three families of non-static universes

In the solutions A and B the curvature of three-dimensional space was necessarily positive, and the mysterious "cosmical constant" λ was also positive. In the non-static solutions this is not so. At first this was not realised. We had become so accustomed to think of λ as an essentially positive quantity, and of a finite world with positive curvature, that the idea of investigating the possibility of solutions with negative or zero values of λ and of the curvature simply did not arise. But when this oversight was corrected, it appeared at once that in the non-static case both λ and the curvature need not be positive, but can be negative or zero quite as well. I will therefore use the value of λ and the sign of the curvature as the principles of classification. The instantaneous state of the universe is characterised by a certain quantity occurring in the equations, which is denoted by the letter R, and which, if there is a curvature, can be interpreted as the radius of curvature, or the "radius" for short. The way in which the universe expands is determined by the variation of this R with the time. There are three types, or families, of non-static universes, which I will call the oscillating universes, and the expanding universes of the first and of the second kind. They are represented graphically in Figure I. The horizontal coördinate is the time, the vertical coördinate is the "radius." Of each type only one example is given in the figure, but it should be realised that each of these is a representative of a family, comprising an infinite number of members differing in size and shape.

In the oscillating universes the "radius" R increases from zero to a certain maximum size, which is different for each member of the family, and then decreases again to zero. The period of oscillation has a certain finite (and rather short) value, different for each member of the family. In the expanding family of the first kind the radius is continually increasing from a certain initial time, when it was zero, to become infinitely large after an infinite time. In the expanding series of the second type the radius has at the initial time a certain minimum value, different for the different members of the family, and increases to become infinite after an infinite time.

If λ is negative, only oscillating universes are possible, whatever the curvature may be, positive, negative or zero. The only choice is between the different members of the oscillating family, and to each special value of λ corresponds one member of the family.

If λ is zero, and the curvature positive, we have still an oscillating universe; if the curvature is zero or negative, the universe is one of the first family of expanding universes.

If λ is positive, then for positive curvatures the three families are possible; for negative and zero curvatures only the first family of expanding universes exists. The different possibilities are given in the following small table.

λ	Curvature		
	Negative	Zero	Positive
Negative	Oscillating	Oscillating	Oscillating
Zero	Expanding I	Expanding I	Oscillating
Positive	Expanding I	Expanding I	Oscillating Expanding I Expanding II

We do not know to which of the three possible families our own universe belongs, and there is nothing in our observational data to guide us in making the choice. And even if we have decided on the family, we have still the freedom to select any particular member of it. This is not because the data are not accurate enough, but because they are deficient in number. The observations give us *two* data, viz. the rate of expansion and the average density, and there are *three* unknowns: the value of λ, the sign of the curvature, and the scale of the figure, i.e. the units of R and of the time. The problem is indeterminate. If we make an hypothesis regarding either λ or the curvature, then we can find the other from the observed data, or rather we would be able to do so, if the data were sufficiently accurate. We might, e.g., decide a priori that the curvature should be zero, i.e. that the three-dimensional space should be euclidean. If we make that hypothesis, we have a sufficient number of data to enable us to determine the value of λ. We might, on the other hand, wish to get rid of the λ, on the ground that it ought never to have been there, being unfortunately introduced into the equations in a past stage of the development of the theory, when we were mistakenly trying to find a static solution. In other words we might make the hypothesis that the true value of λ is zero. In that case the data of observation will allow us, if they are sufficiently accurate, to determine the curvature.

As a matter of fact neither the average density nor the rate of expansion are at the present time known with sufficient accuracy to make an actual *determination* possible, even if an hypothesis of this kind is adopted. All we can say is that, if the curvature is small (as we know it must be, because it is imperceptible by ordinary geometric methods in our neighbourhood), then λ must be small, and *if* the curvature is *very* small, then λ must be very small. On the other hand, *if* λ is very small, or zero, then the curvature must be very small, and may even be zero, for aught we can say at present.

The interpretation of the expanding universe, the making of a mental picture, or a model, of it, which was, or appeared to be, easy when we knew that the universe was finite, is not such a simple matter now that we do not even know whether the curvature is positive, zero or negative, whether the universe is finite or infinite. It sounds rather strange to talk of an infinite universe still expanding. If we were certain that the curvature was negative, we might still, as in the case of positive curvature, replace the phrase "the universe expands" by the equivalent one "the curvature of the universe decreases." But if the curvature is zero, and remains zero throughout, what sort of meaning are we to attach to the "expansion"? The real meaning is, of course, that the mutual distances between the galactic systems, measured in so-called natural measure, increase proportionally to a certain quantity R appearing in the equations, and varying with the time. The interpretation of R as the "radius of curvature" of the universe, though still possible if the universe has a curvature, evidently does not go down to the fundamental meaning of it. The manner in which time and space are bound up with each other in the four-dimensional continuum is variable. It is difficult to express this variability of the cross-connections between space and time in simple language, and different interpretations of it are possible, corresponding to different mathematical transformations of the fundamental line-element, e.g. a different choice of the variable which we interpret as "time." Perhaps the best way we can express it is by saying that the solution of the field-equations of the theory of relativity shows that there is in the universe a tendency to change its scale, which at the present time results in an expansion, but may perhaps at other times become, or have been, a shrinking. This is true of the grand scale model of the universe. If we put in the details, the singularities of the field, viz. the galactic systems and the stars, we find that there is also a tendency, called gravitation, to decrease the mutual distances of these "singularities." At short distances, within the confines of a galactic system, this second tendency is by far the strongest, and the galactic systems retain their size independent of the expansion or contraction of the universe; at large distances, such as those separating one galactic system from the next, the first preponderates.

The interpretation of λ as providing a natural unit of length also, of course, becomes difficult. The radius of curvature, or even the more vaguely defined quantity R, can no longer provide the electron with the means of knowing how large it ought to be, for R changes and the electron does not change its size. The electron could still measure itself in some mysterious way against the square root of the reciprocal of λ—or of *minus* λ, if λ happened to be negative, —but if λ were zero that also would not help it.

Also other interpretations, involving the total mass of, or the number of protons, in the universe, become untenable, if the universe may, for aught we know, be just as well infinite and have no total mass.

The theory of the expanding universe is at the present moment much less definite than we supposed it to be a few months ago, but that does not affect its real significance—it only brings out more clearly what this significance is, and what it is not.

Some consequences of the theory merit a special mention, however briefly.

A question which has long troubled astronomers and physicists is what becomes of the energy that is continually being poured out into space by the sun and the stars. To this question a complete answer is given by the new theory. It is used up, diluted, or degraded, by the expansion of the universe. Just as a man running to catch a bus or a tram-car gets out of breath and spends his energy, or a projectile thrown after a moving train hits it with less force than it would hit a stationary object, so the light travelling through the expanding universe and, so to say, trying to reach a particular star, or stellar system, which is continually receding with great velocity, is losing energy in trying to catch up with it. It is this degradation of the light, technically known as the redshift of the spectral lines, by which we become aware of the receding velocities of the extra-galactic nebulae. It can be shown that the decrease of the total amount of radiant energy in the universe by this degradation exceeds the increase by the radiation of the stars. It would not be correct, however, to conclude that the expansion is caused *by* the energy thus lost by the radiation, any more than it would be correct to say that the tram-car is propelled by the energy expended by the man who runs after it.

There is one very serious difficulty presented by the theory of the expanding universe, which we shall have to face with careful deliberation.

In all solutions there is a certain minimum value of the "radius" R, either zero, or in the expanding family of the second kind a finite value, which the universe had at a definite time in the past. There appears to be a definite "beginning of time," a few thousand million years back in history, as there is a definite "absolute zero" of temperature, corresponding to *minus* 273 degrees on the ordinary scale. What is the meaning of this?

The temptation is strong to identify the epoch of the beginning of the expansion with the "beginning of the world," whatever that may mean. Now astronomically speaking this beginnnig of the expansion took place only yesterday, not much longer ago than the formation of the oldest rocks on the earth. According to all our modern views the evolution of a star, of a double star, or a star cluster, requires intervals of time which are enormously longer. The stars and the stellar systems must be some thousands of times older than the universe!

What must be our attitude with regard to this paradox? It would appear that, if two theories are in contradiction, we must give up either the one or the other. The conflict apparently is between the modern theories of stellar evolution and the dynamical theories of the evolution of double stars and star clusters on the one hand, and the general theory of relativity on the other hand. If this were the real contest, there could be no doubt about the issue: the theory of relativity would come out of the trial victorious, and the theories of evolution would have to be revised. This seems to be Sir Arthur Eddington's standpoint, as he writes: "we must accept this alarmingly rapid dispersal of the nebulae with its important consequences in limiting the time available for evolution." I am afraid, however, that very few astronomers, not to speak of geophysicists, will be prepared to accept this drastic reduction of the time scale.

It is possible to relegate the epoch of the starting of the expansion to minus infinity, e.g. by using instead of the ordinary time the logarithm of the time elapsed since the beginning. But this is only a mathematical trick. We call zero minus infinity, but that only means that we allow the universe an infinite time to get well started on its course of expansion, but it does not make the time during which anything really happens any longer.

Mention should be made of a suggestion that has sometimes been advanced, viz. that the observed shift of the spectral lines towards the red might not indicate a receding motion of the spiral nebulae, but might be accounted for in some different way. In fact, all that the observations tell us is that light coming from great distances—and which therefore has been a long time under way—

is redder when it arrives than when it left its source. *Light is reddened by age,* it loses energy as it gets older, travelling through space. Or expressed mathematically: the wavelength of light is proportional to a certain quantity R, which increases with the time. By the general equations of the theory of relativity the naturally measured distances in a homogeneous and isotropic world are then necessarily proportional to the same quantity R, unless some extraneous cause for the increase of wavelength, or the loss of energy, is present. By extraneous I mean foreign to the theory of relativity and the conception of the nature of light consistent with that theory. Moreover this hypothetical cause should have no other observable consequences, especially it should produce loss of energy without any concomitant dispersion, which would blur the images and make the faint nebulae unobservable. It would require an hypothesis *ad hoc,* and a very carefully framed one too, so as not to overshoot the mark. No such hypothesis, deserving serious consideration, has yet been forthcoming.

It appears to me that there is no way out of the dilemma. It is an unavoidable consequence of the equations that the time taken by the radius to increase from anywhere near its minimum to its present value is of the same order of magnitude as the present radius itself, if we adopt corresponding units of space and time, e.g. years for time and lightyears for space. The scale is determined by the observed rate of expansion. I am afraid all we can do is to accept the paradox and try to accommodate ourselves to it, as we have done to so many paradoxes lately in modern physical theories. We shall have to get accustomed to the idea that the change of the quantity R, commonly called the "radius of the universe," and the evolutionary changes of stars and stellar systems are two different processes, going on side by side without any apparent connection between them. After all the "universe" is an hypothesis, like the atom, and must be allowed the freedom to have properties and to do things which would be contradictory and impossible for a finite material structure. What we observe are the stars and nebulae constituting "our neighbourhood." All that goes beyond that, in time or in space, or both, is pure extrapolation. The conclusions derived about the expanding universe depend on the assumed homogeneity and isotropy, i.e. on the hypothesis that the observed finite material density and expansion of our neighbourhood are not local phenomena, but properties of the "universe." It is not inconceivable that this hypothesis may at some future stage of the development of science have to be given up, or modified, or at least differently interpreted.

Our conception of the structure of the universe bears all the marks of a transitory structure. Our theories are decidedly in a state of continuous, and just now very rapid, evolution. It is not possible to predict how long our present views and interpretations will remain unaltered and how soon they will have to be replaced by perhaps very different ones, based on new observational data and new critical insight in their connection with other data.

Meanwhile the simple workers in science go on quietly, each working at his own particular problem, undisturbed by the many strange and contradictory things that are happening around them and in their own house. And it is on this quiet and unostentatious work that the great advances of science are based. Especially in astronomy two characteristics are common to all data on which the solution of the great problems depends. The first is the extreme minuteness of the quantities to be measured. The determinations of the distances, of the proper motions and radial velocities, which form the materials out of which the great structures of the theories of the galactic system and the universe have been built, all require very accurate measurement of extremely small quantities. It is always a struggle for the last decimal place, and the great triumphs of science are gained when, by new methods or new instruments, the last decimal is made into the penultimate. Thus, as we have seen, new epochs were inaugurated in the beginning of the seventeenth century by the invention of the telescope, and in the last third of the nineteenth by the discovery of photography and spectroscopy.

The other characteristic is that astronomy always requires a very large number of data. As a consequence of the fact that direct experiment is impossible in astronomy, we must have observations of very many stars, and extended to ever fainter objects, in order to enable us to draw reliable conclusions.

These two characteristics of the data that the astronomer requires to build his science on make two things more necessary in astronomy than in any other science: patience and organised coöperation. Patience because many of the phenomena develop so slowly that a long time is necessary for them to become measurable, coöperation because the material is too large and too various to be mastered by one man, or even by one institute. And coöperation not only between different workers and institutions all over the world, but also coöperation with predecessors and successors for the solution of problems that require, by their very nature, more than one man's lifetime. The astronomer—each working at his own task, whether performing long calculations, making theories and hypotheses, or patiently collecting observations in daily routine—is always conscious of belonging to a community, whose members, separated in space and time, nevertheless feel joined by a very real tie, almost of kinship. He does not work for himself alone, he is not guided exclusively, and not even in the first place, by his own insight or preferences, his work is always coördinated with that of others as a part of an organised whole. He knows that, whatever his special work may be it is alway a link in a chain, which derives its value from the fact that there is another link to the left and one to the right of it. It is the chain that is important, not the separate links.

ARTHUR S. EDDINGTON:

Spherical Space

I could be bounded in a nutshell and count myself a king of infinite space.

Hamlet

I

WHEN a physicist refers to curvature of space he at once falls under suspicion of talking metaphysics. Yet space is a prominent feature of the physical world; and measurement of space—lengths, distances, volumes— is part of the normal occupation of a physicist. Indeed it is rare to find any quantitative physical observation which does not ultimately reduce to measuring distances. Is it surprising that the precise investigation of physical space should have brought to light a new property which our crude sensory perception of space has passed over?

Space-curvature is a purely physical characteristic which we may find in a region by suitable experiments and measurements, just as we may find a magnetic field. In curved space the measured distances and angles fit together in a way different from that with which we are familiar in the geometry of flat space; for example, the three angles of a triangle do not add up to two right angles. It seems rather hard on the physicist, who conscientiously meas- ures the three angles of a triangle, that he should be told that if the sum comes to two right angles his work is sound physics, but if it differs to the slightest extent he is straying into metaphysical quagmires.

In using the name "curvature" for this characteristic of space, there is no metaphysical implication. The nomenclature is that of the pure geometers who had already imagined and described spaces with this characteristic before its actual physical occurrence was suspected.

Primarily, then, curvature is to be regarded as the technical name for a property discovered observationally. It may be asked, How closely does "curva- ture" as a technical scientific term correspond to the familiar meaning of the word? I think the correspondence is about as close as in the case of other

From A. S. Eddington, *The Expanding Universe*, Cambridge University Press, 1933, Chapter II. Reprinted by kind permission of the publishers, Cambridge University Press.

familiar words, such as Work, Energy, Probability, which have acquired a specialised meaning in science.

We are familiar with curvature of *surfaces;* it is a property which we can impart by bending and deforming a flat surface. If we imagine an analogous property to be imparted to *space* (three-dimensional) by bending and deforming it, we have to picture an extra dimension or direction in which the space is bent. There is, however, no suggestion that the extra dimension is anything but a fictitious construction, useful for representing the property pictorially, and thereby showing its mathematical analogy with the property found in surfaces. The relation of the picture to the reality may perhaps best be stated as follows. In nature we come across curved surfaces and curved spaces, i.e. surfaces and spaces exhibiting the observational property which has been technically called "curvature". In the case of a surface we can ourselves remove this property by bending and deforming it; we can therefore conveniently describe the property by the operation (bending or curving) which we should have to perform in order to remove it. In the case of a space we cannot ourselves remove the property; we cannot alter space artificially as we alter surfaces. Nevertheless we may conveniently describe the property by the imaginary operation of bending or curving, which would remove it if it could be performed; and in order to use this mode of description a fictitious dimension is introduced which would make the operation possible.

Thus if we are not content to accept curvature as a technical physical characteristic but ask for a picture giving fuller insight, we have to picture more than three dimensions. Indeed it is only in simple and symmetrical conditions that a fourth dimension suffices; and the general picture requires six dimensions (or, when we extend the same ideas from space to space-time, ten dimensions are needed). That is a severe stretch on our powers of conception. But I would say to the reader, do not trouble your head about this picture unduly; it is a stand-by for very occasional use. Normally, when reference is made to space-curvature, picture it as you picture a magnetic field. Probably you do *not* picture a magnetic field; it is something (recognisable by certain tests) which you use in your car or in your wireless apparatus, and all that is needed is a recognised name for it. Just so; space-curvature is something found in nature with which we are beginning to be familiar, recognisable by certain tests, for which ordinarily we need not a picture but a name.

It is sometimes said that the difference between the mathematician and the non-mathematician is that the former can picture things in four dimensions. I suppose there is a grain of truth in this, for after working for some time in four or more dimensions one does involuntarily begin to picture them after a fashion. But it has to be added that, although the mathematician visualises four dimensions, his picture is *wrong* in essential particulars—at least mine is. I see our spherical universe like a bubble in four dimensions; length, breadth, and thickness, all lie in the skin of the bubble. Can I picture this bubble rotating? Why, of course I can. I fix on one direction in the four dimensions as axis, and I see the other three dimensions whirling round it. Perhaps I never actually see more than two at a time; but thought flits rapidly from one pair to another, so that all three seem to be hard at it. Can *you* picture it like that? If you fail, it is just as well. For we know by analysis that a bubble in four dimensions does not rotate that way at all. Three dimensions cannot spin round a fourth. They must rotate two round two; that is to say, the bubble does not rotate about a line axis but about a plane. I know that that is true; but I cannot visualise it.

I need scarcely say that our scientific conclusions about the curvature of space are not derived from the false involuntary picture, but by algebraic working out of formulae which, though they may be to some extent illustrated by such pictures, are independent of pictures. In fact, the pictorial conception of space-curvature falls between two stools: it is too abstruse to convey much illumination to the non-mathematician, whilst the mathematician practically ignores it and relies on the more dependable and more powerful algebraic methods of investigating this property of physical space.

Having said so much in disparagement of the picture of our three-dimensional space contorted by curvature in fictitious directions, I must now mention one application in which it is helpful. We are assured by analysis that in one important respect the picture is not misleading. The curvature, or bending round of space, may be sufficient to give a "closed space"—space in which it is impossible to go on indefinitely getting farther and farther from the starting-point. Closed space differs from an open infinite space in the same way that the surface of a sphere differs from a plane infinite surface.

II

We may say of the surface of a sphere (1) that it is a *curved* surface, (2) that it is a *closed* surface. Similarly we have to contemplate two possible characteristics of our actual three-dimensional space, *curvature* and *closure*. A closed surface or space must necessarily be curved, but a curved surface or space need not be closed. Thus the idea of closure goes somewhat beyond the idea of curvature; and, for example, it was not contemplated in the first announcement of Einstein's general relativity theory which introduced curved space.

In the ordinary application of Einstein's theory to the solar system and other systems on a similar scale the curvature is small and amounts only to a very slight wrinkling or hummocking. The distortion is local, and does not affect the general character of space as a whole. Our present subject takes us much farther afield, and we have to apply the theory to the great super-system of the galaxies. The small local distortions now have cumulative effect. The new investigations suggest that the curvature actually leads to a complete bending round and closing up of space, so that it becomes a domain of finite extent. It will be seen that this goes beyond the original proposal; and the evidence for it is by no means so secure. But all new exploration passes through a phase of insecurity.

For the purpose of discussion this closed space is generally taken to be spherical. The presence of matter will cause local unevenness; the scale that we are now contemplating is so vast that we scarcely notice the stars, but the galaxies change the curvature locally[1] and so pull the sphere rather out of shape. The ideal spherical space may be compared to the geoid used to represent the average figure of the earth with the mountains and ocean beds smoothed away. It may be, however, that the irregularity is much greater, and the universe may be pear-shaped or sausage-shaped; the 150 million light-years over which our observational survey extends is only a small fraction of the whole extent of space, so that we are not in a position to dogmatise as to the actual shape. But we can use the spherical world as a typical model, which will illustrate the peculiarities arising from the closure of space.

In spherical space, if we go on in the same direction continually, we ultimately reach our starting-point again, having "gone round the world." The same thing happens to a traveller on the earth's surface who keeps straight on bearing neither to the left nor to the right. Thus the closure of space may be thought of as analogous to the closure of a surface, and generally speaking it has the same connection with curvature. The whole area of the earth's surface is finite, and so too the whole volume of spherical space is finite. It is "finite but unbounded"; we never come to a boundary, but owing to the re-entrant property we can never be more than a limited distance away from our starting-point.

In the theory that I am going to describe the galaxies are supposed to be distributed throughout a closed space of this kind. As there is no boundary— no point at which we can enter or leave the closed space—this constitutes a self-contained finite universe.

Perhaps the most elementary characteristic of a spherical universe is that at great distances from us there is not so much room as we should have anticipated. On the earth's surface the area within 2 miles of Charing Cross is very nearly 4 times the area within 1 mile; but at a distance of say 4000 miles this simple progression has broken down badly. Similarly in the universe the volume, or amount of room, within 2 light-years of the sun is very nearly 8 times the volume within 1 light-year; but the volume within 4000 million light-years of the sun is considerably less than 8 times the volume within 2000 million light-years. We have no right to be surprised. How could we have expected to know how much room there would be out there without examining the universe to see? It is a common enough experience that simple rules,

---

1. Einstein's law of gravitation connects the various components of curvature of space with the density, momentum, and stress of the matter occupying it. I would again remind the reader that space-curvature is the technical name for an observable physical property, so that there is nothing metaphysical in the idea of matter producing curvature any more than in a magnet producing a magnetic field.

which hold well enough for a limited range of trial, break down when pushed too far. There is no juggling with words in these statements; the meaning of distance and volume in surveying the earth or the heavens is not ambiguous; and although there are practical difficulties in measuring these vast distances and volumes there is no uncertainty as to the ideal that is aimed at. I do not suggest that we have checked by direct measurement the falling off of volume at great distances; like many scientific conclusions, it is a very indirect inference. But at least it has been reached by examining the universe; and, however shaky the deduction, it has more weight than a judgment formed without looking at the universe at all.

Much confusion of thought has been caused by the assertion so often made that we can use any kind of space we please (Euclidean or non-Euclidean) for representing physical phenomena, so that it is impossible to disprove Euclidean space observationally. We can graphically represent (or misrepresent) things as we please. It is possible to represent the curved surface of the earth in a flat space as, for example, in maps on Mercator's projection; but this does not render meaningless the labours of geodesists as to the true figure of the earth. Those who *on this ground* defend belief in a flat universe must also defend belief in a flat earth.

III

There is a widespread impression, which has been encouraged by some scientific writers, that the consideration of spherical space in this subject is an unnecessary mystification, and that we could say all we want to say about the expanding system of the galaxies without using any other conception than that of Euclidean infinite space. It is suggested that talk about expanding space is mere metaphysics, and has no real relevance to the expansion of the material universe itself, which is commonplace and easily comprehensible. This is a mistaken idea. The general phenomenon of expansion, including the explanation provided by relativity theory, can be expounded up to a certain point without any recondite conceptions of space, but there are other consequences of the theory which cannot be dealt with so simply. To consider these we have to change the method, and partly transfer our attention from the properties and behaviour of the material system to the properties and behaviour of the space which it occupies. This is necessary because the properties attributed to the material system by the theory are so unusual that they cannot even be described without self-contradiction if we continue to picture the system in flat (i.e. Euclidean) space. This does not constitute an objection to the theory, for there is, of course, no reason for supposing space to be flat unless our observations show it to be flat; and there is no reason why we should be able to picture or describe the system in flat space if it is not in flat space. It is no disparagement to a square peg to say that it will not fit into a round hole.

I will liken the super-system of galaxies (the universe) to a peg which is fitted into a hole—space. In Chapter 1 we were only concerned with a little bit of the peg (the 150 million light-years surveyed) and the question of fit scarcely arose. When we turn to consider the whole peg we find mathematically that, unless something unforeseen occurs beyond the region explored, it is (for the purposes of this analogy) a square peg. Immediately there is an outcry: "That is an impossible sort of peg—not really a peg at all." Our answer is that it is an excellent peg, as good as any on the market, provided that you do not want to fit it into a round hole. "But holes are round. It is the nature of holes to be round. A Greek two thousand years ago said that they are round." And so on. So whether I want it or not, the argument shifts from the peg to the hole—the space into which the material universe is fitted. It is over the hole that the battle has been fought and won; I think now that every authority admits, if only grudgingly, that the square hole—by which I here symbolise closed space—is a physical possibility.

The issue that I am here dealing with is not whether the theory of a closed expanding universe is right or wrong, probable or improbable, but whether, if we hold the theory, spherical space is necessary to the statement of it. I am not here replying to those who disbelieve the theory, but to those who think its strangeness is due to the mystifying language of its exponents. The following will perhaps show that there has been no gratuitous mystification:

I want you to imagine a system of say a billion stars spread approximately uniformly so that each star has neighbours surrounding it on all sides, the distance of each star from its nearest neighbours being approximately the same everywhere. (Lest there be any doubt as to the meaning of *distance,* I define it as the distance found by parallax observation, or by any other astronomical method accepted as equivalent to actually stepping out the distance.) Can you picture this?

— Yes. Except that you forgot to consider that the system will have a boundary; and the stars at the edge will have neighbours on one side only, so that they must be excepted from your condition of having neighbours on all sides.

— No; I meant just what I said. I want *all* the stars to have neighbours surrounding them. If you picture a place where the neighbours are on one side only—what you call a boundary—you are not picturing the system I have in mind.

— But your system is impossible; there must be a boundary.

— Why is it impossible? I could arrange a billion people on the surface of the earth (spread over the whole surface) so that each has neighbours on all sides, and no question of a boundary arises. I only want you to do the same with the stars.

— But that is a distribution over a surface. The stars are to be distributed in three-dimensional space, and space is not like that.

— Then you agree that if space could be "like that" my system would be quite possible and natural?

— I suppose so. But how could space be like that?

— We will discuss space if you wish. But just now when I was trying to explain that according to present theory space does behave "like that," I was told that the discussion of space was an unnecessary mystification, and that if I would stick to a description of my material system it would be seen to be quite commonplace and comprehensible. So I duly described my material system; whereupon *you* immediately raised questions as to the nature of space.

In the spherical universe the character of the material system is as peculiar as the character of the space. The material system, like the space, exhibits *closure*; so that no galaxy is more central than another, and none can be said to be at the outside. Such a distribution is at first sight inconceivable, but that is because we try to conceive it in flat space. The space and the material system have to fit one another. It is no use trying to imagine the system of galaxies contemplated in Einstein's and Lemaître's theories of the universe, if the only kind of space in our minds is one in which such a system cannot exist.

In the foregoing conversation I have credited the reader with a feeling which instinctively rejects the possibility of a spherical space or a closed distribution of galaxies. But spherical space does not contradict our experience of space, any more than the sphericity of the earth contradicts the experience of those who have never travelled far enough to notice the curvature. Apart from our reluctance to tackle a difficult and unfamiliar conception, the only thing that can be urged against spherical space is that more than twenty centuries ago a certain Greek published a set of axioms which (inferentially) stated that spherical space is impossible. He had perhaps more excuse, but no more reason, for his statement than those who repeat it to-day.

Few scientific men nowadays would reject spherical space as impossible, but there are many who take the attitude that it is an unlikely kind of hypothesis only to be considered as a last resort. Thus, in support of some of the proposed explanations of the motions of the spiral nebulae, it is claimed that they have the "advantage" of not requiring curved space. But what is the supposed disadvantage of curved space? I cannot remember that any disadvantage has ever been pointed out. On the other hand it is well known that the assumption of flat physical space leads to very serious theoretical and logical difficulties.

A closed system of galaxies requires a closed space. If such a system expands, it requires an expanding space. This can be seen at once from the analogy that we have already used, viz. human beings distributed evenly over the surface of the earth; clearly they cannot scatter apart from one another unless the earth's surface expands.

This should make clear how the present theory of the expanding universe stands in relation to (*a*) the expansion of a material system, and (*b*) the expansion of space. The observational phenomenon chiefly concerned (recession of the spiral nebulae) is obviously expansion of a material system; and the onlooker is often puzzled to find theorists proclaiming the doctrine of an expanding space. He suspects that there has been confusion of thought of a rather elementary kind. Why should not the space be there already, and the material system expand into it, as material systems usually do? If the system of galaxies comes to an end not far beyond the greatest distance we have plumbed, then I agree that that is what happens. But the system shows no sign of coming to an end, and, if it goes on much farther, it will alter its character. This change of character is a matter of mathematical computation which cannot be discussed here; I need only say that it is connected with the fact that, if the speed of recession continues to increase outwards, it will ere long approach the speed of light, so that something must break down. The result is that the system becomes a closed system; and we have seen that such a system cannot expand without the space also expanding. That is how *expansion of space* comes in. I daresay that (for historical reasons) expansion of space has often been given too much prominence in expositions of the subject, and readers have been led to think that it is more directly concerned in the explanation of the motions of the nebulae than is actually the case. But if we are to give a full account of the views to which we are led by theory and observation, we must not omit to mention it.

What I have said has been mainly directed towards removing preliminary prejudices against a closed space or a closed system of galaxies. I do not suggest that the reasons for adopting closed space are overwhelmingly strong;[2] but even slight advantages may be of weight when there is nothing to place in the opposite scale. If we adopt open space we encounter certain difficulties (not necessarily insuperable) which closed space entirely avoids; and we do not want to divert the inquiry into a speculation as to the solution of difficulties which need never arise. If we wish to be non-committal, we shall naturally work in terms of a closed universe of finite radius R, since we can at any time revert to an infinite universe by making R infinite.

There is one other type of critic to whom a word may be said. He feels that space is not solely a matter that concerns the physicists, and that by their technical definitions and abstractions they are making of it something different from the common man's space. It would be difficult to define precisely what is in his mind. Perhaps he is not thinking especially of space as a measurable constituent of the physical universe, and is imagining a world order transcending the delusions of our sensory organs and the limitations of our micrometers—a space of "things as they really are." It is no part of my present subject to discuss the relation of the world as conceived in physics to a wider interpretation of our experience; I will only say that that part of our conscious experience representable by physical symbols ought not to claim to be the whole. As a conscious being *you* are not one of my symbols; your domain is

---

2. *Curved* space is fundamental in relativity theory, and the argument for adopting it is generally considered to be overwhelming. It is *closed* space which needs more evidence.

not circumscribed by my spatial measurements. If, like Hamlet, you count yourself king of an infinite space, I do not challenge your sovereignty. I only invite attention to certain disquieting rumours which have arisen as to the state of Your Majesty's nutshell.

IV

The immediate result of introducing the cosmical term into the law of gravitation was the appearance (in theory) of two universes—the Einstein universe and the de Sitter universe. Both were closed spherical universes; so that a traveller going on and on in the same direction would at last find himself back at the starting-point, having made a circuit of space. Both claimed to be static universes which would remain unchanged for any length of time; thus they provided a permanent framework within which the small scale systems—galaxies and stars—could change and evolve. There were, however, certain points of difference between them. An especially important difference, because it might possibly admit of observational test, was that in de Sitter's universe there would be an apparent recession of remote objects, whereas in Einstein's universe this would not occur. At that time only three radial velocities of spiral nebulae were known, and these somewhat lamely supported de Sitter's universe by a majority of 2 to 1. There the question rested for a time. But in 1922 Prof. V. M. Slipher furnished me with his (then unpublished) measures of 40 spiral nebulae for use in my book *Mathematical Theory of Relativity*. As the majority had become 36 to 4, de Sitter's theory began to appear in a favourable light.

The Einstein and de Sitter universes were two alternatives arising out of the same theoretical basis. To give an analogy—suppose that we are transported to a new star, and that we notice a number of celestial bodies in the neighbourhood. We should know from gravitational theory that their orbits must be either ellipses or hyperbolas; but only observation can decide which. Until the observational test is made there are two alternatives; the objects may have elliptic orbits and constitute a permanent system like the solar system, or they may have hyperbolic orbits and constitute a dispersing system. Actually the question whether the universe would follow Einstein's or de Sitter's model depended on how much matter was present in the universe,—a question which could scarcely be settled by theory—and is none too easy to settle by observation.

We have now realised that the changelessness of de Sitter's universe was a mathematical fiction. Taken literally his formulae described a *completely empty* universe; but that was meant to be interpreted generously as signifying that the average density of matter in it, though not zero, was low enough to be neglected in calculating the forces controlling the system. It turned out, however, that the changelessness depended on there being literally no matter present. In fact the "changeless universe" had been invented by the simple expedient of omitting to put into it anything that could exhibit change. We therefore no longer rank de Sitter's as a static universe; and Einstein's is the only form of material universe which is genuinely static or motionless.

The situation has been summed up in the statement that Einstein's universe contains matter but no motion and de Sitter's contains motion but no matter. It is clear that the actual universe containing both matter and motion does not correspond exactly to either of these abstract models. The only question is, Which is the better choice for a first approximation? Shall we put a little motion into Einstein's world of inert matter, or shall we put a little matter into de Sitter's Primum Mobile?

The choice between Einstein's and de Sitter's models is no longer urgent because we are not now restricted to these two extremes; we have available the whole chain of intermediate solutions between motionless matter and matterless motion, from which we can pick out the solution with the right proportion of matter and motion to correspond with what we observe. These solutions were not sought earlier, because their appropriateness was not realised; it was the preconceived idea that a static solution was a necessity in order that everything might be referred to an unchanging background of space. We have seen that this requirement should strictly have barred out de Sitter's solution, but by a fortunate piece of gate-crashing it gained admission; it was the precursor of the other non-static solutions to which attention is now mainly directed.

The deliberate investigation of non-static solutions was carried out by A. Friedmann in 1922. His solutions were rediscovered in 1927 by Abbé G. Lemaître, who brilliantly developed the astronomical theory resulting therefrom. His work was published in a rather inaccessible journal, and seems to have remained unknown until 1930 when attention was called to it by de Sitter and myself. In the meantime the solutions had been discovered for the third time by H. P. Robertson, and through him their interest was beginning to be realised. The astronomical application, stimulated by Hubble and Humason's observational work on the spiral nebulae, was also being rediscovered, but it had not been carried so far as in Lemaître's paper.

The intermediate solutions of Friedmann and Lemaître are "expanding universes." Both the material system and the closed space, in which it exists, are expanding. At one end we have Einstein's universe with no motion and therefore in equilibrium. Then, as we proceed along the series, we have model universes showing more and more rapid expansion until we reach de Sitter's universe at the other end of the series. The rate of expansion increases all the way along the series and the density diminishes; de Sitter's universe is the limit when the average density of celestial matter approaches zero. The series of expanding universes then stops, not because the expansion becomes too rapid, but because there is nothing left to expand.

We can better understand this series of models by starting at the de Sitter end. As explained in Chapter 1 there are two forces operating, the ordinary Newtonian attraction between the galaxies and the cosmical repulsion. In the de Sitter universe the density of matter is infinitely small so that the Newtonian attraction is negligible. The cosmical repulsion acts without check, and we get the greatest possible rate of expansion of the system. When more matter is inserted, the mutual gravitation tends to hold the mass together and opposes the expansion. The more matter put in, the slower the expansion. There will be a particular density at which the Newtonian attraction between the galaxies is just strong enough to counterbalance the cosmical repulsion, so that the expansion is zero. This is Einstein's universe. If we put in still more matter, attraction outweighs repulsion and we obtain a model of a contracting universe.

Primarily this series of models is a series of alternatives, one of which has to be selected to represent our actual universe. But it has a still more interesting application. As time goes on the actual universe travels along the series of models, so that the whole series gives a picture of its life-history. At the present moment the universe corresponds to a particular model; but since it is expanding its density is diminishing. Therefore a million years hence we shall need a model of lower density, i.e. nearer to the de Sitter end of the series.

Tracing this progression as far back as possible we reach the conclusion that the world started as an Einstein universe; it has passed continuously along the series of models having more and more rapid expansion; and it will finish up as a de Sitter universe.

Allusion has been made to the fact that the recession of the galaxies in the present theory of the expanding universe is not precisely the effect foreseen by de Sitter. It may be well to explain the manner of the transition. The phenomenon that is generally called the "de Sitter effect" was a rather mysterious slowing down of time at great distances from the observer; atomic vibrations would be executed more slowly, so that their light would be shifted to the red and imitate the effect of a receding velocity. But besides discovering this, de Sitter examined the equations of motion and noticed that the *real* velocities of distant objects would probably be large; he did not, however, expect these real velocities to favour recession rather than approach. I am not sure when it was first recognised that the complication in the equations of motion was neither more nor less than a repulsive force proportional to the distance; but it must have been before 1922. Summarising the theory at that date, I wrote— "De Sitter's theory gives a double explanation of this motion of recession: first, there is the general tendency to scatter according to the equation $d^2r/ds^2 = \frac{1}{3}\lambda r$; second, there is the general displacement of spectral lines to the red in distant objects due to the slowing down of atomic vibrations which would be erroneously interpreted as a motion of recession."[3] I also pointed out that it was a question of definition whether the latter effect should be regarded as a spurious or a genuine velocity. During the time that its light is travelling to us, the nebula is being accelerated by the cosmical repulsion and acquires an additional outward velocity exceeding the amount in dispute; so that the velocity, which was spurious at the time of emission of the light, has become genuine by the time of its arrival. Inferentially this meant that the slowing

3. *Mathematical Theory of Relativity*, p. 161.

down of time had become a very subsidiary effect compared with cosmical repulsion; but this was not so clearly realised as it might have been. The subsequent developments of Friedmann and Lemaître were geometrical and did not allude to anything so crude as "force"; but, examining them to see what has happened, we find that the slowing down of time has been swallowed up in the cosmical repulsion; it was a small portion of the whole effect (a second-order term) which had been artificially detached by the earlier methods of analysis.

v

An Einstein universe is in equilibrium, but its equilibrium is unstable. The Newtonian attraction and the cosmical repulsion are in exact balance. Suppose that a slight disturbance momentarily upsets the balance; let us say that the Newtonian attraction is slightly weakened. Then repulsion has the upper hand, and a slow expansion begins. The expansion increases the average distance apart of the material bodies so that their attraction on one another is lessened. This widens the difference between attraction and repulsion, and the expansion becomes faster. Thus the balance becomes more and more upset until the universe becomes irrevocably launched on its course of expansion. Similarly if the first slight disturbance were a strengthening of the Newtonian attraction, this would cause a small contraction. The material systems would be brought nearer together and their mutual attraction further increased. The contracting tendency thus becomes more and more reinforced. Einstein's universe is delicately poised so that the slightest disturbance will cause it to topple into a state of ever-increasing expansion or of ever-increasing contraction.

The original unstable Einstein universe might have turned into an expanding universe or into a contracting universe. Apparently it has chosen expansion. The question arises, Can we explain this choice? I do not think it will be any grave discredit if we fail, for I cannot recall any other case in which theory has succeeded in predicting which way an unstable body will fall. However we shall try. We have to consider what kind of spontaneous disturbance could occur in the primordial distribution of matter from which our galaxies and stars have been evolved; for definiteness I picture it as a motionless uniform nebula filling the spherical world. Two kinds of spontaneous change have been suggested:

(1) The matter will form local condensations so as to become unevenly distributed.

(2) Material mass may become converted into radiation, either in the process of building up complex atoms (e.g. the formation of helium from hydrogen) or in the mutual annihilation of electrons and protons.

It can be shown that the conversion of material mass into radiation would start a contraction. Mass for mass, radiation is more effective than matter in exerting gravitational attraction; hence the conversion tips the balance in favour of contraction. Accordingly our hopes of explaining the decision to expand must rest on process (1). The investigation is peculiarly difficult, because it turns out that to a first approximation the redistribution of matter in condensations makes no difference to the balance, and it is necessary to carry the calculation to a high approximation to obtain the deciding term. The problem has been treated by McVittie, McCrae, Lemaître and Sen—not always with accordant results; and I doubt whether I am qualified to judge on so technical a question. I am inclined to think, however, that Lemaître's treatment goes to the root of the matter.[4]

It has been mentioned that, although we often consider models of the universe which are perfectly spherical, the actual universe must be more irregular. A better approximation would be a pimply sphere—the pimples corresponding to the galaxies; for wherever there is matter, the curvature is locally increased. Whilst a "pimply Einstein world" would have approximately the same properties as an ideal Einstein world, it was at first thought that exact equilibrium was only possible for the exact sphere. It is found, however, that a pimply sphere can also be in exact equilibrium and form a static universe. This was pointed out explicitly by Prof. N. K. Sen, who has given a simple and elegant treatment; but it appears to have been implicit in the earlier work of Lemaître.

Suppose for a moment that, when a condensation is formed, the condensation separates completely from the surrounding matter and leaves an empty crack all round. We imagine that a sphere of gas is separated in this way, and continues to condense more and more. Lemaître (by extending a theorem due to Birkhoff) has shown that *after the separation* the gradual condensation of the matter can make no difference whatever to the gravitational force exerted by the sphere on its surroundings; so that, if the universe outside the condensation was originally in equilibrium, its equilibrium will remain undisturbed. In these conditions the formation of condensations will start neither contraction nor expansion of the universe as a whole.

The actual conditions differ from the foregoing in that no empty crack is formed, the condensation merging gradually into its surroundings. The crack, by isolating the condensation from its surroundings, would have prevented any pressure of one on the other; in the absence of a crack there will be a pressure (probably exceedingly small) which will change as the condensation proceeds. It is this change of pressure, neglected in the preceding paragraph, that is the possible cause of expansion or contraction; for we have seen that the mere rearrangement of matter in a more condensed form has no effect. Lemaître describes the change which occurs as a "stagnation" of energy. It is not difficult to see that it is really the converse kind of change to that which occurs when energy of constitution of matter is liberated as radiation; energy

---

4. Lemaître's paper (*Monthly Notices*, vol. xci, p. 490) seems to me very obscure, but I have had the advantage of verbal explanations from the author.

is taken away from the transmissible form (pressure) and immobilised in the constitution of the condensation. Its effect is therefore opposite to that of conversion of material mass into radiation, and it tends to make the universe expand.

Sen's procedure is different. Having found the equations for a "pimply Einstein world" in static equilibrium, he calculates the total mass of such a world, and finds that it is always greater than that of a uniform Einstein world. It follows that, if the matter of the original uniform Einstein world is rearranged in condensations, there is not quite enough mass to form an equilibrium distribution. If we could artificially add a little mass to each condensation, we should obtain one of Sen's pimply spheres in equilibrium; the absence of this mass leaves the gravitational attraction in defect of the amount required to maintain equilibrium. Consequently cosmical repulsion has the upper hand and the universe expands.

Although both Lemaître and Sen agree that expansion (not contraction) ensues, there is a discrepancy between them; for Sen obtains it as the direct result of the rearrangement of matter, whereas Lemaître claims that the direct result is *nil*, and that the expansion is an indirect result dependent on the existence of a small pressure in the primordial nebula. Lemaître's investigation has the advantage that it avoids a very tricky calculation of the mass of the condensations, and seems to offer less likelihood of error.

It is only at the very beginning that we have to look for a cause of expansion or contraction; once started, the expansion or contraction continues and increases automatically. If there were causes of contraction and causes of expansion, victory went to the one which got its shove in first. Thus the formation of condensations must have had the start of the conversion of mass into radiation, since the latter would (as we have seen) have brought about a contracting universe. To my mind this rather suggests that the primordial material consisted of hydrogen (or equivalently free protons and electrons) since there would then be less opportunity for the conversion of mass into radiation than if more complex atoms were present. So long as they are not combined in complex nuclei, protons and electrons are immune from annihilation. The reason for this security is that the photon or quantum of radiation, which results from the annihilation of a proton and electron, has to be provided with momentum, which must be balanced by a recoil momentum. But in hydrogen there is nothing left to recoil. Annihilation of a proton and electron (if it ever occurs) can happen only when they form part of a complex system which will leave a residuum to carry the recoil.[5]

VI

We have been led almost inevitably to the consideration of the beginning of the universe, or at least to the beginning of the present order of physical

5. I am indebted to Sir Alfred Ewing for calling my attention to this.

law. This always happens when we treat of an irreversible one-way process; and the continual expansion of the world raises the same kind of question of an ultimate beginning as has been raised by the continual increase of entropy in the world.

Views as to the beginning of things lie almost beyond scientific argument. We cannot give scientific reasons why the world should have been created one way rather than another. But I suppose that we all have an aesthetic feeling in the matter. The solar system must have started somehow, and I do not know why it should not have been started by projecting nine planets in orbits going in the same direction round the sun. But we have a feeling that that is not the way in which it would naturally be done; and we turn in preference to attempts—none too successful—to account for it by evolution from a nebula. Similarly the theory recently suggested by Einstein and de Sitter, that in the beginning all the matter created was projected with a radial motion so as to disperse even faster than the present rate of dispersal of the galaxies,[6] leaves me cold. One cannot deny the possibility, but it is difficult to see what mental satisfaction such a theory is supposed to afford.

Since I cannot avoid introducing this question of a beginning, it has seemed to me that the most satisfactory theory would be one which made the beginning *not too unaesthetically abrupt*. This condition can only be satisfied by an Einstein universe with all the major forces balanced. Accordingly the primordial state of things which I picture is an even distribution of protons and electrons, extremely diffuse and filling all (spherical) space, remaining nearly balanced for an exceedingly long time until its inherent instability prevails. We shall see later that the density of this distribution can be calculated; it was about one proton and electron per litre. There is no hurry for anything to begin to happen. But at last small irregular tendencies accumulate, and evolution gets under way. The first stage is the formation of condensations ultimately to become the galaxies; this, as we have seen, started off an expansion, which then automatically increased in speed until it is now manifested to us in the recession of the spiral nebulae.

As the matter drew closer together in the condensations, the various evolutionary processes followed—evolution of stars, evolution of the more complex elements, evolution of planets and life. Doubtless in this as in other theories there are serious difficulties of timing, so that one process should not go too fast compared with another. These difficulties of time-scale will be mentioned again later.

Perhaps it will be objected that, if one looks far enough back, this theory does not really dispense with an abrupt beginning; the whole universe must come into being at one instant in order that it may start in balance. I do not regard it in that way. To my mind *undifferentiated sameness* and *nothingness* cannot be distinguished philosophically. The realities of physics are unhomogeneities, happenings, change. Our initial assumption of a homogeneous static

---

6. They do not state this in words, but it is the meaning of their mathematical formulae.

medium is no more than a laying out in order of the conceptions to be used in our analytical description of the distinguishable objects and events whose history we are going to relate. So far as these realities are concerned, the theory achieves its aim of providing an imperceptible and gradual beginning. When at last, by the thermodynamic degradation of energy, the universe with the same gradualness again reaches undifferentiated sameness, that is the end of the physical universe. I do not picture a worn out world careering forlornly through the rest of eternity. What is left is only a few conceptions which we forgot to put away after we had finished using them.

To illustrate the instability of an Einstein universe I will liken it to a pin standing on its point, which may fall either to the left or to the right into two horizontal positions A or B. Position A corresponds to a universe expanded to infinity, and position B to a universe contracted to a point or as nearly to a point as quantum conditions allow. As the only way of avoiding an abrupt beginning, I have supposed the pin to be vertical initially. Its balance then is not quite so precarious as it seems; it would be at the mercy of the slightest disturbance from outside—but there is *nothing* outside. So the fall must come from a slight "decay" in the material of the pin. According to Lemaître and Sen the decay is such as to make it fall towards A, and we now observe it midway in the fall.

If we do not mind a sudden, or even violent, beginning, many other experiments with the pin are possible. We may drop it from an inclined position, or in letting go give it a projection upwards or downwards. Starting from the horizontal position B, we may project it so that it rises and falls again; or, if projected with greater force, it may pass through the vertical position and fall on the other side into position A. Similarly if it is projected from A. The behaviour of a universe is precisely the same; to every adventure of the pin there corresponds a similar adventure of a universe and *vice versa*. These adventures have been treated at length by some writers, and the appropriate formulae calculated. Whilst such a mathematical study is proper in its own sphere, it is liable to give a misleading impression of the complexity of the problem before us. When the different projections are enumerated and presented as though they were all different "theories" of the universe, it looks as though we had come across a bewildering maze of possibilities. But all it amounts to is that the universe is just like any other system that has a position of unstable equilibrium.

At first sight there is a curious difference between the universe and the pin. If the universe has a given mass we cannot project it just how we please; in fact the circumstances of projection determine its mass. But this is explicable when we recollect that energy and mass are equivalent. The total energy of the pin varies according to the way it is projected, and strictly speaking its mass changes in the same way. The mass of the universe behaves analogously. To suppose that velocity of expansion in the (fictitious) radial direction involves kinetic energy, may seem to be taking our picture of spherical space too literally; but the energy is so far real that it contributes to the mass of the universe. In particular a universe projected from B to reach A necessarily has greater mass than one which falls back without reaching the vertical (Einstein) position.

Lemaître does not share my idea of an evolution of the universe from the Einstein state. His theory of the beginning is a *fireworks theory*—to use his own description of it. The world began with a violent projection from position B, i.e. from the state in which it is condensed to a point or atom; the projection was strong enough to carry it past the Einstein state, so that it is now falling down towards A as observation requires. This makes the mass of the universe somewhat greater than in my theory (as explained in the last paragraph); but the change is scarcely important at the present stage of our progress. I cannot but think that my "placid theory" is more likely to satisfy the general sentiment of the reader; but if he inclines otherwise, I would say— "Have it your own way. And now let us get away from the Creation back to problems that we may possibly know something about."

The Einstein configuration was the one escape from an expanding or contracting universe; by proving it to be unstable, we show that it is no more than a temporary escape. Whether the original state was Einstein equilibrium or not, at the date when astronomers arrive on the scene they must be faced with an expanding or contracting universe. This result makes the theory of the expanding universe much more cogent. In 1917 theory was at the crossroads; that is no longer true, and by its own resources it has been guided into the road to a non-static universe. Realising that some degree of expansion (or contraction) is inevitable, we are much more inclined to admit the recession of the spiral nebulae as an indication of its magnitude.[7]

VII

Several counter theories of the observed recession of the nebulae have been proposed and I would like to make clear my general attitude to such theories.

I am a detective in search of a criminal—the cosmical constant. I know he exists, but I do not know his appearance; for instance I do not know if he is a little man or a tall man. Naturally the first move of my chief (de Sitter) was to order a search for footprints on the scene of the crime. The search has revealed footprints, or what look like footprints—the recession of the spiral nebulae. Of course, I am tremendously interested in this possible clue to the criminal. From the length of the stride I calculate the presumed height of the criminal (in approved detective fashion). Having gained this important information as to his appearance, I can now turn to my other clues—in relativity and wave-mechanics—and checking one against the other I think I have now about enough evidence to justify an arrest.

7. I may mention that the proof of the instability of the Einstein configuration was the turning point in my own outlook. Previously the expanding universe (as it appeared in de Sitter's theory) had appealed to me as a highly interesting possibility, but I had no particular preference for it.

It happens that there are other persons interested in the footprints, who are not in the least interested in my criminal. For instance there is a geologist who suggests the theory that they belong to a prehistoric creature. (The counter theories proposed by Einstein and de Sitter and by Milne suppose that the large velocities of the nebulae have existed from the beginning.) Another man thinks they are not footprints at all, but depressions caused by something of unknown nature. To what extent is it incumbent on me to justify myself by criticising these contrary opinions? I do not think they concern me at all closely. Naturally from the beginning I was awake to the possibility that the footprints might not belong to the criminal; the question then to be decided was not whether the clue was sufficient evidence to hang the criminal, but whether it indicated a direction of inquiry which it would be worth while devoting one's energies to following up. Of course, if either the geologist or the depressionist claimed to be able to demonstrate that his idea of the origin of the footprints was correct, I should pay grave attention; for such a demonstration would show that I was altogether on the wrong tack in my own inquiries. But that is not the position; no one claims more for the counter suggestions than that "for all we know, it might be so". That leaves the investigation as open as when we started: footprints have been discovered on the scene of the crime; all sorts of explanations are possible, and it may turn out that they are of little importance; but there is quite a good chance that they were made by the criminal; let us follow up the clue, and try to find out. I am fairly satisfied now that they do belong to the criminal, but that is because by pursuing the clue the further evidence detailed in Chapter iv has come to light.

I have already commented on the theory that the recession of the spiral nebulae is a misinterpretation of the red shift of their light. We may class together the remaining theories which accept the recession of the spiral nebulae as genuine; these accordingly admit the expansion of the universe (perhaps only as a temporary phenomenon) but do not connect it with cosmical repulsion.

The keynote of many of these suggestions seems to be, What is the most general deduction that can be made from our observational knowledge of the positions and motions of the galaxies? I think that those who seek this extreme generality are following a will-o'-the-wisp. The observational data give only the positions and velocities at the present instant; so that it is clear from the start that nothing can definitely be deduced as to the law of force governing the motion. Any *instantaneous* distribution of velocities is compatible with any law of force. If then anyone proposes to treat the problem of the system of galaxies with wider generality than we here attempt—as he would perhaps say, without any preconceptions—we have to ask, What problem? The motions in themselves do not constitute a problem. We have to combine them with other ideas, which we think justified, in order to create a problem at all. It is the preconceptions—imported from other branches of science—that can fertilise an investigation otherwise doomed to barrenness.

Thus I find a difficulty in discussing the proposal of Einstein and de Sitter, and some of de Sitter's separate proposals, because I do not see what are "the rules of the game". These proposals are left as mathematical formulations, all doubtless compatible with what we observe; but there seems nothing to prevent such formulations being indefinitely multiplied. De Sitter has several times emphasised the possibility that the cosmical constant λ might be negative. This gives cosmical attraction instead of cosmical repulsion. Clearly the recession of the nebulae is not evidence in favour of cosmical attraction. The most that can be said is that it is not necessarily fatal evidence against it.

It should not be forgotten that an observational test which is quite inadequate to demonstrate a theory may yet afford welcome confirmation of it. Suppose that by theoretical reasoning we have concluded that the earth is surrounded by a field of force attracting bodies towards it. To test this we are allowed one brief glimpse of what is happening near the earth's surface. Our glimpse may reveal a display of rockets soaring upwards. This is not incompatible with our theory, but it is clearly no confirmation of it. On the other hand we may see a shower of raindrops falling. Nothing can strictly be deduced from this one glimpse; but to observe objects falling to the ground is a tolerable confirmation of the theoretical prediction that there is a force tending to make objects move that way.

E. A. Milne[8] has pointed out that if initially the galaxies, endowed with their present speeds, were concentrated in a small volume, those with highest speed would by now have travelled farthest. If gravitational and other forces are negligible, we obtain in this way a distribution in which speed and distance from the centre are proportional. Whilst accounting for the dependence of speed on distance, this hypothesis creates a new difficulty as to the occurrence of the speeds. To provide a moderately even distribution of nebulae up to 150 million light-years' distance, high speeds must be very much more frequent than low speeds; this peculiar anti-Maxwellian distribution of speeds becomes especially surprising when it is supposed to have occurred originally in a compact aggregation of galaxies.

I might discuss these suggestions more fully if they were likely to be the last. But it would seem that, unless we keep to a defined purpose, an unlimited field of speculation is open; and by the time these remarks are read, some other hypothesis may be in vogue. I define my own purpose as being to find what light (if any) the recession of the spiral nebulae can throw on the problem of the cosmical constant. Having regard to this purpose, it seems sufficient to note that this is not the only direction in which we might look for the explanation of the phenomenon of the nebulae, and then proceed with our task.

---

8. *Nature*, July 2, 1932.

GEORGES LEMAITRE:

The Primeval Atom

Introduction

THE PRIMEVAL ATOM hypothesis is a cosmogonic hypothesis which pictures the present universe as the result of the radioactive disintegration of an atom.

I was led to formulate this hypothesis, some fifteen years ago, from thermodynamic considerations while trying to interpret the law of degradation of energy in the frame of quantum theory. Since then, the discovery of the universality of radioactivity shown by artificially provoked disintegrations, as well as the establishment of the corpuscular nature of cosmic rays, manifested by the force which the Earth's magnetic field exercises on these rays, made more plausible an hypothesis which assigned a radioactive origin to these rays, as well as to all existing matter.

Therefore, I think that the moment has come to present the theory in deductive form. I shall first show how easily it avoids several major objections which would tend to disqualify it from the start. Then I shall strive to deduce its results far enough to account, not only for cosmic rays, but also for the present structure of the universe, formed of stars and gaseous clouds, organized into spiral or elliptical nebulae, sometimes grouped in large clusters of several thousand nebulae which, more often, are composed of isolated nebulae, receding from one another according to the mechanism known by the name of the expanding universe.

For the exposition of my subject, it is indispensable that I recall several elementary geometric conceptions, such as that of the closed space of Riemann, which led to that of space with a variable radius, as well as certain aspects of the theory of relativity, particularly the introduction of the cosmological constant and of the cosmic repulsion which is the result of it.

From Georges Lemaître, *The Primeval Atom: An Essay on Cosmogony*, translated by Betty H. and Serge A. Korff, D. Van Nostrand Co., New York, 1950, Chapter V. Reprinted by kind permission of the publishers, D. Van Nostrand Co., Inc.

Closed Space

All partial space is open space. It is comprised in the interior of a surface, its boundary, beyond which there is an exterior region. Our habit of thought about such open regions impels us to think that this is necessarily so, however large the regions being considered may be. It is to Riemann that we are indebted for having demonstrated that total space can be closed. To explain this concept of closed space, the most simple method is to make a small-scale model of it in an open space. Let us imagine, in such a space, a sphere in the interior of which we are going to represent the whole of closed space. On the rim surface of the sphere, each point of closed space will be supposed to be represented twice, by two points, A and A', which, for example, will be two antipodal points, that is, two extremities of the same diameter. If we join these two points A and A' by a line located in the interior of the sphere, this line must be considered as a closed line, since the two extremities A and A' are two distinct representations of the same, single point. The situation is altogether analogous to that which occurs with the Mercator projection, where the points on the 180th meridian are represented twice, at the eastern and western edges of the map. One can thus circulate indefinitely in this space without ever having to leave it.

It is important to notice that the points represented by the outer surface of the sphere, in the interior of which we have represented all space, are not distinguished by any properties of the other points of space, any more than is the 180th meridian for the geographic map. In order to account for that, let us imagine that we displaced the sphere in such a manner that point A is superposed on B, and the antipodal point A' on B'. We shall then suppose that the entire segment AB and the entire segment A'B' are two representations of a similar segment in closed space. Thus we shall have a portion of space which has already been represented in the interior of the initial sphere which is now represented a second time at the exterior of this sphere. Let us disregard the interior representation as useless; a complete representation of the space in the interior of the new sphere will remain. In this representation, the closed contours will be soldered into a point which is twice represented, namely, by the points B and B', mentioned above, instead of being welded, as they were formerly, to point A and A'. Therefore, these latter are not distinguished by an essential property.

Let us notice that when we modify the exterior sphere, it can happen that a closed contour which intersects the first sphere no longer intersects the second, or, more generally, that a contour no longer intersects the finite sphere at the same number of points. Nevertheless, it is evident that the number of points of intersection can only vary by an even number. Therefore, there are two kinds of closed contours which cannot be continuously distorted within one another. Those of the first kind can be reduced to a point. They do not intersect the outer sphere or they intersect it at an even number of points. The others cannot be reduced to one point, we call them the *odd contours* since they intersect the sphere at an odd number of points.

If, in a closed space, we leave a surface which we can suppose to be horizontal, in going toward the top we can, by going along an odd contour, return to our point of departure from the opposite direction without having deviated to the right or left, backward or forward, without having traversed the horizontal plane passing through the point of departure.

Elliptical Space

That is the essential of the topology of closed space. It is possible to complete these topological ideas by introducing, as is done in a geographical map, scales which vary from one point to another and from one direction to another. That can be done in such a manner that all the points of space and all the directions in it may be perfectly equivalent. Thus, Riemann's homogeneous space, or elliptical space, is obtained. The straight line is an odd contour of minimum length. Any two points divide it into two segments, the sum of which has a length which is the same for all straight lines and which is called the tour of space.

All elliptical spaces are similar to one another. They can be described by comparison with one among them. The one in which the tour of the straight line is equal to $\pi = 3.1416$ is chosen as the standard elliptical space. In every elliptical space, the distances between two points are equal to the corresponding distances in standard space, multiplied by the number R which is called the radius of elliptical space under consideration. The distances in standard space, called space of unit radius, are termed angular distances. Therefore, the true distances, or linear distances, are the product of the radius of space times the angular distances.

Space of Variable Radius

When the radius of space varies with time, space of variable radius is obtained. One can imagine that material points are distributed evenly in it, and that spatio-temporal observations are made on these points. The angular distance of the various observers remains invariant, therefore the linear distances vary proportionally to the radius of space. All the points in space are perfectly equivalent. A displacement can bring any point into the center of the representation. The measurements made by the observers are thus also equivalent, each one of them makes the same map of the universe.

If the radius increases with time, each observer see all points which surround him receding from him, and that occurs at velocities which become greater as they recede further. It is this which has been observed for the extra-galactic nebulae that surround us. The constant ratio between distance and velocity has been determined by Hubble and Humason. It is equal to $T_H = 2 \times 10^9$ years.

If one makes a graph, plotting as abscissa the values of time and as ordinate the value of radius, one obtains a curve, the sub-tangent of which at the point representing the present instant is precisely equal to T_H.

The Primeval Atom

These are the geometric concepts that are indispensable to us. We are now going to imagine that the entire universe existed in the form of an atomic nucleus which filled elliptical space of convenient radius in a uniform manner.

Anticipating that which is to follow, we shall admit that, when the universe had a density of 10^{-27} gram per cubic centimeter, the radius of space was about a billion light-years, that is, 10^{27} centimeters. Thus the mass of the universe is 10^{54} grams. If the universe formerly had a density equal to that of water, its radius was then reduced to 10^{18} centimeters, say, one light-year. In it, each proton occupied a sphere of one angstrom, say, 10^{-8} centimeter. In an atomic nucleus, the protons are contiguous and their radius is 10^{-13}, thus about 100,000 times smaller. Therefore, the radius of the corresponding universe is 10^{13} centimeters, that is to say, an astronomical unit.

Naturally, too much importance must not be attached to this description of the primeval atom, a description which will have to be modified, perhaps, when our knowledge of atomic nuclei is more perfect.

Cosmogonic theories propose to seek out initial conditions which are ideally simple, from which the present world, in all its complexity, might have resulted, through the natural interplay of known forces. It seems difficult to conceive of conditions which are simpler than those which obtained when all matter was unified in an atomic nucleus. The future of atomic theories will perhaps tell us, some day, how far the atomic nucleus must be considered as a system in which associated particles still retain some individuality of their own. The fact that particles can issue from a nucleus, during radioactive transformations, certainly does not prove that these particles pre-existed as such. Photons issue from an atom of which they were not constituent parts, electrons appear there, where they were not previously, and the theoreticians deny them an individual existence in the nucleus. Still more protons or alpha particles exist there, without doubt. When they issue forth, their existence becomes more independent, nevertheless, and their degrees of freedom more numerous. Also, their existence, in the course of radioactive transformations, is a typical example of the degradation of energy, with an increase in the number of independent quanta or increase in entropy.

That entropy increases with the number of quanta is evident in the case of electromagnetic radiation in thermodynamic equilibrium. In fact, in black body radiation, the entropy and the total number of photons are both proportional to the third power of the temperature. Therefore, when one mixes radiations of different temperatures and one allows a new statistical equilibrium to be established, the total number of photons has increased. The degradation of energy is manifested as a pulverization of energy. The total quantity of energy is maintained, but it is distributed in an ever larger number of quanta, it becomes broken into fragments which are ever more numerous.

If, therefore, by means of thought, one wishes to attempt to retrace the course of time, one must search in the past for energy concentrated in a lesser number of quanta. The initial condition must be a state of maximum concentration. It was in trying to formulate this condition that the idea of the primeval atom was germinated. Who knows if the evolution of theories of the nucleus will not, some day, permit the consideration of the primeval atom as a single quantum?

Formation of Clouds

We picture the primeval atom as filling space which has a very small radius (astronomically speaking). Therefore, there is no place for superficial electrons, the primeval atom being nearly an *isotope of a neutron*. This atom is conceived as having existed for an instant only, in fact, it was unstable and, as soon as it came into being, it was broken into pieces which were again broken, in their turn; among these pieces electrons, protons, alpha particles, etc., rushed out. An increase in volume resulted, the disintegration of the atom was thus accompanied by a rapid increase in the radius of space which the fragments of the primeval atom filled, always uniformly. When these pieces became too small, they ceased to break up; certain ones, like uranium, are slowly disintegrating now, with an average life of four billion years, leaving us a meager sample of the universal disintegration of the past.

In this first phase of the expansion of space, starting asymptotically with a radius practically zero, we have particles of enormous velocities (as a result of recoil at the time of the emission of rays) which are immersed in radiation, the total energy of which is, without doubt, a notable fraction of the mass energy of the atoms.

The effect of the rapid expansion of space is the attenuation of this radiation and also the diminution of the relative velocities of the atoms. This latter point requires some explanation. Let us imagine that an atom has, along the radius of the sphere in which we are representing closed space, a radial velocity which is greater than the velocity normal to the region in which it is found. Then this atom will depart faster from the center than the ideal material particle which has normal velocity. Thus the atom will reach, progressively, regions where its velocity is less abnormal, and its proper velocity, that is, its excess over normal velocity, will diminish. Calculation shows that proper velocity varies in this way in inverse ratio to the radius of space. We must therefore look for a notable attenuation of the relative velocities of atoms in the first period of expansion. From time to time, at least, it will happen that, as a result of favorable chances, the collisions between atoms will become sufficiently moderate so as not to give rise to atomic transformations or emissions of radiation, but that these collisions will be elastic collisions, controlled by superficial electrons, so considered in the theory of gases. Thus we shall obtain, at least locally, a beginning of statistical equilibrium, that is, the formation of gaseous clouds. These gaseous clouds will still have considerable velocities, in relation to one another, and they will be mixed with radiations that are themselves attenuated by expansion.

It is these radiations which will endure until our time in the form of cosmic rays, while the gaseous clouds will have given place to stars and to nebulae by a process which remains to be explained.

Cosmic Repulsion

For that explanation, we must say a few words about the theory of relativity. When Einstein established his theory of gravitation, or generalized theory, he admitted, under the name of the principle of equivalence, that the ideas of special relativity were approximately valid in a sufficiently small domain. In the special theory, the differential element of space-time measurements had for its square a quadratic form with four coordinates, the coefficients of which had special constant values. In the generalization, this element will still be the square root of a quadratic form, but the coefficients, designated collectively by the name of *metric tensors,* will vary from place to place. The geometry of space-time is then the general geometry of Riemann at three plus one dimensions. The spaces with variable radii are a particular case in this general geometry, since the theory of spatial homogeneity or of the equivalence of observers is introduced here.

It can be that this geometry differs only apparently from that of special relativity. This is what happens when the quadratic form can be transformed, by a simple change of coordinates, into a form having constant coefficients. Then one says with Riemann that the corresponding variety (that is, space-time) is flat or Euclidian. For that, it is necessary that certain expressions, expressed by components of a tensor with four indices called Riemann's tensor, vanish completely at all points. When it is not so, the tensor of Riemann expresses the departure from flatness. Riemann's tensor is calculated by the average of second derivatives of the metric tensor. Starting with Riemann's tensor with four indices, it is easy to obtain a tensor which has only two indices like the metric tensor; it is called the contracted Riemannian tensor. One can also obtain a scalar, the totally contracted Riemannian tensor.

In special relativity, a free point describes a straight line with uniform motion, that is the principle of inertia. One can also say that, in an equivalent manner, it describes a geodesic of space-time. In the generalization, it is again presumed that a free point describes a geodesic. These geodesics are no longer representable by a uniform, rectilinear motion, they now represent a motion of a point under the action of the forces of gravitation. Since the field of gravitation is caused by the presence of matter, it is necessary that there be a relation between the density of the distribution of matter and Riemann's tensor which expresses the departure from flatness. The density is, in itself, considered as the principal component of a tensor with two indices called the *material tensor;* thus one obtains as a possible expression of the material tensor $T\mu\nu$ as a function of the metric tensor $g\mu\nu$ and of the two tensors of Riemann, contracted to $R\mu\nu$ and totally contracted to R,

$$T_{\mu\nu} = aR_{\mu\nu} + bRg_{\mu\nu} + cg_{\mu\nu}$$

where *a, b,* and *c* are three constants.

But this is not all; certain identities must exist between the components of the material tensor and its derivatives. These identities can be interpreted, for a convenient choice of coordinates, a choice which corresponds, moreover, to the practical conditions of observations, as expressing the principles of conservation, that of energy and that of momentum. In order that such identities may be satisfied, it is no longer possible to choose arbitrarily the values of the three constants. *b* must be taken as equal to—*a/*2. Theory cannot predict either their magnitude or their sign. It is only observation which can determine them.

The constant *a* is linked to the constant of gravitation. In fact, when theory is applied to conditions which are met in the applications (in particular, the fact that astronomical velocities are small in comparison to the speed of light) and when one profits from these conditions by introducing coordinates which facilitate comparison with experiment, one finds that the geodesics differ from rectilinear motion by an acceleration which can be interpreted as an attraction in inverse ratio to the square of the distances, and which is exercised by the masses represented by the material tensor. This is simply the principal effect foreseen by the theory; this theory predicts small departures which, in favorable cases, have been confirmed by observation.

A good agreement with planetary observations is obtained by leaving out the term in *c.* That does not prove that this term may not have experimental consequence. In fact, in the conditions which were employed to obtain Newton's law as an approximation of the theory, the term in *c* would furnish a force varying, not in the inverse square ratio of the distance, but proportionally to this distance. This force could therefore have a marked action at very great distances although, for the distances of the planets, its action would be negligible. Also, the relation *c/a,* designated customarily by the letter *lambda,* is called the cosmological constant. When λ is positive, the additional force proportional to the distance is called *cosmic repulsion.*

The theory of relativity has thus unified the theory of Newton. In Newton's theory, there were two principles posed independently of one another: universal attraction and the conservation of mass. In the theory of relativity, these principles take a slightly modified form, while being practically identical to those of

Newton in the case where these have been confronted with the facts. But universal attraction is now a result of the conservation of mass. The size of the force, the constant of gravitation, is determined experimentally.

The theory again indicates that the constancy of mass has, as a result, besides the Newtonian force of gravitation, a repulsion proportional to the distance of which the size and even the sign can only be determined by observation and by observation requiring great distances.

Cosmic repulsion is not a special hypothesis, introduced to avoid the difficulties which are presented in the study of the universe. If Einstein has re-introduced it in his work on cosmology, it is because he remembered having arbitrarily dropped it when he had established the equations of gravitation. To suppress it amounts to determining it arbitrarily by giving it a particular value: zero.

The Universe of Friedmann

The theory of relativity allows us to complete our description of space with a variable radius by introducing here some dynamic considerations. As before, we shall represent it as being in the interior of a sphere, the center of which is a point which we can choose arbitrarily. This sphere is not the boundary of the system, it is the edge of the map or of the diagram which we have made of it. It is the place at which the two opposite, half-straight lines are soldered into a closed straight line. Cosmic repulsion is manifested as a force proportional to the distance to the center of the diagram. As for the gravitational attraction, it is known that, in the case of distribution involving spherical symmetry around a point, and that is certainly the case here, the regions farther away from the center than the point being considered have no influence upon its motion; as for the interior points, they act as though they were concentrated at the center. By virtue of the homogeneity of the distribution of matter, the density is constant, the force of attraction which results is thus proportional to the distance, just as is cosmic repulsion.

Therefore, a certain density exists, which we shall call the density of equilibrium or the *cosmic density,* for which the two forces will be in equilibrium.

These elementary considerations permit recognition, in a certain measure, of the result which calculation gives and which is contained in Friedmann's equation:

$$\left(\frac{dR}{dt}\right)^2 = -1 + \frac{2M}{R} + \frac{R^2}{T^2}.$$

The last term represents cosmic repulsion (it is double the function of the forces of this repulsion). T is a constant depending on the value of the cosmological constant and being able to replace this. The next-to-last term is double the potential of attraction due to the interior mass. The radius of space R is the distance from the origin of a point of angular distance $\sigma = 1$. If one multiplied the equation by σ^2, one would have the corresponding equation for a point at any distance.

That which is remarkable in Friedmann's equation is the first term -1. The elementary considerations which we have just advanced would allow us to assign it a value which is more or less constant; it is the constant of energy in the motion which takes place under the action of two forces. The complete theory determines this constant and thus links the geometric properties to the dynamic properties.

Einstein's Equilibrium

Since, by virtue of equations, the radius R remains constant, the state of the universe in equilibrium, or Einstein's universe, is reached. The conditions of the universe in equilibrium are easily deduced from Friedmann's equation:

$$R_E = \frac{T}{\sqrt{3}}; \rho_E = \frac{3}{4\pi}\frac{1}{T^2}; M = \frac{T}{\sqrt{3}}.$$

In these formulas, the distances are calculated in light-time, which amounts to taking the velocity of light c as equal to unity, but, in addition, the unit of mass is chosen in such a way that the constant of gravitation may also be equal to unity. It is easy to pass on to the numerical values in C.G.S. by re-establishing in the formulas the constants c and G in such a manner as to satisfy the equations of dimension. In particular, if one takes T as being equal to 2×10^9 years, as we shall suppose in a moment, one finds that the density ρ_E is equal to 10^{-27} gram per cubic centimeter.

These considerations can be extended to a region in which distribution is no longer homogeneous and where even the spherical symmetry is no longer verified, provided that the region under consideration be of small dimension. In fact, it is known that, in a small region, Newtonian mechanics is always a good approximation. Naturally, it is necessary, in applying Newtonian mechanics, to take account of cosmic repulsion but, aside from this easy modification, it is perfectly legitimate to utilize the intuition acquired by the practice of classic mechanics and its application to systems which are more or less complicated. Among other things, it can be noted that the equilibrium of which we have just spoken is unstable and that the equilibrium can even be disturbed in one sense, in one place, and in the opposite sense in another region.

Perhaps it is necessary to mention here that Friedmann's equation is only rigorously exact if the mass M remains constant. While one takes account of the radiation which circulates in space and also of the characteristic velocities of the particles which cross one another in the manner of molecules in a gas and, as in a gas, give rise to pressure, it is necessary to consider the work of this pressure during the expansion of space, in the evaluation of the mass or the energy. But it is apparent that such an effect is generally negligible, as detailed researches elsewhere have shown.

The Significance of Clusters of Nebulae

We are now in a position to take up again the description which we had begun of the expansion of space, following the disintegration of the primeval atom. We had shown how, in a first period of rapid expansion, gaseous clouds must have been formed, animated by great, characteristic velocities. We are now going to suppose that the mass M is slightly larger than $\dfrac{T}{\sqrt{3}}$.

The second member of Friedmann's equation will thus be able to become smaller, but it will not be able to vanish. Thus, we may distinguish three phases in the expansion of space. The first rapid expansion will be followed by a period of deceleration, during the course of which attraction and repulsion will virtually bring themselves into equilibrium. Finally, repulsion will definitely prevail over attraction, and the universe will enter into the third phase, that of the resumption of expansion under the dominant action of cosmic repulsion.

Let us consider the phase of slow expansion in more detail. The gaseous clouds are undoubtedly not distributed in a perfectly uniform manner. Let us consider in a region sufficiently small, and that only from the point of view of classic mechanics, the conflict between the forces of repulsion and attraction which almost produces equilibrium. We easily see that as a result of local fluctuations of density, there will be regions where attraction will finally prevail over repulsion, in spite of the fact that we have supposed that, for the universe in its entirety, it is the contrary which takes place. These regions in which attraction has prevailed will thus fall back upon themselves, while the universe will be entering upon a period of renewed expansion. We shall obtain a universe formed of regions of condensations which are separated from one another. Will not these regions of condensations be elliptical or spiral nebulae? We shall come back to this question in a moment.

Let us note that, although it is of rare occurrence, it will be possible for large regions where the density or the speed of expansion differ slightly from the average to hesitate between expansion and contraction, and remain in equilibrium, while the universe has resumed expansion. Could these regions not be identified with the clusters of nebulae, which are made up of several hundred nebulae located at relative distances from one another, which are a dozen times smaller than those of isolated nebulae? According to this interpretation, these clusters are made up of nebulae which are retarded in the phase of equilibrium; they represent a sample of the distribution of matter, as it existed everywhere, when the radius of space was a dozen times smaller than it is at present, when the universe was passing through equilibrium.

The Findings of De Sitter

This interpretation gives the explanation for a remarkable coincidence upon which De Sitter insisted strongly, in the past. Calculating the radius of the universe in the hypothesis which bears his name, that is, ignoring the presence of matter and introducing into the formulas the value T_H given by the observation of the expansion, he obtained a result which scarcely differs from that which is obtained, in Einstein's totally different hypothesis of the universe, by introducing into the formulas the observed value of the density of matter. The explanation of this coincidence is, according to our interpretation of the clusters of nebulae, that, for a value of the radius which is a dozen times the radius of equilibrium, the last term in Friedmann's formula greatly prevails over the others. The constant T which figures in it is therefore practically equal to the observed value T_H: but since, in addition, the clusters are a fragment of Einstein's universe, it is legitimate to use the relationship existing between the density and the constant T for them. For $T = T_H$ one finds, as we have seen, that the density in the clusters must be 10^{-27} gram per cubic centimeter, which is the value given by observation. This observation is based on counts of nebulae and on the estimate of their mass indicated by their spectroscopic velocity of rotation.

In addition to this argument of a quantitative variety, the proposed interpretation also takes account of important facts of a qualitative order. It explains why the clusters do not show any marked central condensations and have vague forms, with irregular extensions, all things which it would be difficult to explain if they formed dynamic structures controlled by dominant forces, as is manifestly the case for the starclusters or the elliptical and spiral nebulae. It also takes into account a manifest fact which is the existence of large fluctuations of density in the distribution of the nebulae, even outside the clusters. This must be so, in fact, if the universe has just passed through a state of unstable equilibrium, a whole gamut of transition between the properly-termed clusters which are still in equilibrium, while passing through regions where the expansion, without being arrested, has nevertheless been retarded, in such a manner that these regions have a density which is greater than the average.

This interpretation permits the value of the radius at the moment of equilibrium to be determined at a billion light-years, and thus 10^{10} light-years for the present value of the radius. Since American telescopes prospect the universe as far as half a billion light-years, one sees that this observed region already constitutes a sample of a size which is not at all negligible compared to entire space; hence, it is legitimate to hope that the values of the coefficient of expansion T_H and of the density, obtained for this restricted domain, are representative of the whole.

The only indeterminate which exists is that which is relative to the degree of approximation with which the situation of equilibrium has been approached. It is on this value which the estimate of the duration of expansion depends. Perhaps it will be possible to estimate this value by means of statistical considerations regarding the relative frequency of the clusters, compared to the isolated nebulae.

The Proper Motion of Nebulae

Now we must come back to the question of the formation of nebulae from the regions of condensation. We have seen that the characteristic velocities, or the relative velocities of gaseous clouds, which cross one another in the same place, must have been very large. Since certain of them, because of a density which is a little too large, form a nucleus of condensation, they will be able to retain the clouds which have about the same velocity as this nucleus. The proper velocity of the cloud so formed will hence be determined by the velocity of the nucleus of condensation. The nebulae formed by such a mechanism must have large relative velocities. In fact, that is what is observed in the clusters of nebulae. In the one which has been best studied, that of *Virgo,* the dispersion of the velocities about the mean velocity is 650 kilometers per second. The proper velocity must have been the proper velocity of all the nebulae at the moment of passage through equilibrium. For isolated nebulae, this velocity has been reduced to about one-twelfth, as a result of expansion, by the same mechanism which we have explained with reference to the formation of gaseous clouds.

The Formation of Stars

The density of the clouds is, on the average, the density of equilibrium 10^{-27}. For this density of distribution, a mass such as the Sun would occupy a sphere of one hundred light-years in radius. These clouds have no tendency to contract. In order that a contraction due to gravitation can be initiated, their density must be notably increased. This is what can occur if two clouds happen to collide with great velocities. Then the collision will be an inelastic collision, giving rise to ionization and emission of radiation. The two clouds will flatten one another out, while remaining in contact, the density will be easily doubled and condensation will be definitely initiated. It is clear that a solar system or a simple or multiple star may arise from such a condensation, through known mechanisms. That which characterizes the mechanism to which we are led is the greatness of the dimensions of the gaseous clouds, the condensation of which will form a star. This circumstance takes account of the magnitude of the angular momentum, which is conserved during the condensation and whose value could only be nil or negligible if the initial circumstances were adjusted in a wholly improbable manner. The least initial rotation must give rise to an energetic rotation in a concentrated system, a rotation incompatible with the presence of a single body but assuming either multiple stars turning around one another or, simply, one star with one or several large planets turning in the same direction.

The Distribution of Densities in Nebulae

Here is the manner in which we can picture for ourselves the evolution of the regions of condensation. The clouds begin by falling toward the center, and by describing a motion of oscillation following a diameter from one part and another of the center. In the course of these oscillations, they will encounter one another with velocities of several hundreds of kilometers per second and will give rise to stars. At the same time, the loss of energy due to these inelastic collisions will modify the distribution of the clouds and stars already formed in such a manner that the system will be further condensed. It seems likely that this phenomenon could be submitted to mathematical analysis. Certain hypotheses will naturally have to be introduced, in such a way as to simplify the model, so as to render the calculation possible and also so as artificially to eliminate secondary phenomena. There is scarcely any doubt that there is a way of thus obtaining the law of final distribution of the stars formed by the mechanism described above. Since the distribution of brilliance is known for the elliptical nebulae and from that one can deduce the densities in these nebulae, one sees that such a calculation is susceptible of leading to a decisive verification of the theory.

One of the complications to which I alluded, a moment ago, is the eventual presence of a considerable angular momentum. In excluding it, we have restricted the theory to condensations respecting spherical symmetry, that is, nebulae which are spherical or slightly elliptical. It is easy to see what modification will bring about the presence of considerable angular momentum. It is evident that one will obtain, in addition to a central region analogous to the elliptical nebulae, a flat system analogous to the ring of Saturn or the planetary systems, in other words, something resembling the spiral nebulae. In this theory, the spiral or elliptical character of the nebula is a matter of chance; it depends on the fortuitous value of the angular momentum in the region of condensation. It can no longer be a question of the evolution of one type into another. Moreover, the same thing obtains for stars where the type of the star is determined by the accidental value of its mass, that is, of the sum of the masses of the clouds whose encounter produced the star.

Distribution of Supergiant Stars

If the spirals have this origin, it must follow that the stars are formed by an encounter of clouds in two very distinct processes. In the first place, and especially in the central region, the clouds encounter one another in their radial movement, and this is the phenomenon which we have invoked for the elliptical nebulae. Kapteyn's preferential motion may be an indication of it. But besides this relatively rapid process, there must be a slower process of star formation, beginning with the clouds which escaped from the central region as a result of their angular momentum. These will encounter one another in a to-and-fro motion, from one side to another of the plane of the spiral. The existence of these two processes, with different ages, is perhaps the explanation of the fact that supergiant stars are not found in the elliptical nebulae or in the nucleus of spirals, but that one observes them only in the exterior region of the spirals. In fact, it is known that the stars radiate energy which comes from the transformation of their hydrogen into helium. The supergiant stars radiate so much energy that they could only maintain this output during a hundred million years. It should be understood, thus, that, for the oldest stars, the supergiants may be extinct for lack of fuel, whereas they still shine where they have been recently formed.

The Uniform Abundance of the Elements

But it is doubtless not worthwhile to allow ourselves to be prematurely led to the attempted pursuit of the theory in such detail, but rather to restrict ourselves, for the moment, to the more general consequences of the hypothesis of the primeval atom. We have seen that the theory takes account of the formations of stars in the nebulae. It also explains a very remarkable circumstance which could be demonstrated by the analysis of stellar spectra. It concerns the quantitative composition of matter, or the relative abundance of the various chemical elements, which is the same in the Sun, in the stars, on the Earth and in the meteorites. This fact is a necessary consequence of the hypothesis of the primeval atom. Products of the disintegration of an atom are naturally found in very definite proportions, determined by the laws of radioactive transformations.

Cosmic Rays

Finally, we said in the beginning that the radiations produced during the disintegrations, during the first period of expansion, could explain cosmic rays.

These rays are endowed with an energy of several billion electron-volts. We know no other phenomenon currently taking place which may be capable of such effects. That which these rays resemble most is the radiation produced during present radioactive disintegrations, but the individual energies brought into play are enormously greater. All that agrees with rays of superradioactive origin. But it is not only by their quality that these rays are remarkable, it is also by their total quantity. In fact, it is easy, from their observed density which is given in ergs per centimeter, to deduce their density of energy by dividing by c, then their density in grams per cubic centimeter by dividing by c^2. Thus one finds 10^{-34} grams per cubic centimeter, about one ten-thousandth the present density of the matter existing in the form of stars. It seems impossible to explain such an energy which represents one part in ten thousand of all existing energy, if these rays had not been produced by a process which brought into play all existing matter. In fact, this energy, at the moment of its formation, must have been at least ten times greater, since a part of it was able to be absorbed and the remainder has been reduced as a result of the expansion of space. The total intensity observed for cosmic rays is therefore just about that which might be expected.

Conclusion

The purpose of any cosmogonic theory is to seek out ideally simple conditions which could have initiated the world and from which, by the play of recognized physical forces, that world, in all its complexity, may have resulted.

I believe that I have shown that the hypothesis of the primeval atom satisfies the rules of the game. It does not appeal to any force which is not already known. It accounts for the actual world in all its complexity. By a single hypothesis it explains stars arranged in galaxies within an expanding universe as well as those local exceptions, the clusters of nebulae. Finally, it accounts for that mighty phenomenon, the ultrapenetrating rays. They are truly cosmic, they testify to the primeval activity of the cosmos. In their course through wonderfully empty space, during billions of years, they have brought us evidence of the superradioactive age, indeed they are a sort of fossil rays which tell us what happened when the stars first appeared.

I shall certainly not pretend that this hypothesis of the primeval atom is yet proved, and I would be very happy if it has not appeared to you to be either absurd or unlikely. When the consequences which result from it, especially that which concerns the law of the distribution of densities in the nebulae, are available in sufficient detail, it will doubtless be possible to declare oneself definitely for or against.

E. A. MILNE:

The Fundamental Concepts
of Natural Philosophy

FROM THE TIME of Galileo, experiment has been the core of Natural Science. Before him, of course, observation alone had in the development of astronomy played a fundamental part. Besides the great workers of the ancient civilisations, who knew the path of the sun amongst the fixed stars and could predict eclipses, and besides the fruits of Greek astronomy associated with the names of Hipparchus and Ptolemy, the more modern observational work of Tycho Brahe, analysed by Kepler, had vindicated the self-consistency of the Copernican theory of the solar system and had led to its remarkable refinement in the form of Kepler's three quantitative laws—the law of the ellipse, the law of areas, and the law connecting periodic times and major axes. This was a triumphant example of the execution of the programme then being put forward by Francis Bacon for discovering all natural laws—the method of induction from a number of instances. But it was reserved for Galileo to make a start with the process of ascertaining as far as might be, by controlled experiment, the particular nature of *motion*. The *metaphysical* questions associated with motion had not escaped the attention of the Greeks; but Zeno was apparently content with stating paradoxes, and did not resolve them. Galileo, first, experimented with moving bodies; and established that in falling they received equal increments of velocity in equal times—a kinematic theorem, like Kepler's laws. Huyghens was perhaps the first person to establish *dynamical* theorems; that is to say, to infer a kinematic result from a stated physical principle—as, for example, his proof of the approximate isochronism of the pendulum based on the principle of *vis viva*, or, as we should now say, the conservation of energy. Huyghens, together with some of the early Restoration men of science in this country, dealt also with the collisions of bodies. The peerless Newton went further. Assuming outright three primitive "laws of motion," he showed how the results of Galileo, Huyghens, and their contemporaries could be actually deduced; and by the

From *Proceedings of the Royal Society of Edinburgh*, Sec. A., Volume 62, 1943–44, Part I, pp. 10–24. Reprinted by the kind permission of the Royal Society of Edinburgh.

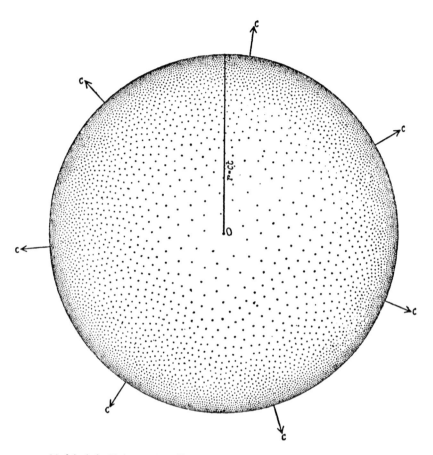

Model of the Universe According to Milne's System of Kinematic Relativity

Diagram representing the expanding universe of nebulae, as made by the observer at O, at a particular epoch in his experience. Each dot represents the nucleus of a nebula, and is in outward motion from O with uniform velocity. At any one epoch, in the experience of O, the velocities are proportional to the distances from O. The points are scattered with increasing density farther and farther away from O, the density approaching infinity near the boundary. The boundary, which is receding from O in every direction with the speed of light, is not itself occupied by points, but the points form an 'open' set of which every point of the boundary is a limiting point. The assembly of moving points has the property that an observer on any point also sees himself as the geometrical centre of the system in his view, with the other points distributed round him with spherical symmetry inside a radius defined by the epoch in his experience to which the diagram refers. For each particle-observer, the system is locally homogeneous near himself, the density increasing outwards, at first slowly, ultimately to infinity. The total population of points is infinite. The particles near the boundary tend towards invisibility, as seen by the central observer, and fade into a continuous background of finite intensity.

(From E. A. Milne, *Relativity Gravitation and World-Structure*, Oxford, 1935.)

addition of a fourth law, the law of universal gravitation, already conjectured by some thinkers, he arrived at the laws of Kepler as inferences. Not only so, but the four highly general and abstract laws introduced by Newton have been found sufficient to deduce an enormous complex of dynamical theorems, to express their relationships in the subsequent beautiful systems of Lagrange and of Hamilton, and to derive all but every detail in the motions both in the solar system and in distant binary stars. The basic principles laid down by Newton remained unaltered till our own day, when Einstein modified simultaneously the laws of motion, the law of gravitation, and the background of space and time which had been explicitly adopted by Newton as the scene in which his laws were to play their parts.

It will be seen that this brief mention of some of the leading names in the evolution of dynamical theory raises immediately a number of fundamental questions which are unsettled to-day. It is impossible even to summarise progress in dynamics without impinging on (i) the relative status of observation, experiment, and inference; (ii) the nature of induction; (iii) the metaphysics of motion; (iv) the status of natural law; (v) the meaning of time and space. There are also raised such more immediately physical questions as the origin of mass and of gravitation, and also (if we look ahead to electrodynamics) the origin of electro-magnetic forces. The Trust Deed of your James Scott Lecture enjoins on the lecturer the duty of lecturing "on the Fundamental Concepts of Natural Philosophy." Natural Philosophy, that splendid term which appertains to so many chairs in Scottish Universities, merges in its more fundamental directions into metaphysics, in the widest sense. Now such progress as has been made in metaphysics since Greek days has been in the direction of logic and epistemology. The more advanced a branch of science, the more it relies on inference and the fewer the independent appeals to experience it contains. As a science progresses, the most diverse phenomenal laws are seen to be deducible from a few general principles, and the number of these, in turn, tends to become smaller and smaller. We deduce the more and more from the less and less, or at least from the simpler and simpler. The rôles of observation and experiment, fundamentally important as they are for discovery and verification, play a less important part in the structure of the finished science, and the rôle of deduction a more important part. The question arises as to whether this process of inferring can come to a stop, and if so, where. Is there an irreducible number of brute facts derived from observation? Or can the complete science be fully constructed by processes of logical inference, without specific appeals to laws based on experience? And if the latter, what is the real basis of argument?

The answer seems to me to be that we can reduce the appeals to *quantitative* experience to zero, provided that, at each stage, in saying exactly what we mean, we state also the precise means by which we become aware of what we are trying to communicate. This is the branch of metaphysics we call epistemology.

The importance of a sound epistemology in physical science was perhaps first realised by Einstein, in his treatment of simultaneity of events. It has been urged also by Eddington, and by Bridgman, who calls its application the "operational" method. What I have to say here is that whilst epistemology, the science of how we know, has been much treated on what I may call its *theoretical* side, it has been less put into practice. Einstein, for example, after his masterly analysis of the meaning of simultaneity for different observers, still relied on an empirical assumption—the constancy of the speed of light—in his derivation of the Lorentz formulæ, not realizing that the same ideas could be developed further so as to dispense with this assumption. I propose to try to show how, when we make actual use of the considerations to which we are led by a proper guidance from epistemology, when we cease from paying it as it were mere lip-service and use it to show how our acquaintance with the external world is synthesised out of units of perception, we are led to quantitative laws relating phenomena in the external world which are inevitable relations between the elements of perception. I think we are thus led to a deeper understanding of some of the questions I have enumerated above, and with that a deeper insight into some of the fundamental concepts of Natural Philosophy.

I propose to illustrate the progress that can be made by examples from my own kinematic studies in relativity, world-structure, dynamics and electrodynamics. I do not deceive myself into thinking that a synthesis of the kind I have mentioned can be made in only one way, or that the structure I have erected will necessarily prove to be the simplest. But I do know that I have been influenced throughout by two principal motives: one is the desire never to introduce, unsuspectedly, any elements of *contingent* law into its construction; the other is the attempt to say exactly what is meant by a quantitative statement in terms of operations that could be actually carried out, and communicated to a distant observer elsewhere in the universe, who could repeat similar observations, on these instructions, himself.

The first of these rules of policy means that when a proposition is such that it might equally well have taken, at that stage, a different form, it must be either denied admittance to the investigation, or, if temporarily admitted, reserved for later rigorous scrutiny in an attempt to put it on a truly deductive footing—that is, to make it depend on results already established.

The second rule of policy is what I mean by practical epistemology. All physical measures—co-ordinates, masses, charges—arise ultimately out of combinations of kinematic measures, which must be made out of time and length measures. These in turn can be reduced to what I call epoch and distance measures, in terms that could be transmitted to and understood by a distant independent observer. This embodies the quintessentials of relativity. For relativity is in essence the science by which one observer, from his own description of a phenomenon, can predict quantitatively an independent observer's description of the same phenomenon.

In putting these rules into practice we must first perform the difficult task of emptying our minds of all our previous knowledge of physics. It is not that we necessarily suspect the existing structure of physics; but we wish to rebuild it from elements consciously and explicitly introduced. In our process of building, in our choice of methods, we cannot help but be guided by our knowledge of the structure with which we wish to compare what we erect; that is to say, with the present accomplished structure of physics—just as an architect in building an edifice is influenced by his knowledge of architectural styles. He is influenced teleologically, but that does not mean that he actually incorporates pieces of other buildings. In the same way we must manufacture our own bricks, and see that each is truly laid. When the building is complete, it is then ready for comparison with the historic structure of physics, built with great charm in what I may call the Gothic manner, with here and there unplanned additions, in changing styles. Gothic buildings are often unnecessarily strong as well as sometimes dangerously weak; the factor of safety of each structural unit is unknown; it cannot be ascertained by measurement of the existing building, without pulling it to pieces, but only by reconstructing a comparison model on engineering lines. The factor of safety of a unit of a building corresponds to the logical status of each separate proposition of physics; its "value" corresponds to the ascertaining of whether each such proposition is consistent with, or inconsistent with, any other given set of physical propositions, and if consistent, whether logically derivable from the latter. In thus synthesising a comparison model we must be prepared to find that to make it hold together with its minimum factors of safety, its internal strutting may have to be different from that of its Gothic prototype.

I do not wish to put forward as a dogma the view that this method is bound to be successful. It is, in J. J. Thomson's words, a policy, not a creed. But we can only find out its limitations by treating it as if it had none, by experimenting with it. This is the true rôle of experiment, to tread on new ground in the faith that can move mountains.

Certain contingent propositions may prove to be necessary before we can start building at all. But I do claim that if we base our structure only on those propositions which assert how I, as an observer, can become aware both of an external world and of other observers in it, capable of description, then an unexpectedly large tract of classical natural philosophy can be covered in quantitative fashion. I have not carried similar studies into the domain of quantum theory, and I will not speak of that. But I believe that in the field of classical mathematical physics there are no inaccessible propositions—not as an *a priori* belief, but as a result of trying. The questions that can be asked can either be answered by sufficient logical pressure, or else are illegitimate questions, whose very putting involves assumptions which are epistemologically unsound.

I have asserted, as an act of faith to be tested, that as soon as we have carefully stated exactly how we become aware of the quantitative aspects of a phenomenon, then automatically we must be able, given the skill, to infer all the relations that exist between those quantitative aspects. This, to my mind, is what we mean by saying that the universe is rational, and so intelligible. This leaves no room for the interposition of magic. Let me now give some examples of the fruits of this policy. Let us take the ideal question of the motion of a free particle.

What are the data of perception relevant to such a question? They are of two distinct kinds: (1) those relating to the position and velocity of the free particle relative to the observer, at the epoch in that observer's experience at which it is being considered; (2) similar data for every other particle in the universe. The reason for the relevance of the latter is that the given "free particle" is in the presence of all these other particles, and is therefore capable of being observed from them. To make progress, the "rest of the universe," *i.e.* the universe apart from the free particle in question, must be given some simple idealised form. At this stage we may admit such a simple idealised model based on observation; later, as I shall show, we scrutinise the status of this model, and derive it from simple ideas. For the moment, we idealise the "rest of the universe" into a system of receding extra-galactic nebulæ—separating island-aggregates, which may in turn be idealised to particles, each provided with a fundamental observer; we adopt a model which we specify. The law of motion of a free particle is then the rule of calculation by which any given observer can predict the observations he would make of the free particle. But, given the rule, any one observer can not only calculate the motion *he* would perceive; he can also calculate from his knowledge of any other observer's motion and from his own rule for determining the motion of the free particle, the rule that that other observer must use for obtaining *his* description of the motion. But one observer is as good as another; the rule for one observer must be the same as the rule for another; so that the second observer's rule, as obtained by *copying* the first observer's rule, must be the same as that obtained by calculating from the first observer's rule, via the motion of the second observer. This sets a restriction on the possible forms that the motion of the free particle may take. Further restrictions are imposed by the circumstance that when we have idealised the "rest of the universe" by a model, that model must effectively exhaust all the other matter in the universe. The final result of all these restrictions is that the motion of the free particle emerges as something quite definite and precise. There is only one form of law of motion of a free particle, in the presence of the idealised model, that satisfies all the conditions.

This is an imperfect attempt to put into words what goes on behind the mathematical technique. But, clumsy as it is, it shows at once why the "rest of the universe" must come into the picture. The "rest of the universe" is the only possible stage for other observers; they must be located on the particles which are the elements of this model, and share their motion; moreover, they must be, in technical language, "equivalent" to one another. The principle that the law of motion of a free particle in the presence of the rest of the universe is determinate by a mere statement of the contents of the rest of the universe, however, does not depend essentially on this notion of *equivalence,* useful as it is; for if the various observers were not equivalent, their non-equivalence would have to be described in specific terms, and the mode of variation of description of the free particle from observer to observer could still be calculated in two ways, and the self-consistency would set a restriction which would fix the motion of the free particle. The rest of the universe is in any case the stage or setting from which observers can view that phenomenon which consists in the possession of a definite equation of motion by the free particle.

Several points now arise. One is that the mode of derivation shows the connection of cosmology with dynamics. No *deduction* of the motion of a single free particle, *per se,* ignoring the rest of the universe, can be legitimate. For the "rest of the universe" is always present, and with it, possible other observers. If we are asked what is the motion of a free particle in "empty space," *i.e.* in the presence of one observer alone, the question is an illegitimate one, and we cannot answer it. This shows in turn why Newton had to *posit* his First Law of Motion. He was not in a position to deduce it, because he left out any mention of the rest of the matter of the universe: empty, infinite space is not a fruitful conception, dynamically, for it provides no *pied-à-terre* for observers.

When we have carried out the above procedure, the equation of motion we obtain may be called the first law of motion, in the presence of that particular model of the universe used, and in terms of those particular measures which were used both for locating the free particle and for the locating of the separate observers by one another. But this form of first law does not coincide as it stands with Newton's form of the first law. This shows two things. First, it guards us against the charge of having tacitly appealed to existing physics; for if we had assumed existing physics, we should inevitably have assumed Newton's First Law, and we ought to have recovered this, short of an actual error. Secondly, it follows that the description of the rest of the universe adopted in the derivation cannot be that appropriate to the measures in which Newtonian dynamics is couched.

As was remarked earlier, co-ordinates are constructed out of observers' observations by definite rules. If we change the rules, we change the coordinates. Now one very general way of changing the rules is to regraduate all clocks used. We must be careful to ensure that the various observers' clocks are regraduated in the same way; for only so can we institute comparisons between different observers' descriptions of the same phenomenon. We shall see later that this is possible. It is now found that if we perform one particular type of regraduation of all observers' clocks, and with it the transformation of all measures derived from these clock readings, then the first law, as deduced above, takes the form of Newton's First Law. This shows that the originally adopted specification of the "rest of the universe," though a very simple one, is not the description of the universe appropriate to the form of Newtonian dynamics.

The original model universe adopted described a uniformly expanding non-homogeneous system of finite extent but of infinite content as regards number of particles. It must be mentioned that this description is that used by the fundamental observer who regards himself as central; but if he shifts the description, by the rules of relativity, to any other fundamental observer, that observer now regards *himself* as the centre. Every fundamental observer is thus central, in his own reckoning. If he now changes his manner of reckoning, by regraduation of his clock in a particular way and, with it, all derived measures, the description of the model changes to that of a stagnant, homogeneous system of infinite extent. This is accordingly the form to which Newton's form of First Law must properly apply.

The form of the first law, in any scale of time, is independent of the scale of density assumed in the model: if we double the density everywhere, we do not alter the form of the motion of a free particle. We can accordingly take the density as small as we please, provided always of course we do not take it absolutely zero. The latter would be to eliminate all observers, and so make the question under discussion meaningless. In fact it would be impossible to take the density as everywhere zero in the original model adopted, for it has a density-singularity at its boundary. Thus the very model insists on the presence of matter to carry observers. In the infinitely extended, homogeneous form of universe, the same difficulty does not arise in the same form; the distinction between an absolutely zero density and a limitingly small density only affects conditions "at infinity." Thus it is reasonable that Newton's form of the First Law should relate, in his system, to empty space; though, in emptying space of matter, we deprive ourselves of the means of deducing the law.

My hearers will ask at this stage, Why all this bother? Why not begin with the homogeneous, infinite, stationary universe and deduce Newton's First Law directly? The answer is that this homogeneous, infinite, stationary type of universe contains in its description a constant, whose origin is apparent only when we return to the non-homogeneous, finite, expanding type of description. In the latter description, the corresponding parameter is the *age* of the system, t, reckoned from its natural zero of time, at the instant in the observer's experience at which the change-over to the homogeneous, infinite, stationary type is made; in the former case the constant corresponds to the *curvature* of the space most conveniently adopted for describing the system. If this curvature is called $-R^2$, then the relation is $R = ct_0$, where c is the originally arbitrary constant which is found subsequently to describe the speed of light, and t_0 is the age at the moment of regraduation of clocks—care being always taken to make the momentary clock-rates agree. This curvature, in the second mode of description, will appear as an unexplained "constant of nature." In the description of the *expanding* model, on the other hand, there is nothing which might be called a "constant of nature"; such constants as occur in its description are conventionally chosen numbers, arbitrarily fixed as the investigation proceeds. The static form of description of the universe, however, necessarily contains a new "constant." The unsuspected presence of this constant, if we try to deduce Newton's First Law outright, makes the deduction extraordinarily difficult. For we should have to introduce this constant without giving a reason for it. It does not appear expressly in the law of free motion of a particle along the geodesic passing through the observer, but it does appear in the law of free motion in cross-country paths.

In the case of the expanding system, which has at any instant its natural age t, t is available for constructing accelerations: apart from refinements, any acceleration must be constructed as a length divided by the square of a time, or a velocity divided by a time, and the distance and velocity of the free particle can at once be divided by appropriate powers of t. Crudely, the answer is that the acceleration of a free particle at position vector $\mathbf{P}$ relative to the observer, and velocity $\mathbf{V}$, is approximately $-(\mathbf{P} - \mathbf{V}t)/t^2$. But no similar quantity corresponding to a time is available in the case of the static form of description, if we start with the latter outright; the desired quantity only appears when we start with the expanding type, and make our transformation of clocks at time $t = t_0$, when t_0 becomes available.

The actual form of transformation of clock-graduations is that a clock formerly graduated to read t, is regraduated to read τ, where

$$\tau = t_0 \log t/t_0 + t_0$$

This makes $d\tau = dt$ at $t = t_0$. The constant-free description in terms of measures made on time-scale t necessarily passes into a description involving the constant t_0; and, as I have said, the transformed description is most conveniently mapped in a space of curvature $-(ct_0)^2$.

I can now illustrate what I have called the illegitimate question in natural philosophy. What is the curvature of the observer's space? The answer is that he may map his observed events in any type of space he chooses; and having chosen a type of space for measures in one scale of time, he may find it convenient, when he changes his scale of time, to adopt a space conformal to the first as his new map. In the present case, in the simplest description of the expanding system, each observer uses a "flat" Euclidean space, of which a portion only is occupied by the system, just as the terrestrial surface can be mapped in two circles of a plane surface. These observers' spaces do not constitute however a "public" space, for the different observers do not agree as to the "distance" between two particles at a common epoch. But when the scale of time is transformed, and relative motion disappears, it becomes possible to choose a "public" space, the whole of which is occupied by the system. The curvature of this space is not a real fact about the system, but only something appertaining to the particular mode of description of the system; t_0 is arbitrary.

I proceed to a second example of the use of deduction and a sound epistemology. What is the origin of the law of gravitation? This, like the First

Law, was by Newton *posited,* not inferred; so it was by Einstein; and no physical discussion of the macro- or micro-structure of matter has ever shed any light on its origin. Yet it is presumably rational, and so accessible to inference. What does our procedure say to it?

Our procedure expects that, as soon as we have said exactly what we mean by *two* free particles, in one another's presence and in the presence of the rest of the universe, then without any empirical assumption, given sufficient logical pressure, we ought to be able to infer their accelerations, and so their future motion. Their accelerations should be determinate and knowable in terms of the data defining them and their environment. The "law of gravitation" is nothing but the pair of equations of motion of *two* free particles in one another's presence. We see at once the connection between the First Law of Motion and the Law of Gravitation; the latter is the next stage of development of the former—the development from one particle to two. It is irrelevant to assume gravitation to be due to any "influence" between the particles; gravitation is the statement of how the pair of particles must move in order to be self-consistently observed by all observers in the universe.

It took me some years to pursue this programme, and eventually I executed it only in a roundabout fashion. I found it easier to discuss the accelerations of a whole class of free particles, and then to simplify the result down to two, but I have little doubt that more powerful analysts will be able to execute it directly. The indirect method was, however, fully worth while, for it threw up what was at the time an unexpected constant, a constant of integration, which proved to represent a new fact about any one of the particles, namely the possibility of the existence of its gravitational mass.

Thus a definite result is actually obtainable. It is capable of stating the "attraction" (carefully defined) between two particles, in terms of the co-ordinates attributed to them by any fundamental observer, at stated epochs, by a rule which is the same for any other observer, in terms of *his* values of the co-ordinates at the same events. The law is thus in strict relativistic form. In fact it is seen to correspond to a "potential" which is the simplest scalar that can be constructed out of the co-ordinates and epochs of the particles and that possesses a singularity when the two particles coalesce. These conditions, indeed, point the way to an alternative derivation of the law, though one not quite so inevitable. Though not taking the form $1/r^2$ for a general observer, it reduces exactly to $1/r^2$ when the observer takes a special position, namely at one of the particles themselves, and calculates the attraction of the other. That is to say, the derivation of the equations of motion of a pair of particles, free in the idealised universe, from the internal properties of the system, results in the putting of the expression $1/r^2$ into proper relativistic form, and so in the relativistic expression of Newton's inverse square law of gravitation. The method succeeds in this object largely because all time co-ordinates are reckoned from the natural zero of time. It was the supposed impossibility of expressing the law $1/r^2$ in relativistic form, within the scope of his early theory of relativity, that led Einstein to his so-called general theory. The present theory is entirely different.

The acceleration of each of a pair of free particles, on the present theory, consists of two components, one of which we may attribute to the second particle, the other to the rest of the universe. The latter only vanishes for a particular choice of the scale of time. It is usual to reserve the name "gravitation" for the former only, but the second component, which we have already encountered as the acceleration of a *single* free particle, may also be described as gravitational: a "gravitational" acceleration is just an acceleration which a particle cannot help possessing. We do not usually call the second component gravitational, because it can be transformed away by a special choice of time-scale. Nevertheless its interpretation as gravitational is extremely fruitful and shows the intimate connection between Newton's First Law and his Fourth, in the more general forms in which they appear here. The two are different aspects of one and the same phenomenon. This also appears in Einstein's theory; but Einstein's theory involves an unexplained empirical constant, the constant of gravitation, which has not so far made any appearance in our theory. We can, however, readily identify it, by this very notion of regarding the first law as essentially gravitational in character.

I have already stated that in the presence of the uniformly expanding universe, as described on the t-scale, t being measured from the natural time-origin, the acceleration of a single free particle takes the approximate form $-(\mathbf{P}-\mathbf{V}t)/t^2$. Here, $\mathbf{P}-\mathbf{V}t$ is the distance of the particle from the apparent centre of the system, in its view, for $\mathbf{P}-\mathbf{V}t$ is its distance from the point $\mathbf{V}t$, which is the position of the fundamental particle relative to which the free particle $(\mathbf{P}, \mathbf{V}, t)$ is momentarily at rest. If we call $(\mathbf{P}-\mathbf{V}t)$ temporarily $\mathbf{r}$, then the acceleration is $-\mathbf{r}/t^2$, directed towards the centre of the system in a frame in which P is momentarily at rest. Compare this with the Newtonian calculation, on the interpretation of this acceleration as gravitational. The Newtonian calculation gives $-\gamma \frac{4}{3}\pi\rho r^3/r^2$, where γ is the Newtonian constant of gravitation and ρ is the density inside the sphere of centre r. Now the particle-density in the expanding universe is, near the "centre," B/c^3t^3. Thus attributing to each "particle" a mass m_0, we get that γ must be defined by

$$-\frac{r}{t^2} = -\frac{\frac{4}{3}\pi\gamma r B m_0}{c^3 t^3}$$

These agree identically in their behaviour with r, and give

$$\gamma = \frac{c^3 t}{M_0}$$

where $M_0 = \frac{4}{3}\pi m_0 B = \frac{4}{3}\pi (ct)^3 . (m_0 B/c^3 t^3)$; M_0 may be regarded as the *apparent* mass of the universe, since it is equal to the mass obtained by filling the sphere of radius ct with matter of a uniform density equal to its central density at time t.

This means that the Newtonian constant of gravitation, the second most famous constant in the world, is expressed in terms of c, the velocity of light, M_0 the apparent mass of the universe and t its age. The values used for these in ordinary physics may be taken as fixing the units of velocity, mass, and time, and are conventional. The derived quantity, γ, is however not conventional, and its numerical value, as calculated from this formula, agrees with the observed value, when t is derived from the nebular recession rate.

This dependence of the constant of gravitation on the age of the universe may at first sight appear very strange. It appears less strange when we find that if we once more transform the scale of time so as to recover the Newtonian measure of time, γ transforms into a constant y_0, given by

$$\gamma_0 = \frac{c^3 t_0}{M_0}$$

where t_0 is the age-parameter at the moment the transformation is made. Thus in the less fundamental Newtonian measure of time, the "constant" of gravitation does masquerade as a constant. A formula similar to the above occurs in Einstein's theory of the spherical universe. Written in the form

$$\frac{\gamma_0 M_0}{c^2} = c t_0 = R$$

it is held to connect the radius of curvature R of this system with its finite mass M_0. This way of looking at it seems to me to be unsatisfactory. If we have to calculate the radius of curvature of the universe from the constant of gravitation, it gives no explanation of either curvature or gravitation. It seems to me preferable to regard γ as fundamentally made up of three more simply explained numbers, c, M_0, and t. M_0 and t can in principle be found by surveying our local cosmic surroundings and determining the local density ρ (which is $m_0 B/c^3 t^3$), and t, which is the "constant" of proportionality in the law of nebular recession $V = P/t$.

The relation $\gamma = c^3 t/M_0$, using the observed value of γ and the only roughly known value of M_0, may be used to calculate t, and so estimate the nebular recession rate. This method has been used by Eddington, but it seems to me, again, to be entirely non-fundamental, inasmuch as it amounts to predicting the moment at which we happen to be observing the universe, which is of course the present moment!

Einstein's, Eddington's, and the present way of regarding the formula are mathematically equivalent; but philosophically they are entirely distinct. Einstein regards the constant of gravitation as a sort of original sin, entering the universe as part of its nature; Eddington similarly takes the constant of gravitation as primitive; the present method regards it as derivative, as constructed out of philosophically simpler numbers, and as varying with the epoch. Einstein's general theory of relativity is expressed wholly in terms of this temporary, Newtonian measure of time, the τ-scale, with its ephemeral constant t_0, and accordingly seems to me to be not part of the *fundamental* concepts of Natural Philosophy.

Gravitation on the present views is an inevitable constituent of external reality. In this way physics comes to have contact with metaphysics. For, to repeat, gravitation is only the name given to the inevitable way in which particles must move in one another's presence and in the presence of the rest of the universe, if they are to move according to the same rules for all equivalent observers in the universe. Observers are an essential element in the situation. There could not be an *observed* universe in which gravitation did not exist; γ could no more be zero than the age of the universe, at the moment at which it is observed, could be zero; in fact γ is zero at the (unobserved) moment of creation, $t = 0$. Instead of the dichotomy of First and Fourth Law, of free motion and unfree motion, in Newton's system, we have a single unity, under which the two laws appear as different aspects of the same phenomenon.

If the "rest of the universe" were suddenly abolished, would the two particles persist in gravitating? This is another example of an illegitimate question, for we cannot within the scope of Natural Philosophy idealise the rest of the universe away. But the indications of our theory are that for two particles "alone in space," gravitation would be a totally different thing. There would be no M_0, and so there could be no γ of the observed order of magnitude. The only way of solving the question would be to put hypothetical observers on the particles, and ask how each would observe the other. I doubt whether anything worth while would emerge: the concept of two particles alone in space is not, fundamentally, a fruitful one.

A kinematic model of the universe implies gravitation. Does it imply the possibility of any other type of interaction between the two particles? The gravitational potential to which it leads, after effects of change of mass with velocity have been separated off, is remarkable as containing only the positions and epochs of the two particles concerned. It does not involve their velocities. By taking more complex forms of scalars of this kind, we build up more and more complicated gravitational situations; but *only* gravitational situations. Is there any other type of potential which could correspond to other types of equation of motion of two particles in one another's presence?

The answer to this question is a rather lengthy investigation, into the details of which I cannot here enter. But, briefly, to introduce a new term into the equations of motion, that is, new forces between particles, one has to consider them as derived by the double differentiation of scalars which may be called *super-potentials*. They must of course be scalars because they must be formed by the same rules by all equivalent observers in the universe, out of their observations, and so must be invariant. But when the simplest such invariants have been formed, they are found to differ essentially from the scalar giving gravitation in that they involve the velocities as well as the coordinates and epochs of the particles. When finally these super-potentials are differentiated down to give equations of motion—and of course they originally grow out of equations of motion—when, I say, the equations of motion become explicit, the new terms correspond to electromagnetic forces between two charged particles in motion. They give, for example, the electrostatic Coulomb field, the magnetic field of a moving charge and the pondero-motive force of Larmor and Lorentz.

The train of thought which leads to these results is logical and mathematical, not physical; it involves no appeals to quantitative experience, or to quantitative empirical laws, save finally when we identify our symbols in nature. We just go on weaving strands of possibility into a texture, and compare the final fabric with nature. As I said earlier, we are guided in our work teleologically by the prospect of reaching something of interest, suitable for comparison with nature. But we do not impose the laws which prior observation of nature had disclosed; and in fact the final laws deduced differ in significant details from the experimental laws. The finally deduced super-potentials prove to be, as it were, source singularities of the next simplest type of singularity to the gravitational type of singularity; that is, they are the next simplest type of singularity which involves velocities as well as positions and epochs.

The significance of this from the point of view of the present lecture is as follows. The electromagnetic laws here found correspond to a second type of interaction between two particles, over and above the gravitational type of interaction. It appears then that there are, *a priori*, at least two types of force possible in the universe; one of these we call gravitational, the other electromagnetic. Each is derived from a point singularity. We may build up more and more complex combinations of such singularities, but essentially they are two. It would be a fruitful subject of research to see if we could go further, and discover the possibility of still further types of force as possible in our abstract scheme, and so to be looked for in nature.

I have already mentioned that the electromagnetic laws inferred by these methods, though coinciding largely with the laws of Faraday and Maxwell, differ from them in detail. The differences become conspicuous at high velocities. The equation of motion of a charged particle in the vicinity of a relatively massive charged particle is found to possess integrals which correspond to energy and angular momentum, but which are not calculable in the same way as energy and momentum. At high speeds, that is, for close proximity of the particles, the differences become so marked that energy and angular momentum can no longer be said to be conserved. When energy and momentum cease to be conserved, the usual dynamical forms of expression cease to be available. Instead of using them, we may use spurious descriptions of the force between the particles arranged so as to give the same spatial orbit. Though the inverse square law of Coulomb is obeyed, in this work, down to indefinitely small distances, the modifications in the high-speed dynamics has the effect of mimicking a force which changes very suddenly at the distance $|e_1 e_2|/mc^2$. This is of course the classical value of the "radius" of the electron. Further, a temporary, unstable, close association of two oppositely charged particles becomes possible, suggesting the neutron, and providing an apparent potential ridge round the one regarded as the nucleus. The further exploration of these ideas is not possible here, but enough has been said to indicate the scope of the field thus opened up.

I have in the course of these three examples—the free particle, the pair of gravitating particles, and the pair of charged particles—touched on cosmology, the curvature of space, time-scales, the natures of gravitation and electromagnetism, and the conservation of energy and momentum. I have not, I hope, given the impression that there is not a great deal more to be done in all these fields. For example, although the notion of gravitational mass emerged of itself, and though in the theory gravitational mass is equal to inertial mass, the idea of inertial mass requires considerable clarification. Further, the exact fulfilment of Newton's Third Law, and the extent to which it is a mere definition of equality of forces, as in Mach's treatment, need a thorough discussion. I propose to pass over these, and deal now with what should logically have been discussed at the beginning, namely the making and combining of the measures by which we can describe these phenomena.

First things first. Before I look out on the world, I am conscious of myself, as thinker and as capable of being a percipient. But "cogito, ergo sum" is mere metaphysics, and takes us nowhere in Natural Philosophy. What is important is that I possess a consciousness, and that my sensations—the events in my consciousness—arrange themselves in a single linear order, which we call a time-order. This is an appeal to a contingent proposition, for sensations might have been capable of a multi-dimensional arrangement. But apart from this initial appeal, we make at present no appeal to our empirical knowledge of the laws obeyed in nature. But the time-order is more than a well-ordered one-dimensional sequence: it is an order with a direction in it. The recognition of this at the outset is represented later by the existence of a natural zero of time when we come to measure time. The time-direction is simply the direction *away from* the natural zero. The zero itself proves to be a singularity, an impassable singularity; we are therefore necessarily on one side of the time zero, and so the time-direction is unique. The fact that all times have later to be measured from this natural zero makes a sharp distinction between the present analysis of time and that, for example, of A. A. Robb. Einstein was the first person to cease discussing Time, with a capital T, and introduce the notion of measuring the passage of time. Kant, for example, in speaking of Time ignored, as Einstein did not, the elements of epistemology at the outset. But even with Einstein, the time co-ordinate is reversible. In the present theory the irreversibility of the passage of time is incorporated throughout.

We cannot rest content with saying that a clock measures time. For this leads at once to the question: What is meant by saying that two similarly named intervals of clock-graduations, like two distinct minutes, are equal? And to the question: If two clocks side by side give different measures, which is to be preferred? We cannot answer such questions outright; when we are synthesising physics from the bottom upwards, we know nothing of uniform clocks. Since the order of events in each individual's consciousness is well-ordered in one dimension, it can be correlated with the order of the real numbers in a multitude of different ways. Any such correlation affords a measure of the position of an elementary event in my consciousness. It suffices to call any such correlation a clock for the time being, and to study the relations between the different clocks so constructed.

This one-dimensional, uniquely directed character of time, which seems to me the most fundamental fact in Natural Philosophy, is intimately connected with the possibility of describing an external world at all; it provides the relation between reality and the ego. It is a sort of window to Plato's cave—we can see nothing save through it. It would be impossible to make a start without it. It is remarkable that the other contingent proposition we are compelled at present to accept on a basis other than that of inference is also connected with dimensionality—the three-dimensionality of what we call space. Whether the number *three* of spatial dimensions is accessible by a process of inference remains for future research. For the moment we notice that a spatial dimension differs from the time-dimension in that it can be traversed in any order. This fundamental distinction between time and space is obscured in the modern habit of speaking of "space-time" as a single entity.

But I am anticipating. So far we have: ego, temporal experience. The next step is for the ego to have something to perceive. Till then, space does not exist. Space is not an object of perception. It is a map I invent for the location in it of objects I perceive. If there are no objects to perceive, there is no space. The unreflecting person tends to understand by space visible emptiness. When we remove all light from it, it becomes invisible darkness; and space then is not. The black-out is inimical to geometry! Now the simplest *modus operandi* of perception is *seeing*. To construct physics out of touching and hearing would be a heroic but unprofitable task. Let us content ourselves with *seeing*. Frequently we see objects continuously, but here we want to build up elements of perception, and therefore we require an object to be seen at a single instant of my consciousness. If, however, an object self-luminously flashes into perception and out again, I have only one datum of observation, namely the instant, by my clock, at which I see it. This affords little basis for further analysis. But if I illuminate the object myself, by uncovering a light at myself, I can make two observations: the instant, by my clock, at which I uncover the light, and the instant at which I see the object flash into view—or the first such insta. :.

The importance of this process is that whilst it assumes nothing of the physics of light, it yet gives me two data on which to work—the instant I seek it and the instant I *ree* it. There is no appeal here to quantitative experience. The use of a source of light is the condition of having anything to discuss at all. Without light the ego would be in dark loneliness.

Out of the two numbers I consider as connected with my momentarily perceived object, I can construct two other numbers in any way whatever, and I call these *co-ordinates*. Conversely, given the co-ordinates I can recover the original data of observation, by reversing the rules. Thus any relations I may subsequently derive between the co-ordinates of different events (as I call a momentarily perceived object) can be immediately translated back into relations between the primitive observations.

Two simple combinations of these primitive observations are sufficiently useful to have names. I call the semi-sum the *epoch* of the event in question, and the semi-difference taken positively, multiplied for conventional reasons by a conventional constant c, the *distance* or *distance* co-ordinate. There is no need for over-anxiety at this stage as to whether these co-ordinates agree with any particular measures according to current standards in physics. Einstein accustomed us to using *any* kind of co-ordinates, though he was not always careful to say what they meant *ab initio*; and there is no objection whatever to our employing co-ordinates defined by means of measures made with an arbitrarily running clock. For these co-ordinates are *operationally* defined, in Bridgman's sense.

In constructing co-ordinates in this way, no assumption whatever is made concerning the speed of light, for the simple reason that until we have set up measures of epoch and distance no meaning attaches to velocity. We have in fact arranged our measures of epoch and distance so that when we come to define velocities, the velocity of light comes out as the constant c. It will be found that if we attempt to define *distance* by some method not involving clock readings, then we need an empirical law about the speed of light in order to connect different observers' observations. This was Einstein's original method—to assume a constant velocity of light as an empirical fact. We make it so by construction—as Mach availed himself of Newton's Third Law.

But there is a further even more fundamental reason for defining distance co-ordinates by time-measures. We need to use *two* time-measures, anyhow, to fix the epoch of a distant event by our clock. Sometimes it is stated that an epoch at a distance can be fixed by reference to a clock at a distance. But to correlate the clock at a distance with our own clock requires precisely the observation of viewing that clock at a distance, at an instant determined by the uncovering of a light at ourselves. Thus we always need ultimately two observations to fix an *epoch-at-a-distance*; and these then always suffice, in addition, to fix a co-ordinate we may call a *distance-at-an-epoch*. To use some different method for fixing the distance of an event when the necessary data have already become available in the course of fixing its epoch, is to throw away the half of our knowledge of it. I apologise for labouring this point, but it has often been almost malevolently misunderstood. The observer in measuring the epoch of a distant event becomes *ipso facto* in a position to assign a measure to its distance.

So far we have introduced myself as conscious observer, light with which to perceive, and an object to be perceived. Having begun by following Descartes, we now follow Leibniz, and introduce a distant monad, a second observer; and we let him also choose any arbitrary correlation between events in his consciousness and the real numbers, and so construct his own clock. We can, if we like, suppose him to provide himself with an actual clock, with its hand running over a graduated dial. This clock I now proceed to read, from a distance, by uncovering a light, as before. I just note the instant at which I uncover my light, the instant I perceive the distant clock, and the particular graduation of the distant clock that its hand is passing at the moment I perceive it—three observations in all.

Reciprocally, my distant friend can now uncover a light at himself, at any moment, and read the clock which I constructed at myself, noting also the time at which he uncovered his light and the time at which he saw my clock. He also makes three observations.

We have now all the elements of a piquant mathematical situation. There are many interesting questions we can now ask, and to which we can find answers. We can correlate the clocks as they stand. We can ask whether a meaning can be attached to regraduating one clock so as to "agree" with the other, in some defined meaning of "agreeing." We can ask for the correlations of the two observers' observations of any event external to both. And we can ask what are the circumstances in which a third, or a fourth, or any number more observers' clocks can all be regraduated so as to "agree" amongst themselves. With the aid of Dr. Whitrow I have been able to find answers to these questions.

The answer to the first of these questions is that, after the clocks have been correlated, it *is* possible to regraduate one of the pair so as to make it "agree," in a well-defined sense, with the other. This does not mean that when A looks at B's clock he sees it reading the same as his own; for if this relation is imposed, it is not true reciprocally. We must have a symmetrical relation between A and B, that is, given A's mode of clock graduation, B must regraduate his so that the totality of observations A makes on B coincide with the totality of observations B can make on A—each to each, as Euclid would say. This is tantamount to giving a definition of *congruence* of clocks separated from one another, and in any relative motion, uniform or non-uniform. The importance of the possibility of a definition of clock congruence by mere superposition of an aggregate of observations is that it allows us subsequently to dispense with a geometrical axiom of congruence, the necessity for which has always been a stumbling-block to geometers. In fact, instead of "applying" lengths to one another, and so assuming the possibility of the transport of rigid bodies, we apply aggregates of time observations to one another, and so obtain an arithmetical definition of clock congruence. When two observers are provided with congruent clocks they are said to be *equivalent* to one another.

The answer to the second question—that of correlating observations of an external event—is a generalisation of Einstein's first relativity theory; it includes what is often called "special relativity," and leads to the Lorentz formulæ as a particular case. The Lorentz formulæ are thus demonstrated without making the physical assumption made by Einstein (constancy of speed of light to all observers in uniform relative motion) or his mathematical assumption, that the formulæ of transformation must be linear.

The answer to the third question—that of the correlating of further observers' clocks with those of the first two—is the theory of linear equivalences.

What Einstein's "special" theory of relativity did was to enable an observer A to calculate from his observations of an event the observations which an observer B would make of the same event, in the circumstances in which A and B were in uniform relative motion. The present theory generalises this to any case of non-uniform relative motion, once the relative motion has been ascertained by either observer's observations of the other. The theory of linear equivalences goes further still.

Uniform relative motions have a certain quality in common: the quality of being uniform; different uniform relative motions match one another, not in the value of the relative velocity, but in the fact of the relative velocity being uniform. The theory of linear equivalences extends this by defining common properties amongst non-uniform motions of any type. It generates classes of non-uniform motions of pairs of particle observers, motions which have a common quality. This quality is the permitting of the carrying of clocks which inside any one linear equivalence are all *congruent* to one another. If we impose the relative motion of A and B arbitrarily, B being provided with a clock congruent to A's, then for any motion of a third observer C we can give C either a clock congruent to A's or a clock congruent to B's; but unless C's motion is of a particular type, depending on the original relative motion of A and B, congruence of C's clock with A's does not in general suffice to ensure congruence of C's clock with B's. In other words, the relations $B \equiv A$, $C \equiv A$ (where $\equiv$ denotes equivalence) are not necessarily transitive, *i.e.* they do not necessarily imply $B \equiv C$. When we require the relations of time-keeping between all observers of a class to be both symmetrical and transitive, the result is a linear equivalence.

The importance of this concept for cosmology is that a world-wide relativity can exist only when we can construct a world-wide class of observers with congruent clocks. They then constitute pencils of linear equivalences. For, for a world-wide relativity, any observer must be able to predict the observations of any other observer provided with similar observing instruments; the simplest similar observing instrument is a congruent clock; and if congruent clocks cannot be constructed everywhere in the universe it can hardly be said that the universe is rational. I am aware that here I infringe on metaphysics; I warned you at the outset that this might be necessary; but it is fully worth while to trace the consequences of this venture.

An important property of the members of an equivalence is that their clocks can all be regraduated so as (1) to remain congruent and (2) at the same time to appear in uniform relative motion. Thus there is essentially only one distinct equivalence; the equivalence is a unique structure, and hence peculiarly appropriate as a representation of the universe. A further property of an equivalence is that if any two members ever coincide in position, then at that instant all coincide in position. This permits complete synchronisation of the zeros of all the clocks concerned, and gives us the natural time zero about which I have spoken previously. Lastly the uniform motion equivalence can be transformed, with introduction of the parameter t_0 I mentioned before, into another particular form of description, namely as a relatively stationary equivalence.

The uniform motion type of equivalence—or rather uniform motion type of description of the unique equivalence—is what I referred to earlier as a simple kinematic model of the expanding universe; elsewhere I have called it the substratum. This system of moving particles, monads, observers, clocks, frames of reference—call them what you will—possesses a great many strange and surprising properties—temporal, geometrical, kinematical, dynamical, gravitational, optical. It seems to me to be as fundamental to these various branches of inquiry which involve the passage of time as is the plane to two-dimensional geometry. Just as the Euclidean plane is the stage, the scene, the background against which the phenomena of geometry—its figures and its theorems— display themselves, so the substratum is the background against which the phenomena of dynamics and gravitation display themselves. The static background of geometry is inadequate for the full description of dynamical phenomena; and no mere extension to four dimensions of a two-dimensional set of points will embody the necessary features. The substratum, containing essentially the idea of a time-origin, embodies the unidirectional aspect of time, which makes irreversible time something much more than an additional dimension. The substratum is essentially a moving three-dimensional set of points. Here we introduce the contingent proposition that space is described by means of three dimensions. As I remarked earlier, it is of great interest that the two contingent propositions we have had to introduce both concern themselves with dimensionality.

I have said that I consider the substratum as one of the fundamental concepts of Natural Philosophy. Just as Euclid's famous "parallels" axiom completes the definition of the Euclidean plane, so that, once this plane has been effectively constructed by the use of this axiom (by denying it curvature) all propositions flow from its properties without further appeals, so once the substratum has been constructed by suitable definitions based on the two contingent dimensional axioms, the one-dimensionality and unidirectional flow of time and the three-dimensionality of space, the properties of all combinations of particles in it and so the relations relating all phenomena taking place in it should flow from its definition, that is, from its description.

The identification of the substratum with the system of receding galaxies is by no means essential to the argument at this stage. That matter will tend to concentrate in regions near, and with motions appropriate to, fundamental particles of the substratum is a later developed gravitational property of the substratum. The theoretical importance of the ideal substratum would remain even if the system of the external galaxies did not form an equivalence. Its *practical* importance, it is true, derives from the fact that the laws of phenomena as we observe them from our position in the universe resemble very closely the laws of phenomena that are predicted as observable from a member of the substratum. But in order to begin the business of predicting laws of nature—of discussing *a priori* the law of motion of a free particle and the law of gravitation—as I did earlier in this lecture, one needs the theoretical, the ideal, the abstract substratum.

The most fundamental property of the substratum is that whilst it appears enclosed within an expanding sphere of radius ct, in the private space of the observer at the centre, when its description is transformed to that of any excentric observer that observer promptly becomes central. Centrality or excentricity is not an intrinsic property of any fundamental observer, but a mode of description. Just as in the Euclidean plane, each particle stands in an identical relation to the aggregate of the other particles. It is for this reason that two observers A, B, members of the substratum, must give the same rule for calculating, for example, the acceleration of a free particle in terms of their own measurements.

The substratum includes an infinite number of members, who constitute an *open* set of moving points. Thus no point or member is situated at its boundary. The boundary is not experienceable; every member of the system is surrounded by other members. It is therefore an illegitimate question to ask— Into what is the system expanding? For no member can ever experience the sensation of passing into "new space." From the point of view of the observer who is central, the external physical space occupied by the enlarging sphere is continually recruited by the gain of space created by the expanding light-wave which started at $t = 0$. Light creates space, as I said before. Can any communication be established with any object "outside" the system? The answer is no, for the infinite number of particle-members between any assigned particle-member and the ideal boundary forms a screen opaque to vision. As far as a sound epistemology is concerned, "space" outside the boundary does not exist, physically. It exists only in the mathematician's imagination. The theoretical boundary is as beyond experience as the theoretical centre at the instant $t = 0$; we cannot experience the universe at the moment of its creation, for we are not yet created ourselves; we are not differentiated as observers.

The properties of the substratum depend essentially on the positive character of t and its positive-increasing character in the experience of each observer. These properties would become totally different if the motion of each particle were reversed in time. "Space" beyond the frontier would become accessible to experience, material objects could either disappear or be crushed by oncoming nebulæ, and in short the system would cease to be rational. It would be irrational because no rules could be given for its prior construction—for the correlation of its clocks, which would have had in their past history no moment of coincidence, and so of synchronisation. These conclusions are confirmed by a study of the irreversibility of the actual equations of motion of a free particle.

The substratum was originally constructed kinematically, *i.e.* its particles were deemed to obey certain prescribed motions. But it also has the property that these motions will continue of themselves; it constitutes also a dynamical system. This possession of this and other dynamical and indeed also gravitational properties markedly differentiates it from the assemblages of points considered in geometry alone.

We saw earlier that we have to identify the time-variable τ, not t, with Newtonian time. This allows us to identify co-ordinate distance, on the τ-scale, with distance as measured by means of our ordinary physical methods. A further consequence of this is that a length which is constant on the τ-scale will appear to be expanding on the t-scale, at a uniform rate. This is of course a necessary condition for self-consistency, since any length whatever must be entirely included in the substratum, and so must have been very small near $t = 0$. We may suppose that the material metre-rod is constant on the τ-scale, and so expanding on the t-scale; that means that if two nebulæ were tied together by means of a chain of rigid metre-links, the expansion of the metre-links would just keep pace with the mutual recession of the nebulæ. This problem, of the "real" behaviour of the metre-rod, is not an easy one. It has ramifications in the field of electrodynamics, and it makes it doubtful whether the Lorentz transformation in its usual form applies to our ordinary physical measures. It must be remembered that the deviation between τ-time and t-time amounts to only one part in 2×10^9 per year, as given by the present age of the universe, so that experimental tests of the applicability of the Lorentz formulæ to ordinary physical measures would need to be extremely precise before decisions could be reached. I have in fact worked out the possible modification to the Lorentz formulæ to make them appropriate to τ-measures, though they have not yet been published.

I do not know whether the account given in this lecture in any way conveys the feeling of excitement and high adventure that accompanies researches by kinematic methods into the nature of dynamics, gravitation, and electromagnetism. One starts from so little and ends with so much. The tempo, the pulse of the investigations is of a different quality from those of more usual mathematical physics. One breathes a different air at the outset; the metabolism of the research is different. The possibility of making progress by these methods seems to rest on the execution of elementary acts of perception which could actually be carried out. The structure is therefore secure epistemologically. Every symbol which occurs in the analysis has an immediate operational meaning. Moreover, the observer is always explicitly present in the analysis; and different observers' observations of any event are related to one another by the analysis of their observations *of one another,* observations which could actually be carried out. The totality of possible observers forms a substratum, whose simplest mode of description is the uniformly expanding substratum.

Once this has been constructed, no further appeals to the existence of each observer's temporal experience are needed, for the substratum embodies the totality of all observers' temporal experiences. The properties of the substratum now flow from its definition. The substratum itself has no size and no age—only a size and an age when a particular observer contained in it has been mentioned. The uniformly expanding type of substratum, containing no "constants of nature" in its description, yields theorems which may be considered as fundamental. The properties described in these theorems transcend the mere kinematical properties of a set of moving particles, and include results which are usually separated out as dynamical, gravitational, or electromagnetic. It is for this reason that I have ventured to give some account of these properties in a lecture on the Fundamental Concepts of Natural Philosophy.

H. P. ROBERTSON:

Geometry as a Branch of Physics

IS SPACE REALLY CURVED? That is a question which, in one form or another, is raised again and again by philosophers, scientists, T. C. Mits and readers of the weekly comic supplements. A question which has been brought into the limelight above all by the genial work of Albert Einstein, and kept there by the unceasing efforts of astronomers to wrest the answer from a curiously reluctant Nature.

But what is the meaning of the question? What, indeed, is the meaning of each word in it? Properly to formulate and adequately to answer the question would require a critical excursus through philosophy and mathematics into physics and astronomy, which is beyond the scope of the present modest attempt. Here we shall be content to examine the rôles of deduction and observation in the problem of physical space, to exhibit certain high points in the history of the problem, and in the end to illustrate the viewpoint adopted by presenting a relatively simple caricature of Einstein's general theory of relativity. It is hoped that this, certainly incomplete and possibly naïve, description will present the essentials of the problem from a neutral mathematico-physical viewpoint in a form suitable for incorporation into any otherwise tenable philosophical position. Here, for example, we shall not touch directly upon the important problem of form versus substance—but if one wishes to interpret the geometrical substratum here considered as a formal backdrop against which the contingent relations of nature are exhibited, one should be able to do so without distorting the scientific content.

First, then, we consider geometry as a deductive science, a branch of mathematics in which a body of theorems is built up by logical processes from a postulated set of axioms (not "self-evident truths"). In logical position geometry differs not in kind from any other mathematical discipline—say the theory of numbers or the calculus of variations. As mathematics, it is not the science of measurement, despite the implications of its name—even though it did, in keeping with the name, originate in the codification of rules for land surveying. The principal criterion of its validity as a mathematical discipline

From *Albert Einstein: Philosopher-Scientist* (Library of Living Philosophers, Volume VII), 1949, Evanston, Illinois, pp. 313–32. Reprinted with the kind permission of The Library of Living Philosophers, Inc. and the Tudor Publishing Company.

is whether the axioms as written down are self-consistent, and the sole criterion of the truth of a theorem involving its concepts is whether the theorem can be deduced from the axioms. This truth is clearly relative to the axioms; the theorem that the sum of the three interior angles of a triangle is equal to two right angles, true in Euclidean geometry, is false in any of the geometries obtained on replacing the parallel postulate by one of its contraries. In the present sense it suffices for us that geometry is a body of theorems, involving among others the concepts of point, angle and a unique numerical relation called distance between pairs of points, deduced from a set of self-consistent axioms.

What, then, distinguishes Euclidean geometry as a mathematical system from those logically consistent systems, involving the same category of concepts, which result from the denial of one or more of its traditional axioms? This distinction cannot consist in its "truth" in the sense of observed fact in physical science; its truth, or applicability, or still better appropriateness, in this latter sense is dependent upon observation, and not upon deduction alone. The characteristics of Euclidean geometry, as mathematics, are therefore to be sought in its internal properties, and not in its relation to the empirical.

First, Euclidean geometry is a *congruence geometry*, or equivalently the space comprising its elements is *homogeneous and isotropic*; the intrinsic relations between points and other elements of a configuration are unaffected by the position or orientation of the configuration. As an example, in Euclidean geometry all intrinsic properties of a triangle—its angles, area, etc.,—are uniquely determined by the lengths of its three sides; two triangles whose three sides are respectively equal are "congruent"; either can by a "motion" of the space into itself be brought into complete coincidence with the other, whatever its original position and orientation may be. These motions of Euclidean space are the familiar translations and rotations, use of which is made in proving many of the theorems of Euclid. That the existence of these motions (the axiom of "free mobility") is a desideratum, if not indeed a necessity, for a geometry applicable to physical space, has been forcibly argued on *a priori* grounds by von Helmholtz, Whitehead, Russell and others; for only in a homogeneous and isotropic space can the traditional concept of a rigid body be maintained.[1]

But the Euclidean geometry is only one of several congruence geometries; there are in addition the "hyperbolic" geometry of Bolyai and Lobachewsky, and the "spherical" and "elliptic" geometries of Riemann and Klein. Each of these geometries is characterized by a real number K, which for the Euclidean geometry is zero, for the hyperbolic negative, and for the spherical

---

1. Technically this requirement, as expressed by the axiom of free mobility, is that there exist a motion of the 3-dimensional space into itself which takes an arbitrary configuration, consisting of a point, a direction through the point, and a plane of directions containing the given direction, into a standard such configuration. For an excellent presentation of this standpoint see B. A. W. Russell's *The Foundations of Geometry* (Cambridge, 1897), or Russell and A. N. Whitehead's article "Geometry VI: Non-Euclidean Geometry" 11th Ed. *Encyclopædia Britannica*.

and elliptic geometries positive. In the case of 2-dimensional congruence spaces, which *may* (but need not) be conceived as surfaces embedded in a 3-dimensional Euclidean space, the constant K may be interpreted as the *curvature* of the surface into the third dimension—whence it derives its name. This name and this representation are for our purposes at least psychologically unfortunate, for we propose ultimately to deal exclusively with properties intrinsic to the space under consideration—properties which in the later physical applications can be measured within the space itself—and are not dependent upon some extrinsic construction, such as its relation to an hypothesized higher dimensional embedding space. We must accordingly seek some determination of K—which we nevertheless continue to call curvature—in terms of such inner properties.

In order to break into such an intrinsic characterization of curvature, we first relapse into a rather naïve consideration of measurements which may be made on the surface of the earth, conceived as a sphere of radius R. This surface is an example of a 2-dimensional congruence space of positive curvature $K = 1/R^2$ on agreeing that the abstract geometrical concept "distance" r between any two of its points (not the extremities of a diameter) shall correspond to the lesser of the two distances *measured on the surface* between them along the unique great circle which joins the two points.[2] Consider now a "small circle" of radius r (measured on the surface!) about a point P of the surface; its perimeter L and area A (again measured on the surface!) are clearly less than the corresponding measures $2\pi r$ and πr^2 of the perimeter and area of a circle of radius r in the Euclidean plane. An elementary calculation shows that for sufficiently small r (i.e., small compared with R) these quantities on the sphere are given approximately by:

$$L = 2\pi r \, (1 - Kr^2/6 + \ldots),$$

(1)

$$A = \pi r^2 \, (1 - Kr^2/12 + \ldots).$$

Thus, the ratio of the area of a small circle of radius 400 miles on the surface of the earth to that of a circle of radius 40 miles is found to be only 99.92, instead of 100.00 as in the plane.

Another consequence of possible interest for astronomical applications is that in spherical geometry the sum σ of the three angles of a triangle (whose sides are arcs of great circles) is *greater* than 2 right angles; it can in fact be shown that this "spherical excess" is given by

(2) $\sigma - \pi = K\delta,$

where δ is the area of the spherical triangle and the angles are measured in radians (in which $180° = \pi$). Further, each full line (great circle) is of finite

---

2. The motions of the surface of the earth into itself, which enable us to transform a point and a direction through it into any other point and direction, as demanded by the axiom of free mobility, are here those generated by the 3-parameter family of rotations of the earth about its center (not merely the 1-parameter family of diurnal rotations about its "axis."!).

length $2\pi R$, and any two full lines meet in two points—there are no parallels!

In the above paragraph we have, with forewarning, slipped into a non-intrinsic quasi-physical standpoint in order to present the formulae (1) and (2) in a more or less intuitive way. But the essential point is that these formulae are in fact independent of this mode of presentation; they are relations between the mathematical concepts distance, angle, perimeter and area which follow as logical consequences from the axioms of this particular kind of non-Euclidean geometry. And since they involve the space-constant K, this "curvature" may in principle at least be determined *by measurements made on the surface*, without recourse to its embedment in a higher dimensional space.

Further, these formulae may be shown to be valid for a circle or triangle in the hyperbolic plane, a 2-dimensional congruence space for which $K < 0$. Accordingly here the perimeter and area of a circle are *greater*, and the sum of the three angles of a triangle *less*, than the corresponding quantities in the Euclidean plane. It may also be shown that each full line is of infinite length, that through a given point outside a given line an infinity of full lines may be drawn which do not meet the given line (the two lines bounding the family are said to be "parallel" to the given line), and that two full lines which meet do so in but one point.

The value of the intrinsic approach is especially apparent in considering 3-dimensional congruence spaces, where our physical intuition is of little use in conceiving them as "curved" in some higher-dimensional space. The intrinsic geometry of such a space of curvature K provides formulae for the surface area S and the volume V of a "small sphere" of radius r, whose leading terms are

$$S = 4\pi r^2 \ (1 - Kr^2/3 + \ldots),$$

(3)

$$V = 4/3\pi r^3 \ (1 - Kr^2/5 + \ldots).$$

It is to be noted that in all these congruence geometries, except the Euclidean, there is at hand a natural unit of length $R = 1/|K|^{\frac{1}{2}}$; this length we shall, without prejudice, call the "radius of curvature" of the space.

So much for the congruence geometries. If we give up the axiom of free mobility we may still deal with the geometry of spaces which have only limited or no motions into themselves.[3] Every smooth surface in 3-dimensional Euclidean space has such a 2-dimensional geometry; a surface of revolution has a 1-parameter family of motions into itself (rotations about its axis of symmetry), but not enough to satisfy the axiom of free mobility. Each such surface has at a point $P(x, y)$ of it an intrinsic "total curvature" $K(x, y)$, which will in general vary from point to point; knowledge of the curvature at all points essentially determines all intrinsic properties of the surface.[4] The deter-

3. We are here confining ourselves to metric (Riemannian) geometries, in which there exists a differential element ds of distance, whose square is a homogeneous quadratic form in the co-ordinate differentials.

4. That is, the "differential," as opposed to the "macroscopic," properties. Thus the Euclidean plane and a cylinder have the same differential, but not the same macroscopic, structure.

mination of $K(x, y)$ by measurements on the surface is again made possible by the fact that the perimeter L and area A of a closed curve, every point of which is at a given (sufficiently small) distance r from $P(x, y)$, are given by the formulae (1), where K is no longer necessarily constant from point to point. Any such variety for which $K = 0$ throughout is a ("developable") surface which may, on ignoring its macroscopic properties, be rolled out without tearing or stretching onto the Euclidean plane.

From this we may go on to the contemplation of 3- or higher dimensional ("Riemannian") spaces, whose intrinsic properties vary from point to point. But these properties are no longer describable in terms of a single quantity, for the "curvature" now acquires at each point a directional character which requires in 3-space 6 components (and in 4-space 20) for its specification. We content ourselves here to call attention to a single combination of the 6, which we call the "mean curvature" of the space at the point $P(x, y, z)$, and which we again denote by K—or more fully by $K(x, y, z)$; it is in a sense the mean of the curvatures of various surfaces passing through P, and reduces to the previously contemplated space-constant K when the space in question is a congruence space.[5] This concept is useful in physical applications, for the surface area S and the volume V of a sphere of radius r about the point $P(x, y, z)$ as center are again given by formulae (3), where now K is to be interpreted as the mean curvature $K(x, y, z)$ of the space at the point P. In four and higher dimensions similar concepts may be introduced and similar formulae developed, but for them we have no need here.

We have now to turn our attention to the world of physical objects about us, and to indicate how an ordered description of it is to be obtained in accordance with accepted, preferably philosophically neutral, scientific method. These objects, which exist for us in virtue of some pre-scientific concretion of our sense-data, are positioned in an extended manifold which we call physical space. The mind of the individual, retracing at an immensely accelerated pace the path taken by the race, bestirs itself to an analysis of the interplay between object and extension. There develops a notion of the permanence of the object and of the ordering and the change in time—another form of extension, through which object and subject appear to be racing together—of its extensive relationships. The study of the ordering of actual and potential relationships, the physical problem of space and time, leads to the consideration of geometry and kinematics as a branch of physical science. To certain aspects of this problem we now turn our attention.

We consider first that proposed solution of the problem of space which is based upon the postulate that space is an *a priori* form of the understanding. Its geometry must then be a congruence geometry, independent of the physical

5. The quantities here referred to are the six independent components of the Riemann-Christoffel tensor in 3 dimensions, and the "mean curvature" here introduced (not to be confused with the mean curvature of a surface, which is an extrinsic property depending on the embedment) is $K = -R'/6$, where R' is the contracted Ricci tensor. I am indebted to Professor Herbert Busemann, of the University of Southern California, for a remark which suggested the usefulness for my later purposes of this approach. A complete exposition of the fundamental concepts involved is to be found in L. P. Eisenhart's *Riemannian Geometry* (Princeton 1926).

content of space; and since for Kant, the propounder of this view, there existed but one geometry, space must be Euclidean—and the problem of physical space is solved on the epistemological, pre-physical, level.

But the discovery of other congruence geometries, characterized by a numerical parameter K, perforce modifies this view, and restores at least in some measure the objective aspect of physical space; the *a posteriori* ground for this space-constant K is then to be sought in the contingent. The means for its intrinsic determination is implicit in the formulae presented above; we have merely (!) to measure the volume V of a sphere of radius r or the sum σ of the angles of a triangle of measured area δ, and from the results to compute the value of K. On this modified Kantian view, which has been expounded at length by Russell,[6] it is inconceivable that K might vary from point to point—for according to this view the very possibility of measurement depends on the constancy of space-structure, as guaranteed by the axiom of free mobility. It is of interest to mention in passing, in view of recent cosmological findings, the possibility raised by A. Calinon (in 1889!) that the space-constant K might vary with time.[7] But this possibility is rightly ignored by Russell, for the same arguments which would on this *a priori* theory require the constancy of K in space would equally require its constancy in time.

In the foregoing sketch we have dodged the real hook in the problem of measurement. As physicists we should state clearly those aspects of the physical world which are to correspond to elements of the mathematical system which we propose to employ in the description ("realisation" of the abstract system). Ideally this program should prescribe fully the operations by which numerical values are to be assigned to the physical counterparts of the abstract elements. How is one to achieve this in the case in hand of determining the numerical value of the space-constant K?

Although K. F. Gauss, one of the spiritual fathers of non-Euclidean geometry, at one time proposed a possible test of the flatness of space by measuring the interior angles of a terrestrial triangle, it remained for his Göttingen successor K. Schwarzschild to formulate the procedure and to attempt to evaluate K on the basis of astronomical data available at the turn of the century.[8] Schwarzschild's pioneer attempt is so inspiring in its conception and so beautiful in its expression that I cannot refrain from giving here a few short extracts from his work. After presenting the possibility that physical

---

6. In the works already referred to in footnote 1 above.

7. "Les espaces géometriques," *Revue Philosophique*, vol. 27, pp. 588–595 (1889). The possibilities at which Calinon arrives are, to quote in free translation:

"1. Our space is and remains rigorously Euclidean;

"2. Our space realizes a geometrical space which differs very little from the Euclidean, but which always remains the same;

"3. Our space realizes successively in time different geometrical spaces; otherwise said, our spatial parameter varies with the time, whether it departs more or less away from the Euclidean parameter or whether it oscillates about a definite parameter very near to the Euclidean value."

8. "Ueber das zulässige Krümmungsmaass des Raumes," *Vierteljahrsschrift der astronomischen Gesellschaft*, vol. 35, pp. 337–347 (1900). The *annual parallax*, as used in practice, is one-half that defined below.

space may, in accordance with the neo-Kantian position outlined above, be non-Euclidean, Schwarzschild states (in free translation):

One finds oneself here, if one but will, in a geometrical fairyland, but the beauty of this fairy tale is that one does not know but what it may be true. We accordingly bespeak the question here of how far we must push back the frontiers of this fairyland; of how small we must choose the curvature of space, how great its radius of curvature.

In furtherance of this program Schwarzschild proposes:

A triangle determined by three points will be defined as the paths of light-rays from one point to another, the lengths of its sides a, b, c, by the times it takes light to traverse these paths, and the angles α, β, γ will be measured with the usual astronomical instruments.

Applying Schwarzschild's prescription to observations on a given star, we consider the triangle ABC defined by the position A of the star and by two positions B, C of the earth—say six months apart—at which the angular positions of the star are measured. The base $BC = a$ is known, by measurements within the solar system consistent with the prescription, and the interior angles β, γ which the light-rays from the star make with the base-line are also known by measurement. From these the *parallax* $p = \pi - (\beta + \gamma)$ may be computed; in Euclidean space this parallax is simply the inferred angle α subtended at the star by the diameter of the earth's orbit. In the other congruence geometries the parallax is seen, with the aid of formula (2) above, to be equal to

$$(2') \qquad p = \pi - (\beta + \gamma) = \alpha - K\delta,$$

where α is the (unknown) angle at the star A, and δ is the (unknown) area of the triangle ABC. Now in spite of our incomplete knowledge of the elements on the far right, certain valid conclusions may be drawn from this result. First, if space is hyperbolic $(K < 0)$, for distant stars (for which $\alpha \sim 0$), the parallax p will remain positive; hence if stars are observed whose parallax is zero to within the errors of observation, this estimated error will give an upper limit to the absolute value $- K$ of the curvature. Second, if space is spherical $(K > 0)$, for a sufficiently distant star (more distant than one-quarter the circumference of a Euclidean sphere of radius $R = 1/K^{\frac{1}{2}}$, as may immediately be seen by examining a globe) the sum $\beta + \gamma$ will exceed two right angles; hence the parallax p of such a star should be negative, and if no stars are in fact observed with negative parallax, the estimated error of observation will give an upper limit to the curvature K. Also, in this latter case the light sent out by the star must return to it after traversing the full line of length $2\pi R$, (πR in elliptic space), and hence we should, but for absorption and scattering, be able to observe the returning light as an anti-star in a direction opposite to that of the star itself! On the basis of the evidence then available, Schwarzschild concluded that if space is hyperbolic its radius of curvature $R = 1/(-K)^{\frac{1}{2}}$ cannot be less than 64 light-years (i.e., the distance light travels in 64 years), and that if the space is elliptic its radius of curvature $R = 1/K^{\frac{1}{2}}$ is at least 1600 light-years. Hardly imposing figures for us today, who believe on other astronomical grounds that objects as distant as 500 million light-years have been sighted in the Mt. Wilson telescope, and who are expecting to find objects at twice that distance with the new Mt. Palomar mirror! But the value for us of the work of Schwarzschild lies in its sound operational approach to the problem of physical geometry—in refreshing contrast to the pontifical pronouncement of H. Poincaré, who after reviewing the subject stated:[9]

> If therefore negative parallaxes were found, or if it were demonstrated that all parallaxes are superior to a certain limit, two courses would be open to us; we might either renounce Euclidean geometry, or else modify laws of optics and suppose that light does not travel rigorously in a straight line.
>
> It is needless to add that all the world would regard the latter solution as the more advantageous.
>
> The Euclidean geometry has, therefore, nothing to fear from fresh experiments. [!]

So far we have tied ourselves into the neo-Kantian doctrine that space must be homogeneous and isotropic, in which case our proposed operational approach is limited in application to the determination of the numerical value of the space-constant K. But the possible scope of the operational method is surely broader than this; what if we do apply it to triangles and circles and spheres in various positions and at various times and find that the K so determined is in fact dependent on position in space and time? Are we, following Poincaré, to attribute these findings to the influence of an external force postulated for the purpose? Or are we to take our findings at face value, and accept the geometry to which we are led as a natural geometry for physical science?

The answer to this methodological question will depend largely on the *universality* of the geometry thus found—whether the geometry found in one situation or field of physical discourse may consistently be extended to others—and in the end partly on the predilection of the individual or of his colleagues or of his times. Thus Einstein's special theory of relativity, which offers a physical kinematics embracing measurements in space and time, has gone through several stages of acceptance and use, until at present it is a universal and indispensable tool of modern physics. Thus Einstein's general theory of relativity, which offers an extended kinematics which includes in its geometrical structure the universal force of gravitation, was long considered by some contemporaries to be a *tour de force*, at best amusing but in practice useless. And now, in extending this theory to the outer bounds of the observed universe, the kind of geometry suggested by the present marginal data seems to many so repugnant that they would follow Poincaré in postulating some *ad hoc* force, be it a double standard of time or a secular change in the velocity of light or Planck's constant, rather than accept it.

---

9. *Science and Hypothesis*, p. 81; transl. by G. B. Halsted (Science Press 1929).

But enough of this general and historical approach to the problem of physical geometry! While we should like to complete this discussion with a detailed operational analysis of the solution given by the general theory of relativity, such an undertaking would require far more than the modest mathematical background which we have here presupposed. Further, the field of operations of the general theory is so unearthly and its *experimenta crucis* so delicate that an adequate discussion would take us far out from the familiar objects and concepts of the workaday world, and obscure the salient points we wish to make in a welter of unfamiliar and esoteric astronomical and mathematical concepts. What is needed is a homely experiment which could be carried out in the basement with parts from an old sewing machine and an Ingersoll watch, with an old file of *Popular Mechanics* standing by for reference! This I am, alas, afraid we have not achieved, but I do believe that the following example of a simple theory of measurement in a heat-conducting medium is adequate to expose the principles involved with a modicum of mathematical background. The very fact that it will lead to a rather bad and unacceptable physical theory will in itself be instructive, for its very failure will emphasize the requirement of universality of application—a requirement most satisfactorily met by the general theory of relativity.

The background of our illustration is an ordinary laboratory, equipped with Bunsen burners, clamps, rulers, micrometers and the usual miscellaneous impedimenta there met—at the turn of the century, no electronics required! In it the practical Euclidean geometry reigns (hitherto!) unquestioned, for even though measurements are there to be carried out with quite reasonable standards of accuracy, there is no need for sophisticated qualms concerning the effect of gravitational or magnetic or other general extended force-fields on its metrical structure. Now that we feel at home in these familiar, and disarming, surroundings, consider the following experiment:

Let a thin, flat metal plate be heated in any way—just so that the temperature T is not uniform over the plate. During the process clamp or otherwise constrain the plate to keep it from buckling, so that it can reasonably be said to remain flat by ordinary standards. Now proceed to make simple geometrical measurements on the plate with a short metal rule, which has a certain coefficient of expansion c, taking care that the rule is allowed to come into thermal equilibrium with the plate at each setting before making the measurement. The question now is, what is the geometry of the plate *as revealed by the results of these measurements?*

It is evident that, unless the coefficient of expansion c of the rule is zero, the geometry will not turn out to be Euclidean, for the rule will expand more in the hotter regions of the plate than in the cooler, distorting the (Euclidean) measurements which would be obtained by a rule whose length did not change according to the usual laboratory standards. Thus the perimeter L of a circle centered at a point at which a burner is applied will surely turn out to be greater than π times its measured diameter $2r$, for the rule will expand in measuring through the hotter interior of the circle and hence give a smaller reading than if the temperature were uniform. On referring to the first of formulae (1) above it is seen that the plate would seem to have a negative curvature K at the center of the circle—the kind of structure exhibited by an ordinary twisted surface in the neighborhood of a "saddle-point." In general the curvature will vary from point to point in a systematic way; a more detailed mathematical analysis of the situation shows that, on removing heat sources and neglecting radiation losses from the faces of the plate, K is everywhere negative and that the "radius of curvature" $R = 1/(-K)^{\frac{1}{2}}$ at any point P is inversely proportional to the rate s at which heat flows past P. (R is in fact equal to k/cs, where k is the coefficient of heat conduction *of the plate* and c is as before the coefficient of expansion *of the rule*.) The hyperbolic geometry is accordingly realized when the heat flow is constant throughout the plate, as when the long sides of an elongated rectangle are kept at different fixed temperatures.[10]

And now comes the question, what is the true geometry of the plate? The flat Euclidean geometry we had uncritically agreed upon at the beginning of the experiment, or the un-Euclidean geometry revealed by measurement? It is obvious that the question is improperly worded; the geometry is determinate only when we prescribe the method of measurement, i.e., when we set up a correspondence between the physical aspects (here readings on a definite rule obtained in a prescribed way) and the elements (here distances, in the abstract sense) of the mathematical system. Thus our original common-sense requirement that the plate not buckle, or that it be measured with an invar rule (for which $c \sim o$), leads to Euclidean geometry, while the use of a rule with a sensible coefficient of expansion leads to a locally hyperbolic type of Riemannian geometry, which is in general not a congruence geometry.

There is no doubt that anyone examining this situation will prefer Poincaré's common-sense solution of the problem of the physical geometry of the plate—i.e., to attribute to it Euclidean geometry, and to consider the measured deviations from this geometry as due to the action of a force (thermal stresses in the rule). Most compulsive to this solution is the fact that this disturbing force lacks the requirement of universality; on employing a brass rule in place of one of steel we would find that the local curvature is trebled—and an ideal rule ($c = o$) would, as we have noted, lead to the Euclidean geometry.

In what respect, then, does the general theory of relativity differ in principle from this geometrical theory of the hot plate? The answer is: *in its universality*; the force of gravitation which it comprehends in the geometrical structure acts equally on all matter. There is here a close analogy between the

---

10. This case, in which the geometry is that of the Poincaré half-plane, has been discussed in detail by E. W. Barankin "Heat Flow and Non-Euclidean Geometry," *American Mathematical Monthly*, vol. 49, pp. 4–14 (1942). For those who are numerically-minded it may be noted that for a steel plate ($k = 0.1$ cal/cm deg) 1 cm thick, with a heat flow of 1 cal/cm^2 sec, the natural unit of length R of the geometry, as measured by a steel rule ($c = 10^{-5}$/deg), is 10^6 cm $\sim$ 328 feet!

gravitational mass M of the field-producing body (Sun) and the inertial mass m of the test-particle (Earth) on the one hand, and the heat conduction k of the field (plate) and the coefficient of expansion c of the test-body (rule) on the other. *The success of the general relativity theory of gravitation as a physical geometry of space-time is attributable to the fact that the gravitational and inertial masses of any body are observed to be rigorously proportional for all matter.* Whereas in our geometrical theory of the thermal field the ratio of heat conductivity to coefficient of expansion varies from substance to substance, resulting in a change of the geometry of the field on changing the test-body.

From our present point of view the great triumph of the theory of relativity lies in its absorbing the universal force of gravitation into the geometrical structure; its success in accounting for minute discrepancies in the Newtonian description of the motions of test-bodies in the solar field, although gratifying, is nevertheless of far less moment to the philosophy of physical science.[11] Einstein's achievements would be substantially as great even though it were not for these minute observational tests.

Our final illustration of physical geometry consists in a brief reference to the cosmological problem of the geometry of the observed universe as a whole—a problem considered in greater detail elsewhere in this volume. *If* matter in the universe can, taken on a sufficiently large scale (spatial gobs millions of light-years across), be considered as uniformly distributed, and if (as implied by the general theory of relativity) its geometrical structure is conditioned by matter, then to this approximation our 3-dimensional astronomical space must be homogeneous and isotropic, with a spatially-constant K which may however depend upon time. Granting this hypothesis, how do we go about measuring K, using of course only procedures which can be operationally specified, and to which congruence geometry are we thereby led? The way to the answer is suggested by the second of the formulae (3), for if the nebulae are by-and-large uniformly distributed, then the number N within a sphere of radius r must be proportional to the volume V of this sphere. We have

---

11. Even here an amusing and instructive analogy exists between our caricature and the relativity theory. On extending our notions to a 3-dimensional heat-conducting medium (without worrying too much about how our measurements are actually to be carried out!), and on adopting the standard field equation for heat conduction, the "mean curvature" introduced above is found at any point to be $-(cs/k)^2$, which is of second order in the characteristic parameter c/k. (The case in which the temperature is proportional to $a^2 - r^2$, which requires a continuous distribution of heat sources, has been discussed in some detail by Poincaré, *Loc. cit.* pp. 76–78, in his discussion of non-Euclidean geometry.) The field equation may now itself be given a geometrical formulation, at least to first approximation, by replacing it by the requirement that the mean curvature of the space *vanish* at any point at which no heat is being supplied to the medium—in complete analogy with the procedure in the general theory of relativity by which the classical field equations are replaced by the requirement that the Ricci contracted curvature tensor vanish. Here, as there, will now appear certain deviations, whose magnitude here depends upon the ratio c/k, between the standard and the modified theories. One curious consequence of this treatment is that on solving the modified field equation for a spherically-symmetric source (or better, sink) of heat, one finds precisely the same spatial structure as in the Schwarzschild solution for the gravitational field of a spherically-symmetric gravitational mass—the correspondence being such that the geometrical effect of a sink which removes 1 calorie per second from the medium is equivalent to the gravitational effect of a mass of 10^{23} gm, e.g., of a chunk of rock 200 miles in diameter!

then only to examine the dependence of this number N, as observed in a sufficiently powerful telescope, on the distance r to determine the deviation from the Euclidean value. But how is r operationally to be defined?

If all the nebulae were of the same intrinsic brightness, then their apparent brightness as observed from the Earth should be an indication of their distance from us; we must therefore examine the exact relation to be expected between apparent brightness and the abstract distance r. Now it is the practice of astronomers to assume that brightness falls off inversely with the square of the "distance" of the object—as it would do in Euclidean space, if there were no absorption, scattering, and the like. We must therefore examine the relation between this astronomer's "distance" d, as inferred from apparent brightness, and the distance r which appears as an element of the geometry. It is clear that *all* the light which is radiated at a given moment from the nebula will, after it has traveled a distance r, lie on the surface of a sphere whose area S is given by the first of the formulae (3). And since the practical procedure involved in determining d is equivalent to assuming that all this light lies on the surface of a Euclidean sphere of radius d, it follows immediately that the relationship between the "distance" d used in practice and the distance r dealt with in the geometry is given by the equation

$$4\pi d^2 = S = 4\pi r^2 (1 - Kr^2/3 + \ldots);$$

whence, to our approximation

(4)
$$d = r(1 - Kr^2/6 + \ldots), \text{ or}$$
$$r = d(1 + Kd^2/6 + \ldots).$$

But the astronomical data give the number N of nebulae counted out to a given inferred "distance" d, and in order to determine the curvature from them we must express N, or equivalently V, to which it is assumed proportional, in terms of d. One easily finds from the second of the formulae (3) and the formula (4) just derived that, again to the approximation here adopted,

(5)
$$V = 4/3\pi d^3 (1 + 3/10Kd^2 + \ldots).$$

And now on plotting N against inferred "distance" d and comparing this empirical plot with the formula (5), it should be possible operationally to determine the "curvature" K.[12]

The search for the curvature K indicates that, after making all known corrections, the number N seems to increase faster with d than the third power, which would be expected in Euclidean space, hence K is *positive*. The space

---

12. This is, of course, an outrageously over-simplified account of the assumptions and procedures involved. All nebulae are *not* of the same intrinsic brightness, and the modifications required by this and other assumptions tacitly made lead one a merry astronomical chase through the telescope, the Earth's atmosphere, the Milky Way and the Magellanic Clouds to Andromeda and our other near extra-galactic neighbors, and beyond. The story of this search has been delightfully told by E. P. Hubble in his *The Realm of the Nebulae* (Yale 1936) and in his *Observational Approach to Cosmology* (Oxford 1937), the source of the data mentioned below.

implied thereby is therefore bounded, of finite total volume, and of a present "radius of curvature" $R = 1/K^{\frac{1}{2}}$ which is found to be of the order of 500 million light-years. Other observations, on the "red-shift" of light from these distant objects, enable us to conclude with perhaps more assurance that this radius is increasing in time at a rate which, if kept up, would double the present radius in something less than 2000 million years.

With this we have finished our brief account of Geometry as a branch of Physics, a subject to which no one has contributed more than Albert Einstein, who by his theories of relativity has brought into being physical geometries which have supplanted the tradition-steeped *a priori* geometry and kinematics of Euclid and Newton.

GEORGE GAMOW:

Modern Cosmology

THE SUBJECT of cosmology is the study of our Universe's general features, its extension in space and its duration in time. With the great 200-inch telescope on Palomar Mountain man today can look over two billion light-years into space and see nearly a billion galaxies, spread more or less uniformly through that vast volume.

It is important to realize that we are looking not only far into distance but also far back in time. For instance, the present-day photograph of the great Andromeda Nebula shows that group of stars as it looked about two million years ago, for it has taken this time for its light to reach us. The most distant galaxies detected by the 200-inch are seen by us in the state in which they were more than two billion years ago.

The view of the Universe that we are seeing at this instant can be represented schematically by a cone-shaped diagram that takes the time factor into account [*see illustration on page* 394]. At the apex of the cone is our own galaxy as it is now; down the surface of the cone are the other galaxies photographed by our telescopes as they were at dates in the past corresponding to their distance from us. A horizontal cross section through the cone would show the Universe as it was at a given date; this is known as a world map.

Theoretical cosmology attempts to correlate the observed facts about the Universe at large with known physical laws and to draw a consistent picture of the Universe's structure in space and its changes in time. In studying the structure we must accept the Copernican point of view and deny to man the honor of a privileged position in the Universe; in other words, we must assume that the structure of space is very much the same in distant regions as it is in the part we can observe. We cannot suppose that our particular neighborhood is specially adorned with beautiful spiral galaxies for the enjoyment of professional and amateur astronomers.

From *Scientific American*, March 1954, Volume 190, No. 3, pp. 55–63. Reprinted by the kind permission of the publishers, The Scientific American.

The Paradox of Finite Light

To make clear the nature of the problems with which cosmologists must deal, let us begin with a paradox first pointed out by the German astronomer Heinrich Olbers more than a century ago. If stars are distributed uniformly through space, and if space is infinite, why, he asked, are we not blinded by their light? (Nowadays we must think of space as filled with galaxies, then unknown, but that does not affect the question.) Olbers' argument goes as follows: Suppose

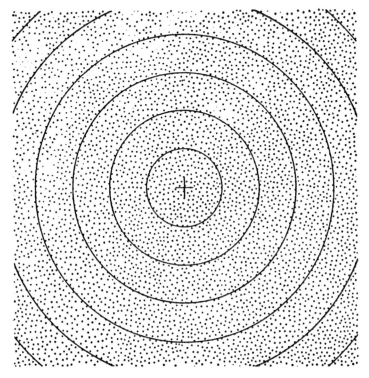

Olbers' paradox supposed an even distribution of stars in the Universe and concentric layers of equal thickness. The modern paradox substitutes galaxies for the stars.

we think of space as a series of concentric spheres with ourselves at the center—imagine it as having the structure of an infinitely big onion. Each sphere is larger in radius than the next smaller one by a certain fixed amount; that is, the thickness of the onion layers, or shells, is uniform [*see illustration above*]. Now the volume of each successive shell is greater than that of the next smaller one in proportion to the square of the increase in radius, and the number of galaxies in the shell is larger in the same proportion. On the other hand, the light reaching the center from galaxies farther and farther away decreases in proportion to the square of the increase in radius. Hence the two opposing factors—the increase in the number of galaxies and the reduction in light from each galaxy—cancel out, and we should expect the center to receive the same amount of light from every shell, no matter how near or how far. Therefore in an infinite universe any given point theoretically should receive an infinite amount of light!

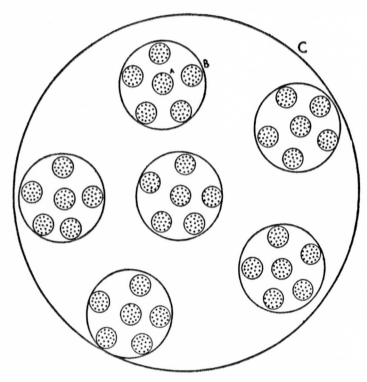

Charlier's Universe supposed galaxies formed into clusters formed into larger clusters and so on. The dots are galaxies; the circles, clusters of galaxies or of clusters.

Actually, of course, the light sources partly screen one another and as a consequence of this interference the illumination could not exceed the surface brightness of an individual star. But this means that our night sky would be as bright as the sun's disk from horizon to horizon! In daytime the sun itself would be practically unnoticeable against the shining background of the galaxies in the heavens.

What is wrong with this reasoning? Early in this century the Swedish astronomer C. V. L. Charlier proposed an ingenious answer to Olbers' paradox. The visible stars in the Milky Way system altogether occupy so negligible a fraction of the sky area, and our galaxy itself is so limited—a droplet in the vast reaches of space—that all the Milky Way's starlight scarcely illuminates the earth at all. And the distances between galaxies are far greater than those between stars in our galaxy. Because of their distance from us and their great dilution in space, the total illumination of our night sky from the billion galaxies within the range of the 200-inch telescope is only a small percentage of the faint light we get from the Milky Way. This still does not invalidate Olbers' argument, if we assume that space is filled with the same density of galaxies for an indefinite distance beyond the range of our telescopes. But Charlier suggested that there may be a limit to this population: that we may be part of a giant cluster of galaxies which is surrounded by empty space at some distance beyond our telescopic range. If this is so, the total illuminiation at the earth from the cluster would indeed be negligible.

Of course we cannot stop there; we have to assume that there are other giant clusters, and that they are combined in superclusters, and these in turn in super-superclusters, and so on without end. It is apparent, however, that as we take in larger and larger volumes of space, the mean number of galaxies per unit of space becomes smaller and smaller, because of the increasingly large portions of empty space between the clusters and combinations of clusters [*see illustration, page* 392]. Since Olbers' paradox rested on the assumption that the number of galaxies per space unit remains the same no matter how large a volume is considered, Charlier's "hierarchy Universe" neatly solved the puzzle.

Expanding Space

Today we have a more direct answer to Olbers' paradox: namely, the shift toward the red end of the spectrum in the light reaching us from distant galaxies, which weakens or "dims" their light in proportion to their distance from us. The discovery of the red shift has had far more important consequences, however, than merely the solving of old puzzles; it has profoundly changed man's thinking about the cosmos. The chief change was to introduce the notion of the expanding Universe—an idea which has now become firmly established. One should remember that the expanding Universe theory finds support not only in the red shift but also in classical Newtonian mechanics. Because of the gravitational forces between the galaxies, the cosmic system cannot be expected to remain static, just as a tennis ball cannot hang motionless in midair. The system must either contract, under the forces of gravitational attraction, or expand, as the result of some dispersing force overcoming the attraction.

From the observed red shift one can calculate that the galaxies are fleeing from one another with a kinetic energy which is about 50 times as great as the

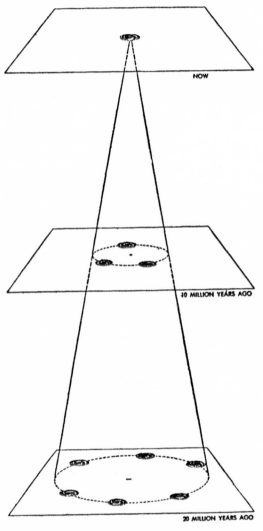

Cone-shaped map depicts our historical view of the Universe. At the vertex is our galaxy. The galaxies within 10 million light-years appear as they did 10 million years ago, and so on. The relative size of these galaxies has been greatly exaggerated for clarity.

potential energy of gravitational attraction between them. This means that the present expansion of the Universe will never stop, or, in mathematical language, that the expansion of our Universe is hyperbolic. Further, from the observed recession velocity and the distances between the galaxies one can also compute how long ago the Universe began to expand from its original compressed state.

On this basis Edwin P. Hubble and Milton L. Humason calculated a quarter of a century ago that the age of the Universe was 1.8 billion years. Until recently that estimate stood in serious contradiction to the estimates of geologists and astrophysicists, who calculated from the decay of radioactive materials in the earth and from the rate of burning of nuclear fuel by the stars that the Universe must be five billion years old. But the discrepancy was eliminated a little over a year ago when Walter Baade of the Mount Wilson and Palomar Observatories discovered that as a result of new observations the distances between galaxies, and therefore the age calculated on the basis of the red shift, must be multiplied by a factor of 2.8. This correction (2.8 times 1.8) raises the expansion age to five billion years, in perfect agreement with the geological and astrophysical estimates!

Curved Space

So far we have been discussing the properties of the Universe without reference to the so-called relativistic cosmology based on Einstein's general theory of relativity. The essential point of Einstein's general theory is the introduction of the notion of curved space, and the identification of the effect of gravitational forces with the change of free motion of material bodies in a curved non-Euclidean space. After the great success of his theory in predicting the deflection of light rays by the gravitational field around the sun, Einstein proceeded to apply the theory to the Universe as a whole. According to the cosmological principle of uniformity, one should assume that the curvature of space is the same throughout the Universe; in terms of a two-dimensional analogy, our Universe should be round like the surface of a basketball. There are two possible types of curvature for a curved surface: positive and negative. Positive curvature turns inward, like the surface of a ball; negative curvature turns outward, like a western saddle. Between these of course lies the surface of zero curvature, which is perfectly flat.

In complete analogy with these two-dimensional examples, three-dimensional space can be curved positively or negatively, with zero curvature representing ordinary Euclidean space. In Euclidean space the volume of a sphere increases as the cube of its radius. But in a positively curved space the volume increases at a less rapid rate, while in a negatively curved space it increases more rapidly.

Now if the space of our Universe is curved either way, in principle it should be possible to find that out observationally by counting the galaxies within volumes of space of successively greater radius from us. If the number of galaxies increases more slowly or more rapidly than the cube of the distance, this would indicate a positive or a negative curvature. During the past two decades the late E. P. Hubble carried out such counts at the Mount Wilson and Palomar Observatories, but unfortunately with very indefinite results.

The difficulty is that we can expect to find noticeable curvature effects only at very great distances, and we cannot make reliable distance estimates for galaxies so far away. The only way we can judge their distance is by the faintness of their light. But we must also remember that we are looking far back in time. The intrinsic brightness of galaxies may change with time. Consequently we cannot be sure that a distant galaxy which is fainter than another is farther away; it may instead be at a different stage of evolution. Until we know more about evolutionary changes in galaxies, we shall not be able to reach any definite conclusions about the curvature of space from counts of the nebulae.

Models of the Universe

Relativistic cosmology went through several interesting stages. Due to an algebraic error in his calculations, Einstein concluded that the Universe must be static. (This was, of course, before the discovery of the red shift.) The only way to make such an idea work was to introduce some kind of repulsive force to counteract the gravitational one. This force, in contrast to any other known in physics, would have to be assumed to increase with distance. Einstein met this dilemma by introducing into his equations of general relativity the so-called "cosmological term," and that led to the famous spherical Universe—a finite cosmos closed on itself. But Einstein's static model failed to agree with astronomical observations: it was too small to represent the actual Universe.

Soon afterward the Dutch astronomer Willem de Sitter found another possible solution. His model of the Universe, however, turned out to be even less acceptable than Einstein's. It satisfied the equations only if one assumed that space was completely empty and there was no matter whatsoever!

Then in the early 1920s the Russian mathematician Alexander Friedmann noticed the error in Einstein's computation. He showed that with this correction one could get solutions of basic relativistic equations which yielded models of a Universe that changed with time. The matter was further developed by the Belgian cosmologist Georges Lemaître. Associating Friedmann's dynamic Universe with the new red-shift observations of Hubble and Humason, he formulated the theory of the expanding Universe in the form in which we know it now.

Einstein's original equations of a static Universe related its curvature to the mean density of matter in space and to an *ad hoc* cosmological constant. The present dynamic equations of the expanding Universe connect its curvature with two directly observable quantities: the mean density of matter and the rate of expansion. With the observed value of these two quantities one can calculate that the curvature of our Universe is negative, so that space is open and infinite. It bends in the way a western saddle does. The radius of curvature comes out as five billion light-years.

About five years ago an entirely new idea was introduced into theoretical cosmology by the British mathematicians Herman Bondi and Thomas Gold. They started from the assumption that if the Universe is homogeneous in space, it must also be homogeneous in time. This would mean that any region of the Universe must always have looked in the past, and will always look in the future, essentially the same as it looks now. The only way to reconcile this postulate with the well established movement of the galaxies away from one another was to assume that new galaxies are continuously being formed to compensate for the dispersal of the older ones. If new galaxies are being formed, then new matter must be continuously created throughout space. Bondi and Gold calculated that the creation of new matter must proceed at the rate of one hydrogen atom per hour per cubic mile in intergalactic space. This idea of Bondi and Gold was soon extended by the British astronomer Fred Hoyle, who modified the original Einstein equations of general relativity so that they would permit the continuous creation of matter in space.

Besides circumventing the philosophical question as to the "beginning" of the Universe, the Bondi-Gold-Hoyle theory claimed to dispose of the painful discrepancy in the estimates of the age of the Universe that was still troubling astronomers at the time. If new galaxies were continuously being created, the Universe must be populated with galaxies of all ages, from babies to oldsters living on borrowed time. Bondi, Gold and Hoyle assumed that the average age of the population was about one third of the figure of 1.8 billion years that Hubble had arrived at for the total age of the Universe, that is, 600 million years. According to this point of view, since our own galaxy is estimated to be several billion years old, we are living in a rather elderly member of the population.

The recent revision of distances that eliminated the age discrepancy and placed the age of the Universe at five billion years does not disprove the Bondi-Gold-Hoyle theory of a steady-state Universe; it merely raises the average age of galaxies to about 1.7 billion years and makes our own galaxy three times instead of nine times as old as the average. Nevertheless the elimination of the discrepancy does deprive the steady-state idea of its main support. As far as observations go, the weight of the evidence at present is definitely in favor of the idea of an evolving Universe rather than a steady-state one such as is envisioned by Bondi, Gold and Hoyle.

One of the most important recent pieces of evidence was a discovery made in 1948 by the U. S. astronomers Joel Stebbins and Albert E. Whitford. Using light filters of different colors, they measured the reddening of light from distant galaxies, and to their own and everyone else's great surprise they found the reddening to be about 50 per cent greater than could be accounted for by the red shift, or Doppler effect, due to the galaxies' movement away from us.

It is well known that light rays may be reddened by dust, which scatters and screens out the blue part of the light; the dust in the atmosphere is what makes the sun look red at sunrise and sunset. The excess reddening of the distant galaxies was therefore attributed at first to dust floating in intergalactic space. But when the investigators calculated how much dust would have to be present there to produce the observed reddening, they got an astounding result: the dust in intergalactic space would add up to 100 times as much matter as the total amount concentrated in the galaxies themselves! This finding not only contradicted all accepted views about the distribution of matter in the Universe but also played havoc with astronomers' distance scales and the theory of the curvature of space.

Fortunately further studies by Whitford extricated the astronomers from this impasse. To understand them we must look into the composition of galaxies. Baade has shown that there are two kinds of stellar populations: Population I consists predominantly of blue stars, with great clouds of dust and gas floating among them; Population II, mainly redder stars, has no dust or gas whatever. The spiral galaxies are made up largely of Population I, the elliptical ones of Population II. Since interstellar dust and gas afford material for forming new stars, one may assume that spiral galaxies are in a more or less steady state, with newborn stars replacing old ones that are slowly fading out. On the other hand, elliptical galaxies, lacking dust and gas, are producing no new stars to replace the dying population. The two types of communities may be likened respectively to the population of Cambridge, Mass., a dynamic community in which new births replace those who die, and to the Harvard University alumni of the class of 1925, a declining population.

Now Stebbins and Whitford had limited their original observations to elliptical galaxies. In his new studies Whitford included spiral galaxies, located in the same clusters. And he found that the light from the spirals showed no excess reddening! Thus the extra reddening of the light from the elliptical galaxies could not be due to any factor (such as dust) affecting it in its travel through space, since the elliptical and spiral galaxies observed lay side by side the same distance from us. The only possible conclusion was that the excessive redness of the distant elliptical galaxies is due to the fact that we see them now as they were in a distant past when they were intrinsically redder. This finding is one of the strongest evidences in favor of the idea that galaxies are evolving and against the theory of a steady-state Universe.

There are other arguments against that theory. For instance, if our own galaxy is older than the average, we should expect the stars in our system to be older and different from most of those in neighboring galaxies, but no such general difference has been observed. On the whole it appears that the steady-Universe theory, attractive as it may seem from certain philosophical points of view, is neither necessary nor correct.

The Evolving Universe

Returning now to Lemaître's theory of the expanding Universe, let us try to explain how the cosmos evolved from the original highly compressed, very hot gas to stars, galaxies and matter as we know it today. First of all we must consider the relation between matter (represented by particles such as protons, neutrons and electrons) and radiation (represented by light quanta). In classical physics it was customary to regard matter as ponderable and radiation as imponderable, but we know now that radiant energy has mass, which is calculated, according to Einstein's basic law, by dividing the quantity of energy by the square of the velocity of light. On the earth the weight of radiant energy is negligibly small compared to that of matter: the total mass of all the light quanta passing through the atmosphere on a bright, sunny day is less than a thousandth of a millionth of a millionth of a millionth of the weight of the air. Heat radiation is slightly heavier than light, but its weight amounts to only one microgram per 10 billion tons of air in the atmosphere!

AGE	0-5 MINUTES	5-30 MINUTES	250 MILLION YEARS	1 BILLION YEARS	5 BILLION YEARS
ORGANIZATION					

Evolution of the Universe is symbolized at five stages. During the first five minutes of its expansion photons (*wavy lines*) outweighed solitary particles of matter such as protons (*black circles*), neutrons (*larger white circles*) and electrons (*smaller white circles*). Between five and 30 minutes matter had gained the upper hand and the fundamental particles had begun to coalesce into more complex nuclei such as those of deuterium (*proton and neutron*) and helium (*two protons and two neutrons*). After 250 million years the primordial gas began to break up into huge protogalaxies. After a billion years the matter in the protogalaxies had condensed into stars and planets. The present epoch is characterized here by the presence of life on at least one planet.

In interstellar and intergalactic space the ratio is not so large: the mass of matter there is only about 1,000 times the mass of the stellar radiation. Still, in the Universe as we know it today matter is everywhere more massive than radiation. But it need not always have been so. During the early stages of the Universe's evolution the mass density of radiation must have exceeded that of ordinary matter. The reason for that conclusion, which the writer first suggested several years ago, lies in the different behavior of matter and radiation. Imagine two cylinders, one filled with a material gas, the other a vacuum containing only thermal radiation. Both cylinders are sealed and thermally insulated, and the one containing the radiation has its inner walls made of an ideal mirror which does not absorb radiation. The cylinders have movable pistons. Now we pull the pistons, increasing the volume of space in each cylinder. In the cylinder filled with material gas, the density of the gas will be reduced in direct proportion to the increase in volume. But in the other cylinder, the mass density of the radiation will fall off more sharply, because the energy (and consequently the mass) of each quantum of radiation will be reduced by reflection from the receding piston. The laws of physics tell us the radiation's mass density will decrease in the ratio of 4/3 to the increase in volume.

Applying similar considerations to the Universe as a whole, we arrive at the conclusion that once upon a time in the distant past radiant energy had the upper hand over ordinary matter; there must have been pounds and pounds of light quanta for every ounce of atoms.

A Universe filled almost entirely with thermal radiation presents a rather simple case in the relativistic theory of expanding space. One can show that, starting from the time of maximum compression, all distances will increase in proportion to the square root of the elapsed time, and that the temperature of the radiation will decrease in inverse proportion to the square root of the elapsed time. The temperature of the Universe at any date is equal to 15 billion degrees absolute (degrees Centigrade above absolute zero) divided by the square root of its age expressed in seconds. Thus we get a chronological picture of the "changing climate" of our Universe: at the age of five minutes its mean temperature was about one billion degrees absolute; at one day it was about 40 million degrees (comparable to the temperature at the center of the sun or of an atomic bomb); at 300,000 years it was 6,000 degrees (the temperature at the surface of the sun) and at 10 million years it was 300 degrees (about room temperature).

Computing the mass densities of radiation and of matter at various epochs, we can find the date of the great event when matter took over from radiation, i.e., surpassed it in mass density. The date was about the year 250,000,000 A.B. (After the Beginning). The temperature of space was then about 170 degrees absolute, and the density both of radiation and of matter was comparable with the present density of interstellar gas. The Universe, in short, was dark and cool.

The Genesis of Galaxies

The transition from the reign of thermal radiation to the reign of matter must have been characterized by a very important event: formation of giant gaseous clouds. From these "protogalaxies" the galaxies of today must have developed, somewhat later, by the condensation of gas into individual stars. During the period when matter had played only a secondary role in the infinite ocean of thermal radiation, it had had, so to speak, no will of its own; the particles of matter were "dissolved" in the thermal radiation, much as molecules of salt are dissolved in water. As soon as matter took the upper hand, however, the forces of gravity acting between the particles must have caused a growing inhomogeneity of the matter in space. The English astronomer James Jeans showed more than half a century ago that the size of the clouds into which a gas of particles will be collected by gravitational forces can be calculated from the density and temperature of the spread-out gas. Using Jeans's formula and the transition-period temperature and density values given above, we find that the primordial gas clouds must have been about 40,000 light-years across, and each cloud must have had a total mass about 200 million times that of our sun. These figures, derived purely from theory, are in quite reasonable agreement with the observed figures for the average dimensions and mass of the present galaxies. (The Milky Way and the Andromeda Nebula are considerably larger than the average galaxy.)

The protogalaxies were pulled apart by the general expansion process. Their material later condensed into billions of stars, presumably by repetition on a smaller scale of Jeans's accretion process. Planets were formed, and the Universe again became brightly illuminated, as a result of nuclear reactions taking place in the interiors of the stars. But these "secondary" processes are a topic in themselves, which we shall not discuss in detail here.

The Beginning

Let us now go back to the beginning—the earliest stages of expansion. According to our calculations, when our Universe was five minutes old its temperature was a billion degrees, and it must have been still higher before that. At such temperatures particles move with energies of millions of electron volts—energies comparable with those in modern atom-smashing accelerators. This means that nuclear reactions must have been going on at a high rate all through the matter of the Universe. It is natural to conclude that the chemical elements were formed, in the relative abundances that were to make up the Universe we know, during that early stage of evolution. This assumption is strengthened by the fact that the natural radioactive elements are calculated today, from the extent of decay, to be about five billion years old.

During the first few minutes of the Universe's existence matter must have consisted only of protons, neutrons and electrons, for any group of particles that combined momentarily into a composite nucleus would immediately have dissociated into its components at the extremely high temperature. One can call the mixture of particles *ylem* (pronounced eelem)—the name that Aristotle gave to primordial matter. As the Universe went on expanding and the temperature of ylem dropped, protons and neutrons began to stick together, form-ing deuterons (nuclei of heavy hydrogen), tritons (still heavier hydrogen), helium and heavier elements.

On the basis of what we know about the behavior of nuclear particles and of the assumptions about the rate of temperature and density changes in the expanding Universe, one can calculate the net result of all the possible nuclear reactions that must have taken place during those early minutes of the Universe's history. The time available for the formation of the elements must have been very short, for two reasons: (1) the free neutrons in the original ylem would have decayed rapidly, and (2) the temperature quickly dropped below the level at which nuclear reactions could take place. The mean life of a neutron is known to be only about 12 minutes; hence half an hour after the expansion had started there would have been practically no neutrons left if they had not been combined in atomic nuclei. Favorable temperature conditions lasted about the same length of time. Thus all the chemical elements must have been formed in that half-hour.

Many people would argue that it makes no physical sense to talk about half an hour which took place five billion years ago. To answer that criticism, let us consider a site, somewhere in Nevada where an atomic bomb was set off several years ago. The site is still "hot" with long-lived fission products. It took only about one microsecond for the nuclear explosion to produce all the fission products. And simple arithmetic will show that a period of several years stands in the same ratio to one microsecond as five billion years do to a half-hour!

Early Matter

Calculations of the rate at which elementary atomic nuclei would have been synthesized under the assumed conditions were carried out by the writer a number of years ago and were later extended by Enrico Fermi and Anthony L. Turkevich. The composite nuclei whose production was estimated were deuterium (a combination of a proton and a neutron), tritium (one proton and two neutrons) and two isotopes of helium. By the end of 30 minutes free neutrons would practically have disappeared, and after that there would be no further change in the relative abundances of these elementary nuclei. At that time the Universe, according to these calculations, would have consisted of roughly equal amounts of hydrogen and helium, and about 1 per cent of the original ylem would have been converted into rare isotopes of hydrogen and helium which could combine to form the nuclei of heavier elements.

Now the fact that the amounts of hydrogen and helium come out approximately equal in these calculations is highly gratifying, because this is just about the relative abundance of these two elements in the Universe today. The results of the calculations gives good support to the expansion theory, for it can be shown that the Universe would have emerged from the ylem state consisting practically entirely of hydrogen or entirely of helium if conditions had been much different from those postulated.

Beyond these first two elements, however, the theory runs into a serious and as yet unresolved difficulty. The theory assumes that the light elements combined in successive steps to form the heavier ones. Helium consists of four nucleons (nuclear particles); that is, its atomic mass is four. The next nucleus should have the atomic mass five, but the fact is that no nucleus of mass five exists; at least, none of any appreciable length of life is known. For some reason five nucleons simply do not hold together. After helium 4 the next nucleus is an isotope of lithium of mass six. One must therefore assume that helium was built up to the next nucleus either by the simultaneous capture of two neutrons (an extremely unlikely event) or by fusion with a tritium nucleus. But the rate at which such fusions could have occurred under the given conditions is much too low to account for the amount of heavier elements that was actually produced. No likely reaction that bridges the gap at mass five has yet been found.

Beyond mass five there is little or no trouble; once that gap has been bridged, one can account quite satisfactorily for the relative abundances of the elements from lithium up through the periodic table to uranium, as has been shown by calculations carried out by the writer, Ralph A. Alpher and Robert C. Herman. If no way is found to bridge the gap, we may have to conclude that the main bulk of the heavier elements was formed not in the early stages of the Universe's expansion but some time later, perhaps in the interiors of fantastically hot stars.

The Explosion

A theory which suggests that our Universe started from an extremely compressed concentration of matter and radiation naturally raises the question: How did it get into that state, and what made it expand? In his original version of the expanding Universe Lemaître visualized the beginning as a giant "primordial atom" which exploded because of violent radioactive decay processes. But this conception is quite out of keeping with the picture of early evolution that we have arrived at. The young Universe must have consisted almost exclusively of high-temperature thermal radiation, and atoms, radioactive or not, could have played only a negligible role in its behavior.

A much more satisfactory answer can be obtained by considering the operation in reverse of those same relativistic formulae that we have used to describe the expansion process. The formulae tell us that various parts of the Universe are flying apart with an energy exceeding the forces of Newtonian attraction between them. Extrapolating these formulae to the period before the Universe reached the stage of maximum contraction, we find that the Universe must then have been collapsing, with just as great speed as it is now expanding!

Thus we conclude that our Universe has existed for an eternity of time, that until about five billion years ago it was collapsing uniformly from a state of infinite rarefaction; that five billion years ago it arrived at a state of maximum compression in which the density of all its matter may have been as great as that of the particles packed in the nucleus of an atom (*i.e.,* 100 million million times the density of water), and that the Universe is now on the rebound, dispersing irreversibly toward a state of infinite rarefaction.

Such motion is hyperbolic; it can be compared with the motion of a comet, which does not revolve around the sun as planets do but comes in from the infinity of space (in certain cases), sails around the sun in a bent path, developing a beautiful tail, and vanishes into infinity again without promise of return.

Before the Beginning

Any inquisitive person is bound to ask: "What was the Universe like while it was collapsing?" One might give a metaphysical answer in the words of Saint Augustine of Hippo, who wrote in his *Confessions:* "Some people say that before He made Heaven and Earth, God prepared Gehenna for those who have the hardihood to inquire into such high matters."

More recently a mathematical-physical answer was given by the Japanese physicist Chushiro Hayashi, and his idea has been elaborated by Alpher, Herman and James W. Follin of the Applied Physics Laboratory at The Johns Hopkins University. Considering the known facts about the behavior of fundamental particles, they came to the conclusion that the present chemical composition of the Universe is quite independent of its constitution before the state of maximum collapse. Transformations of particles must have occurred so rapidly during that state that the outcome was determined entirely by the conditions at the time rather than by what had gone on before.

Thus from the physical point of view we must forget entirely about the pre-collapse period and try to explain all things on the basis of facts which are no older than five billion years—plus or minus five per cent.

HERMANN BONDI:

Theories of Cosmology

COSMOLOGY is the name of the subject that deals with the large scale properties of the universe as a whole. At first sight it may seem surprising that it should be at all possible to deal with such a subject in a scientific manner. For the essence of science is the possibility of observational disproof and it might be argued that no observational knowledge of the entire universe can be at our disposal for a long time to come, if, indeed, ever. The main purpose of this article is to correct this attitude, to show that cosmology can be treated scientifically and to indicate the main lines of thought now current in the subject.

Apart from some valuable considerations due to Newton, the first serious step in scientific cosmology was taken by the Hamburg astronomer Olbers about 130 years ago. In many ways Olbers' argument is the basis of all modern cosmology and it will accordingly be presented fully.

Olbers contemplated the fact that one saw in the sky a few really bright stars, a larger number of medium bright ones, and great numbers of faint ones. He argued that persumably, in general, the brightest stars were close to us, explaining thereby both their appearance of brightness and their small number, since there is relatively little space close to us. Similarly, he supposed the medium bright stars to be somewhat further away, accounting both for their smaller luminosity and their greater number. Finally, the faint stars would be really far away and hence be faint and numerous. But what about yet more distant regions of space? Should one not expect them to appear to be populated with exceedingly numerous but individually extremely faint stars? Olbers thought that this was so and accordingly he expected these immensely distant regions to contribute a faint background brightness to the night sky. He expected stars in the depths of space to provide an almost uniform glow in the sky, individual stars being too faint to be visible but so numerous as to provide, in the aggregate, such a glow.

Olbers was not content with this thought, but proceeded to calculate the background brightness to be expected. In this he followed the established

From *The Advancement of Science*, Volume XII, No. 45, 1955, pp. 33–38. Reprinted with the kind permission of The British Association for the Advancement of Science.

scientific procedure of working out the observable consequences of hypotheses so as to be able to check them observationally. Since the individual distant stars were supposed to be too faint to be seen, Olbers could not rely on astronomical knowledge of the distant regions of space but had to make assumptions about them, assumptions that would enable him to infer the background light of the night sky and so to check the correctness of his assumptions. The assumptions Olbers made concerning the constitution of the distant parts of the universe were as follows:

(i) The distant regions are essentially like our astronomical neighbourhood. The average distance between stars and the average luminosity of each star are more or less the same throughout the universe. In other words, the universe is homogeneous when viewed on a large scale.

(ii) The general character of the universe is not only the same at all places but also at all times. (This assumption is needed since, owing to the finite velocity of light, we see distant regions not as what they are now, but what they were like a long time ago.) In other words the universe is unchanging in time when viewed on a large scale.

(iii) On an average the relative velocity of any two stars vanishes, so that there are no major systematic motions in the universe.

(iv) The laws of physics as derived from our terrestrial experience apply throughout the universe.

These assumptions seemed to Olbers, as they seem now, to be the first and most obvious ones to come to mind. They are sufficient to deduce the corresponding background brightness of the sky. The actual derivation requires a simple and brief mathematical argument.

Imagine spheres to be drawn with the earth at the centre and of radii $a, a + h, a+2h, a + 3h$ and so on, where a is much greater than h and h is a very large distance. The choice of h and a is determined by the need for every one of the spherical shells bounded by two successive spheres to be so large that it contains large numbers of stars (or, more correctly in the light of present knowledge, numerous galaxies). The number should be so large that average properties of luminosity and spacings can be used without serious error. Since the thickness of each shell is the same, the volume of each shell will be proportional to r^2, where r is the radius of the shell. (It is irrelevant whether the inner or outer radius is used since the thickness h is far smaller than r.) Accordingly, the number of stars in each shell is proportional to r^2 and so is the total rate at which stars in each shell emit light, since the average spacing and luminosity of stars have been assumed to be the same everywhere at all times. How much of this light reaches us? All the stars in the shell of radius r are at approximately the same distance r from us. Accordingly, the intensity of light received from any one of these stars is its rate of sending out light, divided by $4\pi r^2$. Therefore the intensity of light received from all stars in a shell is the rate at which all stars in the shell emit light, divided by $4\pi r^2$. Since this rate is proportional to r^2, it follows that the light received from a shell is independent of its radius (and does not vanish). Each shell contributes the same finite amount of light here. Since shells can be added without limit, the total amount of light received from all shells is infinite.

This result is not, however, quite correctly derived. True infinities rarely if ever arise in physical arguments. The discussion omitted to account for the finite size of stars which implies that each star not only sends out light, but intercepts the light of the stars behind it. This consideration does indeed prevent the sum becoming infinite, but, since stars send out so much light from a relatively small surface, the sum remains large. It turns out to be equal to the intensity of light on the surface of an average star which is about 40,000 times as strong as sun light when the sun is in the zenith.

This important result also follows from the consideration that, in Olbers' system, in whichever direction one looks, one's line of sight will eventually intercept a stellar surface. We are, therefore, entirely surrounded by stellar sur-faces and so are at the same temperature and hence in the same radiation field. Alternatively, one can argue that the space between the stars forms a thermo-dynamic system that, by assumptions (ii) and (iii), has had time to settle down to equilibrium and hence the temperature everywhere in it is the same as that of the boundary, the surfaces of the stars.

The result derived from these assumptions is patently wrong. The earth is not immersed in a diffuse bath of light 40,000 times as strong as sunlight. Olbers' paradox, as it may be called, is in striking disagreement with observation. The derivation of the paradox is logically sound and so the assumptions cannot be valid.[1] This is a remarkable result. The assumptions made concerning the nature of the universe have been disproved by observation. The possibility of observational disproof is the decisive characteristic of science. Olbers' paradox, in demonstrating the possibility of the observational disproof of a set of cosmological assumption, shows cosmology to be a science.

But the utility of Olbers' paradox does not end with this achievement. The set of four assumptions cannot be valid, but might some of them at least be maintained? Which of them can be spared most easily? There is no point in dropping assumption (iv) concerning the validity of the physical laws, since to throw away all our knowledge would seem to offer little hope of progress. Assumption (i) (Uniformity) has considerable direct and indirect observational evidence in its favour, so that the two assumptions remaining must be

---

1. The only serious attempt at getting round the argument was made by Olbers himself who suggested that absorbing matter in space was responsible for our dark night sky. This suggestion is, however, incorrect, since absorption would raise the temperature of the absorber until eventually the absorber attained a temperature so high that it would radiate as much as absorb and so would be useless. By assumption (ii) it has had time to reach this state. Nor is it possible to resolve Olbers' paradox by supposing the geometry of space to be non-Euclidean. For all that is implied by this mathematical term is that the surface of a sphere of radius r *is not* $4\pi r^2$, but some other function of r. As Olbers' result only depends on the ratio of the surfaces of two spheres of equal radius being unity, the change is irrelevant. Similarly, even if the universe were finite, light could still encircle it arbitrarily often. The night sky would be illuminated by the indefinitely repeated images of a finite number of stars and so would be just as bright as in the Euclidean case.

considered first. Can Olbers' paradox be resolved by dropping either of these?

It is clear that this can be achieved by dropping assumption (ii). For if the universe is not unchanging, then one can postulate that the stars began to shine only a finite and not too long time ago. In this case we would not receive any light from distant shells since the stars there would not have been radiating at the time light would have had to leave them to get here now. This way of resolving the paradox may be stated briefly in the phrase 'The universe is young.'

The paradox can also be resolved if assumption (ii) is retained, but assumption (iii) is dropped. This is not quite so easy to see, though.

First, one must examine what motions exist that are compatible with assumptions (i) and (ii). It is not obvious that there are such motions, and some serious mathematics is required to find the answer. It turns out that the only motions compatible with assumption (i), *i.e.* the only motions of a homogeneous system preserving its homogeneity, are[2] those in which the relative motion of any two particles is along the line joining them, has a velocity proportional to the distance between the particles (if the velocity so derived is so large as to be comparable with the velocity of light, a more complicated formula has to be used) and is either inward or outward throughout. In the first case the entire system is contracting, in the second it is expanding. The rate of expansion (or contraction) is the ratio of velocity to distance, which is the same for all particles. If assumption (ii) (unchanging character) is made this rate must be constant, otherwise it may vary in time.

Having now found the motions that become possible by rejecting only Olbers' assumption (iii), we have to see whether either of them can resolve Olbers' paradox. Fortunately, terrestrial physics readily supplies the answer. It is known that light from a receding source appears to be redder and weaker than if the source were static, that these effects become large as the velocity of the source approaches the velocity of light, and that light from an approaching source is similarly bluer and stronger. If the universe is an expanding system, then the stars of the distant shells (in Olbers' argument) are receding from us, and the far distant shells are receding very fast. The light from these shells will therefore be considerably weaker than according to Olbers' argument, and the intensity of the background light will be less than calculated before. Accordingly, if the rate of expansion is sufficiently high the background light of the sky will be as faint as it actually is, and so Olbers' paradox will be resolved. If the universe were contracting then the light from distant shells would be enhanced and the paradox would be made worse.

It follows then that the darkness of the night sky, together with assumptions (i) and (iv), implies that the universe is either expanding or it is young or both these statements apply.

---

2. The additional, probably not very important, assumption of isotropy has also been made to simplify the result.

It is probably advantageous to discuss the modern observations at this stage before describing the current theories. A big telescope shows that the stars we see form an organised group, our galaxy. This is a disc-shaped object with a radius of about 40,000 light years (the average distance between stars is a few light years, which is a unit of distance equal to six million million miles) and a thickness of a few thousand light years. Looking out in the plane of the disc we see vast numbers of stars forming the phenomenon of the Milky Way. There are perhaps 100,000 million stars in our galaxy.

Looking further into space one sees, well beyond the confines of our galaxy, other similar groups of stars, other galaxies more or less similar to our own. These galaxies frequently occur in clusters containing anything from a few to a few thousand member galaxies. The average distance between galaxies (outside clusters) is towards a million light years, about thirty times the radius of an average galaxy. With modern telescopes galaxies can be observed even at distances of many hundreds of millions of light years, and hence vast numbers of galaxies are known. Allowing for clusters, the distribution appears to be reasonably uniform and so lends support to Olbers' assumption (i). A very difficult examination of the light of distant galaxies shows that the spectral lines are shifted to the red. The only plausible inference that can be drawn is that this reddening is due to a velocity of recession. It turns out that the red shift varies between different galaxies so as to be proportional to their distances from us (as inferred by the faintness of the light received) so that the velocity of recession is proportional to the distance. It was pointed out before that expansion with such a velocity-distance law is (apart from contraction) the only possible type of motion compatible with the assumption of uniformity. The fact that the observed motion is just of this type is a strong indication that assumption (i) (uniformity) is correct. Incidentally, the velocities measured by the red shift are very large (up to one-fifth of the velocity of light).

This is the setting in which the principal current theories of cosmology must be appreciated. The first of these gives up both assumptions (ii) (unchanging aspect of the universe) and (iii) (no motion), but bases itself firmly on assumption (i) (uniformity) and (iv) (validity of the terrestrially discovered laws of nature). Since the best formulation for the behaviour of matter-in-the-large is the general theory of relativity, this cosmological theory is based on it and is known as relativistic cosmology. General relativity and uniformity do not define a unique model of the universe, but a whole range of them. However, one of these, the model discovered by G. Lemaître, is generally thought to agree best with our universe. This is an evolving, changing model. The model starts off with all the matter of the universe in a highly condensed hot nuclear state, Lemaître's 'primeval atom.' The elements are made from the primitive hydrogen in a kind of nuclear explosion that leads to violent expansion. Under the influence of gravitation the rate of expansion slows down until an almost static state is reached in which the primeval uniform distribution of matter condenses into galaxies. A repulsive force that arises naturally in relativistic cosmology finally asserts itself sufficiently to start off an accelerating period of expansion, in which we now live.

The other current theory bases itself firmly on Olbers' assumptions (i) and (ii). Because of the stress it lays on the unchanging aspect of the universe it is called the steady-state theory. The reason for this stress is found in a critical examination of assumption (iv). The laws of nature, on which relativistic cosmology is based, are a summary of our terrestrial experiences gained in what is, on the cosmical scale, an extremely small region of space and an extremely brief period of time. It is impossible, on the basis of such limited experience, to judge which features are permanent and which are only temporary, or of only local significance. In particular, if the universe varied in space or time it would be likely that some features of our physical laws (especially the so-called constants of nature) should vary with it. To put it differently, in such a varying universe there is no reason to expect our experiences to be typical, no reason to expect our summaries of these experiences, our laws of physics, to apply elsewhere. In these circumstances cosmology would be an exceptionally difficult subject in which every conceivable variation of the 'laws of nature' should be allowed for. The only possibility of avoiding these difficulties would arise if the universe were uniform in space and time, *i.e.* if Olbers' first two assumptions were valid. Then our experiences would not refer to any special place and time, but to typical conditions and hence would be applicable everywhere at all times. The assumption that this is so (*i.e.* (i) and (ii) valid) is known as the *perfect cosmological principle* as opposed to the so-called *cosmological principle* which states merely that the universe is, on the large scale, uniform in space, though not necessarily in time. It is the purpose of the steady-state theory to draw inferences from the perfect cosmological principle, inferences that can be checked observationally, so confirming or disproving the principle.

In the discussion of Olbers' paradox it was shown that if (i) and (ii) are retained, then the resolution of the paradox requires the universe to be expanding. In other words, since in the steady-state theory the aspect of the universe is unchanging, the resolution of the paradox given by the statement 'the universe is young' is inadmissible, and so the expansion follows. On the basis of the perfect cosmological principle the recession of the galaxies follows from the observation that it is dark at night, and the results of intricate astronomical work confirms this implication.

The next point is a little more difficult. If the universe is unchanging in the large, then all its large-scale physical characteristics (such as the density of matter) must be constant. But if the galaxies are receding then the intergalactic distances are increasing, matter is moving away from matter and so it would appear, by the law of conservation of matter, that the average density must be diminishing. This is in clear-cut contradiction to the perfect cosmological principle. The only way out of the contradiction is to suppose the law of conservation of matter to be incorrect to the extent that matter is being continually created. Although this may sound outrageous and contrary to experience, it is necessary to ask what the rate of this creation process is before ruling out this possibility. Now the mean density of matter in the universe is so low and the time scale of the expansion so large, that on calculation the mean rate of creation required to compensate for the effects of the expansion turns out to be one atom of hydrogen (or an equivalent mass) per quart volume every few thousand million years. This rate is obviously too small by many orders of magnitude to conflict with the experience on which the law of conservation of matter is based. There is therefore no need whatever to reject the idea of continual creation on the grounds that it disagrees with known evidence. It only disagrees with a mathematical extrapolation from evidence. Changing this may be inconvenient, but there is no reason to suppose the world to be arranged to suit the convenience of mathematicians.

In what form is this new matter created? This question is closely concerned with the problem of evolution. Every closed system is known to go through irreversible changes. Cosmically the most important of these is probably the conversion of hydrogen into helium, which takes place in every star, the excess energy being radiated away into space. Each system, each galaxy, is therefore ageing. How can the overall aspect of the universe remain unchanging, if every galaxy is evolving irreversibly? Only if new galaxies are being born, and old ones drift out of the range of telescopes through the expansion of the universe. The newly created matter must therefore stand at the beginning of the evolutionary chain, and, according to current astrophysics, this is cold diffuse hydrogen. The creation process must therefore imply the (presumably random) creation of hydrogen atoms of low velocity at a uniform rate.

The universe of the steady-state theory may be compared with a stationary human population. Each individual is born, grows up, ages and dies, but the overall aspect of the population is unchanging. This requires not only that new individuals arise to take the place of the ones who have died, but that these new individuals should be babies, so as to stand at the beginning of the evolution of the individual. Similarly, the model of the universe is inhabited by galaxies drifting apart and ageing. In the growing spaces between the galaxies, continual creation leads to the existence of vast clouds of gas that condense into new galaxies which in turn age and drift away. All the large-scale average properties (density, distance between galaxies, size of galaxies, etc.) stay constant, though each member passes through irreversible changes.

The general description of the model of the steady-state theory leads up to the essential purpose of the theory, the discovery of crucial observations. There are several of these, but only a few can be discussed here.

(i) The light that reaches us now from distant galaxies left them a long time ago. If, as in Lemaître's model, all galaxies were formed at more or less the same time, then these distant galaxies must (on an average) have been younger when they sent out the light now received than near galaxies. On the steady-state theory, however, time does not matter. The average age of galaxies a long time ago was just the same as it is now.

The observations in question consist therefore in seeing whether there are any systematic changes in the characteristics of galaxies (intrinsic colour, shape, size, etc.) with distance. If there are any such changes, the steady-state theory will have been disproved; if there are none, it will be strengthened and relativistic cosmology will be weakened.

It is not easy to separate the effects of the red shift and of the faintness from changes of intrinsic characteristics. This difficulty does not allow definite conclusions to be drawn from existing observations. There is good reason to believe that further observational work will lead to a conclusive answer in a few years.

(ii) In the example of the human population it is clear that if the population is to remain stationary, the age distribution must follow a definite pattern. Correspondingly, the age distribution of galaxies must follow a certain pattern (with young galaxies predominating) if the steady-state theory applies. In Lemaître's model, on the other hand, the range of ages is quite limited, with no galaxy less than a certain age. Once it becomes possible to judge the age of a galaxy from its appearance, this test should be powerful. This possibility should arise before long.

(iii) and (iv) Tests relating to the origin of galaxies and of heavy elements. These will be discussed in D. W. Sciama's article.

Whatever may be thought of current theories, it is clear that they are not idle speculations, but serious attempts to discern crucial observational tests. The discovery of the cosmological significance of the darkness of the night sky made cosmology a science. Current theories, in asking for more intricate but entirely possible observations, are continuing this scientific path.

D. W. SCIAMA:

Evolutionary Processes
in Cosmology

A GREAT COMPLEXITY exists in the world. This complexity greatly complicates the physicist's task. Instead of matter consisting of a simple substance like hydrogen, it manifests itself in all the complex elements—about a hundred of them. And instead of being distributed uniformly it is clumped together into galaxies and systems of galaxies separated by distances large compared with their size. In many scientific problems such complications are a nuisance, but to the cosmologist they may provide valuable clues to the history and structure of the universe. The two examples I have given here have been the most analysed from this point of view. I want, therefore, to discuss how these complexities may have arisen.

The simplest possibility to state is that they came into being as such just as the rest of the universe did; there is nothing to explain. Such an answer is logically possible, as is the suggestion that the whole universe has just been created, our memories and all. However, it is more fruitful to suppose that the complexities have developed from something essentially simpler. In particular, I want to explore the possibility that the complex elements originated as hydrogen, from which they developed as a result of nuclear reactions induced by high temperatures; and that the galaxies have developed from some simpler distribution of matter owing to the action of gravitational forces.

This does not mean that I must restrict my considerations to a universe that was once nothing but hydrogen uniformly distributed, as is supposed in Lemaître's theory of an evolving universe. For example, in the steady-state theory of the universe, the newly created matter begins life as uniformly distributed hydrogen and evolves in a complex way because of the action on it of already existing complexities. Thus, in this theory the complexities are self-propagating. I should now like to discuss in a little more detail the processes envisaged by the two theories and I shall begin with the formation of the heavy elements.

From *The Advancement of Science*, Volume XII, No. 45, 1955, pp. 38–42. Reprinted by the kind permission of The British Association for the Advancement of Science.

A good deal of painstaking work has gone into building up our present picture of the abundance of the different elements in different parts of the universe. Studies have been made of the Earth, of meteorites, the Sun, the stars, interstellar gas and dust, and other galaxies. Objects which we cannot handle reveal themselves largely through their spectra; the abundance of an element can be deduced from the strength of its characteristic contribution to the total spectrum. Unfortunately, this deduction is not a very certain one because one does not usually have a sufficiently detailed knowledge of the factors besides abundance which affect a particular element's contribution. Despite these difficulties, a fairly definite picture is beginning to emerge. Its most remarkable feature is that the abundance distribution is more or less the same everywhere; the surface of the Earth, the meteorites, the Sun, the stars, interstellar gas and distant galaxies all give very similar results when allowance is made for the special properties of each region. For instance, the surface of the Earth does not have its fair share of hydrogen, as gravity on the Earth is not strong enough to retain such a light gas.

The standard pattern is dominated by hydrogen which accounts for more than 95 per cent of all the atoms in the universe—this is a further reason for starting everything off as hydrogen—the heavy elements which seem so important to us on the Earth are really only a slight impurity produced by special circumstances. Most of the remaining 5 percent is helium, such as can be formed from hydrogen in the interiors of normal stars. As we go to heavier and heavier elements the abundances drop very rapidly until we reach an atomic weight of about 100. After this the abundances stay about the same up to the heaviest elements. This regular general behaviour, and, indeed, the detailed small deviations from it, show that the abundances are related to the nuclear properties of the atoms and not to their chemical properties. Hence, it is in the field of nuclear reactions that we must seek the key to these patterns.

As I have already said, a high temperature is needed for nuclear reactions to take place; only then do the particles collide with sufficient speed for transmutation to occur. Hydrogen gets converted into helium at the centre of the Sun and in similar stars, but elements heavier than helium need a temperature of at least fifty million degrees, which is more than three times the Sun's central temperature, and the whole range of elements need temperatures up to a thousand million degrees.

If now a critic insists that the stars are not hot enough to make heavy elements, I would reply with Eddington, 'Go and find me a hotter place.'

In the last few years Eddington's advice has been taken and two main candidates for the hotter place have been proposed: the universe itself in its first quarter of an hour, and certain special stars, particularly supernovae. The first possibility has led to what is known as the α-β-γ theory of element formation after its propounders Alpher and Gamow, who for obvious reasons implicated a well-known nuclear physicist called Bethe. This theory is set in a model of an evolving universe which started with an explosion and in its early stages had a high density and temperature. Owing to its high temperature, the contents of the universe at this time were mainly radiation, the relatively small amount of matter present being in the form of neutrons. Radiation is not dominant at the present time because as the universe expands the density of radiation decreases faster than the density of matter. About a million years after the explosion the densities of the two were equal. Now after five thousand million years the density of radiation is only one millionth of the density of matter. Although the early neutrons have little influence on the universe in its extreme youth, their behaviour then determines the present abundances of heavy elements.

In order to determine how the neutrons behave, the α-β-γ theory uses the well-established results of nuclear physics. These show that the first step consists of many of the neutrons decaying into protons and electrons. These protons then absorb neutrons to form deuterons. The deuterons then absorb neutrons and so on. In this way heavy elements are built up, though it must be admitted that there is an obstacle to this process, which I shall mention later. When the nuclei become too neutron-rich, they adjust themselves by β-decay. Thus the build-up process consists of neutron capture followed from time to time by β-decay. In such a process the amount of any particular nucleus will be mainly determined by the ease with which it captures a further neutron, the more readily it does so the fewer of that type of nucleus will be left. In this way the theory provides a simple explanation of the observed distribution that I mentioned earlier in which the abundance falls rapidly as the atomic weight increases until a weight of about 100, after which the abundance remains constant. This is just what we should expect from experimental results on the ease with which different nuclei capture neutrons. After about fifteen minutes from the beginning of the explosion, the density of free neutrons has dropped so much that the whole buildup process comes to an end. So the observed abundance distribution now after five thousand million years corresponds to the state of affairs after only a quarter of an hour.

Those who advocate the steady-state universe have to look in special localities for their hot place, since the universe as a whole was never hotter than it is now. Indeed, they can only appeal to high-temperature processes that can now be observed.

The only regions known in which sufficiently high temperatures are generated are the supernovae—stars which explode with such violence that they often outshine the entire galaxy in which they are situated. The central temperature at the beginning of the explosion is believed to be about a thousand million degrees, which is of the required order of magnitude. The rate of occurrence of supernovae is roughly one per galaxy per century, so that as many as ten or a hundred million supernovae may have exploded in the Milky Way since it was formed. The exploding stars distribute their material throughout large regions of the galaxy, so that both the production and the distribution of the heavy elements can be accounted for. Furthermore, the explosion is so rapid that the abundances of the elements will correspond more to the initial high temperature than to the lower temperatures that occur subsequently.

A detailed study of the problem has recently been made by Hoyle, who has shown that a combination of the theory of stellar structure, of observations of stars, and of nuclear physics gives the right total amounts of heavy elements and the detailed way in which abundances vary with atomic weight. A striking success of this work was the prediction of a particular energy level of the C^{12} nucleus which was needed to obtain agreement with the observed abundances. Nuclear physicists at the California Institute of Technology were thereby provoked into seeking this level, which was, indeed, found at the predicted energy.

Now I want to discuss the way in which the two theories attempt to account for the other evolutionary process I have mentioned, namely the formation of galaxies. These objects seem to be the largest units from which the universe is built; each one contains a hundred million to ten thousand million stars. These galaxies are spread out through space separated by distances of about a hundred times their own size. This distribution is itself far from regular, clusters of galaxies which appear to be gravitationally bound together being a prominent feature of the distribution.

Now, how are these galaxies formed? The basic mechanism is presumably gravitation holding the material together or perhaps collecting it from an initially more widely distributed configuration. However, there are also opposing tendencies at work in the universe. Gas pressure is one, and another is the tendency for the whole system to expand. In order that galaxies should form, gravitation must overcome these expansive tendencies in localised regions, but not in the system as a whole. The two cosmological theories have characteristically different ways of achieving this.

The basic idea of the evolving universe theory is that there was one special interval of time when gravitation could win. The importance of gravitation relative to the expansive tendencies decreases as the density of matter decreases, so that at a later stage in the history of the universe galaxies could no longer form. This is the state we are in to-day. On the other hand, at a very early stage of the universe radiation was completely dominant and its dispersive effect on matter prevented the formation of galaxies. But it is suggested that there was a compromise time about a million years after the beginning of the universe when the density of radiation was sufficiently small and the density of matter sufficiently great. At this stage the gas is supposed to have broken up into blobs which remained the same size but which separated from one another, so that they are now much further apart. Since the present age of the Milky Way can be estimated independently, one can calculate how far apart the galaxies now ought to be, and one does indeed get an answer of the right order. The theory also gives a rough estimate for the mass of a galaxy in agreement with observation.

In the steady-state theory there is no moment of time with special properties; galaxies must be forming at all times. The density of matter between galaxies is not too low for this because the creation of matter keeps it at a steady value despite the expansion. In this theory there is no first galaxy: any particular galaxy is born into a universe already full of galaxies. It is the galaxies existing at any time which enable all the dispersive forces to be overcome for the ones about to form. For the gravitational action of the galaxies on the gas around them perturbs it so as to produce concentrations which can then collapse under their own gravitational action. Thus these complexities are self-propagating. Obviously we require the children to have the same general properties as their parents, or the universe would not be in a steady state. This requirement that it must be an accurately self-propagating system is so stringent that it enables one to fix the properties of the galaxies entirely theoretically— their mass, their size, their distances apart, and their clustering tendencies with results in general agreement with observation.

I should like to end by making my own comparison between these two attempts to account for the composition and distribution of matter in the universe. I favour the steady-state theory for the following reasons. The α-β-γ theory of the heavy elements faces a severe difficulty. In the building up of the elements by neutron capture and subsequent β-decay there is a bottleneck: for there is no stable nucleus of weight 5, which means that the building up process stops at weight 4. The same difficulty would occur at weight 8. This difficulty was discovered by the proponents of the theory, who state that no way through this bottleneck has yet been found.

This difficulty does not exist in the supernova theory, where the density is sufficiently high for three-body collisions to be fairly frequent—these break through the bottleneck.

Furthermore, the steady-state theory only appeals to processes that can now be observed. These must in any case be taken into account, and if they suffice to explain the observed abundances any other substantial source of heavy elements is ruled out.

As regards galaxies, it has never been properly shown that they would form in the conditions envisaged by the α-β-γ theory, and owing to technical difficulties no detailed predictions have been made about their properties and distribution. In the steady-state theory this can be done, and general agreement with observation is then obtained.

But to my mind there is a more important reason for preferring the steady-state theory. For in theories which start from an explosion the initial properties of the universe are entirely arbitrary. Thus it is possible to find an initial temperature which is favourable for making heavy elements, and then one simply has to assume that this was the initial temperature although the general theory would be equally compatible with any other initial temperature. This means that in this type of theory the laws of physics do not specify the contents of the universe, but only show how one state of the universe follows from another. The steady-state theory opens up the exciting possibility that the laws of physics may indeed determine the contents of the universe through the requirement that all features of the universe be self-propagating. This has already been done for the distribution of galaxies, and there seems to be no obstacle to extending this principle to other properties. The requirement of self-propagation is thus a powerful new principle with whose aid we see for the first time the possibility of answering the question why things are as they are without merely saying: it is because they were as they were.

FRED HOYLE:

Continuous Creation
and the Expanding Universe

. . . LOOK OUT at the heavens on a clear night; if you
want a really impressive sight do so from a steep mountainside or from a ship
at sea. . . . By looking at any part of the sky that is distant from the Milky Way
you can see right out of the disk that forms our Galaxy. What lies out there?
Not just scattered stars by themselves, but in every direction space is strewn
with whole galaxies, each one like our own. Most of these other galaxies—
or extra-galactic nebulae as astronomers often call them—are too faint to be
seen with the naked eye, but vast numbers of them can be observed with a
powerful telescope. When I say that these other galaxies are similar to our
Galaxy, I do not mean that they are exactly alike. Some are much smaller than
ours, others are not disk-shaped but nearly spherical in form. The basic
similarity is that they are all enormous clouds of gas and stars, each one with
anything from 100,000,000 to 10,000,000,000 or so members.

Although most of the other galaxies are somewhat different from ours, it
is important to realize that some of them are indeed very like our Galaxy even
so far as details are concerned. By good fortune one of the nearest of them,
only about 700,000 light years away, seems to be practically a twin of our
Galaxy. You can see it for yourself by looking in the constellation of
Andromeda. With the naked eye it appears as a vague blur, but with a powerful telescope it shows up as one of the most impressive of all astronomical
objects. On a good photograph of it you can easily pick out places where there
are great clouds of dust. These clouds are just the sort of thing that in our
own Galaxy produces troublesome fog. . . . It is this fog that stops us
seeing more than a small bit of our own Galaxy. If you want to get an idea
of what our Galaxy would look like if it were seen from outside, the best
way is to study this other one in Andromeda. If the truth be known I expect
that in many places there living creatures are looking out across space at our

From Fred Hoyle, *The Nature of the Universe*, Harper & Bros., 1950, Chapter 6, originally
entitled "The Expanding Universe." Reprinted with the kind permission of the publishers, Harper
and Bros.

Galaxy. They must be seeing much the same spectacle as we see when we look at their galaxy.

It would be possible to say a great deal about all these other galaxies: how they are spinning round like our own; how their brightest stars are supergiants, just like those of our Galaxy; and how in those where supergiants are common, wonderful spiral patterns are found. . . . We can also find exploding stars in these other galaxies. In particular, supernovae are so brilliant that they show up even though they are very far off. Now the existence of supernovae in other galaxies has implications for our cosmology. . . . The basic requirement of the process for the coming into being of planetary systems is the supernova explosion. So we can conclude, since supernovae occur in the other galaxies, planetary systems must exist there just as in our own. Moreover, by observing the other galaxies we get a far better idea of the rate at which supernovae occur than we could ever get from our Galaxy alone. A general survey by the American observers Baade and Zwicky has shown that on the average there is a supernova explosion every four or five hundred years in each galaxy. . . . On the average each galaxy must contain more than 1,000,000 planetary systems.

How many of these gigantic galaxies are there? Well, they are strewn through space as far as we can see with the most powerful telescopes. Spaced apart at an average distance of rather more than 1,000,000 light years, they certainly continue out to the fantastic distance of 1,000,000,000 light years. Our telescopes fail to penetrate further than that, so we cannot be certain that the galaxies extend still deeper into space, but we feel pretty sure that they do. One of the questions we shall have to consider later is what lies beyond the range of our most powerful instruments. But even within the range of observation there are about 100,000,000 galaxies. With upward of 1,000,000 planetary systems per galaxy the combined total for the parts of the Universe that we can see comes out at more than a hundred million million. I find myself wondering whether somewhere among them there is a cricket team that could beat the Australians.

We now come to the important question of where this great swarm of galaxies has come from. . . . In the space between the stars of our Galaxy there is a tenuous gas, the interstellar gas. At one time our Galaxy was a whirling disk of gas with no stars in it. Out of the gas, clouds condensed, and then in each cloud further condensations were formed. This went on until finally stars were born. Stars were formed in the other galaxies in exactly the same way. But we can go further than this and extend the condensation idea to include the origin of the galaxies themselves. Just as the basic step in explaining the origin of the stars is the recognition that a tenuous gas pervades the space within a galaxy, so the basic step in explaining the origin of the galaxies is the recognition that a still more tenuous gas fills the whole of space. It is out of this general background material, as I shall call it, that the galaxies have condensed.

Here now is a question that is important for our cosmology. What is the present density of the background material? The average density is so low that a pint measure would contain only about one atom. But small as this is, the total amount of the background material exceeds about a thousandfold the combined quantity of material in all the galaxies put together. This may seem surprising but it is a consequence of the fact that the galaxies occupy only a very small fraction of the whole of space. You see here the characteristic signature of the New Cosmology. . . . Inside our Galaxy the interstellar gas outweighs the material in all the stars put together. Now we see that the background material outweighs by a large margin all the galaxies put together. And just as it is the interstellar gas that controls the situation inside our Galaxy, so it is the background material that controls the Universe as a whole. This will become increasingly clear as we go on.

The degree to which the background material has to be compressed to form a galaxy is not at all comparable with the tremendous compression necessary to produce a star. This you can see by thinking of a model in which our Galaxy is represented by a fifty-cent piece. Then the blob of background material out of which our Galaxy condensed would be only about a foot in diameter. This incidentally is the right way to think about the Universe as a whole. If in your mind's eye you take the average galaxy to be about the size of a bee—a small bee, a honeybee, not a bumblebee—our Galaxy, which is a good deal larger than the average, would be roughly represented in shape and size by the fifty-cent piece, and the average spacing of the galaxies would be about three yards, and the range of telescopic vision about a mile. So sit back and imagine a swarm of bees spaced about three yards apart and stretching away from you in all directions for a distance of about a mile. Now for each honeybee substitute the vast bulk of a galaxy and you have an idea of the Universe that has been revealed by the large American telescopes.

Next I must introduce the idea that this colossal swarm is not static: it is expanding. There are some people who seem to think that it would be a good idea if it was static. I disagree with the idea, if only because a static universe would be very dull. To show you what I mean by this I should like to point out that the Universe is wound up in two ways—that is to say, energy can be got out of the background material in two ways. Whenever a new galaxy is formed, gravitation supplies energy. For instance, gravitation supplies the energy of the rotation that develops when a galaxy condenses out of the background material. And gravitation again supplies energy during every subsequent condensation of the interstellar gas inside a galaxy. It is because of this energy that a star becomes hot when it is born. The second source of energy lies in the atomic nature of the background material. It seems likely that this was originally pure hydrogen. This does not mean that the background material is now entirely pure hydrogen, because it gets slightly adulterated by some of the material expelled by the exploding supernovae. As a source of energy hydrogen does not come into operation until high temperatures develop

—and this only arises when stars condense. It is this second source of energy that is more familiar and important to us on the Earth.

Now, why would a Universe that was static on a large scale, that was not expanding in fact, be uninteresting? Because of the following sequence of events. Even if the Universe were static on a large scale it would not be locally static: that is to say, the background material would condense into galaxies, and after a few thousand million years this process would be completed— no background would be left. Furthermore, the gas out of which the galaxies were initially composed would condense into stars. When this stage was reached hydrogen would be steadily converted into helium. After several hundreds of thousands of millions of years this process would be everywhere completed and all the stars would evolve toward the black dwarfs. . . . So finally the whole Universe would become entirely dead. This would be the running down of the Universe that was described so graphically by Jeans.

One of my main aims will be to explain why we get a different answer to this when we take account of the dynamic nature of the Universe. You might like to know something about the observational evidence that the Universe is indeed in a dynamic state of expansion. Perhaps you've noticed that a whistle from an approaching train has a higher pitch, and from a receding train a lower pitch, than a similar whistle from a stationary train. Light emitted by a moving source has the same property. The pitch of the light is lowered, or as we usually say reddened, if the source is moving away from us. Now we observe that the light from the galaxies is reddened, and the degree of reddening increases proportionately with the distance of a galaxy. The natural explanation of this is that the galaxies are rushing away from each other at enormous speeds, which for the most distant galaxies that we can see with the biggest telescopes become comparable with the speed of light itself.

My nonmathematical friends often tell me that they find it difficult to picture this expansion. Short of using a lot of mathematics I cannot do better than use the analogy of a balloon with a large number of dots marked on its surface. If the balloon is blown up the distances between the dots increase in the same way as the distances between the galaxies. Here I should give a warning that this analogy must not be taken too strictly. There are several important respects in which it is definitely misleading. For example, the dots on the surface of a balloon would themselves increase in size as the balloon was being blown up. This is not the case for the galaxies, for their internal gravitational fields are sufficiently strong to prevent any such expansion. A further weakness of our analogy is that the surface of an ordinary balloon is two dimensional—that is to say, the points of its surface can be described by two co-ordinates; for example, by latitude and longitude. In the case of the Universe we must think of the surface as possessing a third dimension. This is not as difficult as it may sound. We are all familiar with pictures in perspective—pictures in which artists have represented three-dimensional scenes on two-dimensional canvases. So it is not really a difficult conception to imagine the three dimensions of space as being confined to the surface of a balloon. But then what does the radius of the balloon represent, and what does it mean to say that the balloon is being blown up? The answer to this is that the radius of the balloon is a measure of time, and the passage of time has the effect of blowing up the balloon. This will give you a very rough, but useful, idea of the sort of theory investigated by the mathematician.

The balloon analogy brings out a very important point. It shows we must not imagine that we are situated at the center of the Universe, just because we see all the galaxies to be moving away from us. For, whichever dot you care to choose on the surface of the balloon, you will find that the other dots all move away from it. In other words, whichever galaxy you happen to be in, the other galaxies will appear to be receding from you.

Now let us consider the recession of the galaxies in a little more detail. The greater the distance of a galaxy the faster it is receding. Every time you double the distance you double the speed of recession. The speeds come out as vast beyond all precedent. Near-by galaxies are moving outward at several million miles an hour, whereas the most distant ones that can be seen with our biggest telescopes are receding at over 200,000,000 miles an hour. This leads us to the obvious question: If we could see galaxies lying at even greater distances, would their speeds be still vaster? Nobody seriously doubts that this would be so, which gives rise to a very curious situation that I will now describe.

Galaxies lying at only about twice the distance of the furthest ones that actually can be observed with the new telescope at Mount Palomar would be moving away from us at a speed that equalled light itself. Those at still greater distances would have speeds of recession exceeding that of light. Many people find this extremely puzzling because they have learned from Einstein's special theory of relativity that no material body can have a speed greater than light. This is true enough in the special theory of relativity which refers to a particularly simple system of space and time. But it is not true in Einstein's general theory of relativity, and it is in terms of the general theory that the Universe has to be discussed. The point is rather difficult, but I can do something toward making it a little clearer. The further a galaxy is away from us the more its distance will increase during the time required by its light to reach us. Indeed, if it is far enough away the light never reaches us at all because its path stretches faster than the light can make progress. This is what is meant by saying that the speed of recession exceeds the velocity of light. Events occurring in a galaxy at such a distance can never be observed at all by anyone inside our Galaxy, no matter how patient the observer and no matter how powerful his telescope. All the galaxies that we actually see are ones that lie close enough for their light to reach us in spite of the expansion of space that's going on. But the struggle of the light against the expansion of space does show itself, as I said before, in the reddening of the light.

As you will easily guess, there must be intermediate cases where a galaxy is at such a distance that, so to speak, the light it emits neither gains ground nor loses it. In this case the path between us and the galaxy stretches at just such a rate as exactly compensates for the velocity of the light. The light gets lost on the way. It is a case, as the Red Queen remarked to Alice, of "taking all the running you can do to keep in the same place." We know fairly accurately how far away a galaxy has to be for this special case to occur. The answer is about 2,000,000,000 light years, which is only about twice as far as the distances that we expect the giant telescope at Mount Palomar to penetrate. This means that we are already observing about half as far into space as we can ever hope to do. If we built a telescope a million times as big as the one at Mount Palomar we could scarcely double our present range of vision. So what it amounts to is that owing to the expansion of the Universe we can never observe events that happen outside a certain quite definite finite region of space. We refer to this finite region as the observable Universe. The word "observable" here does not mean that we actually observe, but what we could observe if we were equipped with perfect telescopes.

So far we have been entirely concerned with the rich fruits of twentieth century observational astronomy and in particular with the results achieved by Hubble and his colleagues. We have seen that all space is strewn with galaxies, and we have seen that space itself is continually expanding. Further questions come crowding in: What causes the expansion? Does the expansion mean that as time goes on the observable Universe is becoming less and less occupied by matter? Is space finite or infinite? How old is the Universe? To settle these questions we shall now have to consider new trains of thought. These will lead us to strange conclusions.

First I will consider the older ideas—that is to say, the ideas of the nineteen-twenties and the nineteen-thirties—and then I will go on to offer my own opinion. Broadly speaking, the older ideas fall into two groups. One of them is distinguished by the assumption that the Universe started its life a finite time ago in a single huge explosion. On this supposition the present expansion is a relic of the violence of this explosion. This big bang idea seemed to me to be unsatisfactory even before detailed examination showed that it leads to serious difficulties. For when we look at our own Galaxy there is not the smallest sign that such an explosion ever occurred. This might not be such a cogent argument against the explosion school of thought if our Galaxy had turned out to be much younger than the whole Universe. But this is not so. On the contrary, in some of these theories the Universe comes out to be younger than our astrophysical estimates of the age of our own Galaxy. Another really serious difficulty arises when we try to reconcile the idea of an explosion with the requirement that the galaxies have condensed out of diffuse background material. The two concepts of explosion and condensation are obviously contradictory, and it is easy to show, if you postulate an explosion of sufficient violence to explain the expansion of the Universe, that condensations looking at all like the galaxies could never have been formed.

And so we come to the second group of theories that attempt to explain the expansion of the Universe. These all work by monkeying with the law of gravitation. The conventional idea that two particles attract each other is only accepted if their distance apart is not too great. At really large distances, so the argument goes, the two particles repel each other instead. On this basis it can be shown that if the density of the background material is sufficiently small, expansion must occur. But once again there is a difficulty in reconciling all this with the requirement that the background material must condense to form the galaxies. For once the law of gravitation has been modified in this way the tendency is for the background material to be torn apart rather than for it to condense into galaxies. Actually there is just one way in which a theory along these lines can be built so as to get round this difficulty. This is a theory worked out by Lemaître which was often discussed by Eddington in his popular books. But we now know that on this theory the galaxies would have to be vastly older than our astrophysical studies show them actually to be. So even this has to be rejected.

I should like now to approach more recent ideas by describing what would be the fate of our observable universe if any of these older theories had turned out to be correct. According to them every receding galaxy will eventually increase its distance from us until it passes beyond the limit of the observable universe—that is to say, they will move to a distance beyond the critical limit of about 2,000,000,000 light years that I have already mentioned. When this happens they will disappear—nothing that then occurs within them can ever be observed from our Galaxy. So if any of the older theories were right we should end in a seemingly empty universe, or at any rate in a universe that was empty apart perhaps from one or two very close galaxies that became attached to our Galaxy as satellites. Nor would this situation take very long to develop. Only about 10,000,000,000 years—that is to say, about a fifth of the lifetime of the Sun—would be needed to empty the sky of the 100,000,000 or so galaxies that we can now observe there.

My own view is very different. Although I think there is no doubt that every galaxy we observe to be receding from us will in about 10,000,000,000 years have passed entirely beyond the limit of vision of an observer in our Galaxy, yet I think that such an observer would still be able to see about the same number of galaxies as we do now. By this I mean that new galaxies will have condensed out of the background material at just about the rate necessary to compensate for those that are being lost as a consequence of their passing beyond our observable universe. At first sight it might be thought that this could not go on indefinitely because the material forming the background would ultimately become exhausted. The reason why this is not so, is that new material appears to compensate for the background material that is constantly being condensed into galaxies. This is perhaps the most surprising of all the conceptions of the New Cosmology. For I find myself forced to assume that the nature of the Universe requires continuous creation—the perpetual bringing into being of new background material.

The idea that matter is created continuously represents our ultimate goal in this book. It would be wrong to suppose that the idea itself is a new one. I know of references to the continuous creation of matter that go back more than twenty years, and I have no doubt that a close inquiry would show that the idea, in its vaguest form, goes back very much further than that. What is new about it is this: it has now been found possible to put a hitherto vague idea in a precise mathematical form. It is only when this has been done that the consequences of any physical idea can be worked out and its scientific value assessed. I should perhaps explain that besides my personal views, which I shall now be putting forward, there are two other lines of thought on this matter. One comes from the German scientist P. Jordan, whose views differ from my own by so wide a gulf that it would be too wide a digression to discuss them. The other line of attack has come from the Cambridge scientists H. Bondi and T. Gold, who, although using quite a different form of argument from the one I adopted, have reached conclusions almost identical with those I am now going to discuss.

The most obvious question to ask about continuous creation is this: Where does the created material come from? It does not come from anywhere. Material simply appears—it is created. At one time the various atoms composing the material do not exist, and at a later time they do. This may seem a very strange idea and I agree that it is, but in science it does not matter how strange an idea may seem so long as it works—that is to say, so long as the idea can be expressed in a precise form and so long as its consequences are found to be in agreement with observation. Some people have argued that continuous creation introduces a new assumption into science—and a very startling assumption at that. Now I do not agree that continuous creation is an additional assumption. It is certainly a new hypothesis, but it only replaces a hypothesis that lies concealed in the older theories, which assume, as I have said before, that the whole of the matter in the Universe was created in one big bang at a particular time in the remote past. On scientific grounds this big bang assumption is much the less palatable of the two. For it is an irrational process that cannot be described in scientific terms. Continuous creation, on the other hand, can be represented by precise mathematical equations whose consequences can be worked out and compared with observation. On philosophical grounds too I cannot see any good reason for preferring the big bang idea. Indeed it seems to me in the philosophical sense to be a distinctly unsatisfactory notion, since it puts the basic assumption out of sight where it can never be challenged by a direct appeal to observation.

Perhaps you may think that the whole question of the creation of the Universe could be avoided in some way. But this is not so. To avoid the issue of creation it would be necessary for all the material of the Universe to be infinitely old, and this it cannot be for a very practical reason. For if this were so, there could be no hydrogen left in the Universe. . . . Hydrogen is being steadily converted into helium throughout the Universe and this conversion is a one-way process—that is to say, hydrogen cannot be produced in any appreciaole quantity through the breakdown of the other elements. How comes it then that the Universe consists almost entirely of hydrogen? If matter were infinitely old this would be quite impossible. So we see that the Universe being what it is, the creation issue simply cannot be dodged. And I think that of all the various possibilities that have been suggested, continuous creation is easily the most satisfactory.

Now what are the consequences of continuous creation? Perhaps the most surprising result of the mathematical theory is that the average density of the background material must stay constant. The new material does not appear in a concentrated form in small localized regions but is spread throughout the whole of space. The average rate of appearance of matter amounts to no more than the creation of one atom in the course of about a year in a volume equal to that of a moderate-sized skyscraper. As you will realize, it would be quite impossible to detect such a rate of creation by direct experiment. But although this seems such a slow rate when judged by ordinary ideas, it is not small when you consider that it is happening everywhere in space. The total rate for the observable universe alone is about a hundred million, million, million, million, million tons per second. Do not let this surprise you because, as I have said, the volume of the observable universe is very large. Indeed I must now make it quite clear that here we have the answer to our question, Why does the Universe expand? For it is this creation that drives the Universe. The new material produces an outward pressure that leads to the steady expansion. But it does much more than that. With continuous creation the apparent contradiction between the expansion of the Universe and the requirement that the background material shall be able to condense into galaxies is completely overcome. For it can be shown that once an irregularity occurs in the background material a galaxy must eventually be formed. Such irregularities are constantly being produced by the gravitational effect of the galaxies themselves. For the gravitational field of the galaxies disturbs the background material and causes irregularities to form within it. So the background material must give a steady supply of new galaxies. Moreover, the created material also supplies unending quantities of atomic energy, since by arranging that newly created material should be composed of hydrogen we explain why in spite of the fact that hydrogen is being consumed in huge quantities in the stars, the Universe is nevertheless observed to be overwhelmingly composed of it.

We must now leave this extraordinary business of continuous creation for a moment to consider the question of what lies beyond the observable part of the Universe. In the first place you must let me ask, Does this question have any meaning? According to the theory it does. Theory requires the galaxies to go on forever, even though we cannot see them. That is to say, the galaxies are expanding out into an infinite space. There is no end to it all. And what is more, apart from the possibility of there being a few freak galaxies, one bit of this infinite space will behave in the same way as any other bit.

The same thing applies to time. You will have noticed that I have used the concepts of space and time as if they could be treated separately. According to the relativity theory this is a dangerous thing to do. But it so happens that it can be done with impunity in our Universe, although it is easy to imagine other universes where it could not be done. What I mean by this is that a division between space and time can be made and this division can be used throughout the whole of our Universe. This is a very important and special property of our Universe, which I think it is important to take into account in forming the equations that decide the way in which matter is created.

Perhaps you will allow me a short diversion here to answer the question: How does the idea of infinite space fit in with the balloon analogy that I mentioned earlier? Suppose you were blowing up a balloon that could never burst. Then it is clear that if you went on blowing long enough you could make its size greater than anything I cared to specify, greater for instance than a billion billion miles or a billion billion billion miles and so on. This is what is meant by saying that the radius of the balloon tends to infinity. If you are used to thinking in terms of the balloon analogy, this is the case that gives you what we call an infinite space.

Now let us suppose that a film is made from any space position in the Universe. To make the film, let a still picture be taken at each instant of time. This, by the way, is what we are doing in our astronomical observations. We are actually taking the picture of the Universe at one instant of time—the present. Next, let all the stills be run together so as to form a continuous film. What would the film look like? Galaxies would be observed to be continually condensing out of the background material. The general expansion of the whole system would be clear, but though the galaxies seemed to be moving away from us there would be a curious sameness about the film. It would be only in the details of each galaxy that changes would be seen. The overall picture would stay the same because of the compensation whereby the galaxies that were constantly disappearing through the expansion of the Universe were replaced by newly forming galaxies. A casual observer who went to sleep during the showing of the film would find it difficult to see much change when he awoke. How long would our film show go on? It would go on forever.

There is a complement to this result that we can see by running our film backward. Then new galaxies would appear at the outer fringes of our picture as faint objects that come gradually closer to us. For if the film were run backward the Universe would appear to contract. The galaxies would come closer and closer to us until they evaporated before our eyes. First the stars of a galaxy would evaporate back into the gas from which they were formed. Then the gas in the galaxy would evaporate back into the general background from which it had condensed. The background material itself would stay of constant density, not through matter being created, but through matter disappearing. How far could we run our hypothetical film back into the past? Again according to the theory, forever. After we had run backward for about 5,000,000,000 years our own Galaxy itself would disappear before our eyes. But although important details like this would no doubt be of great interest to us there would again be a general sameness about the whole proceeding. Whether we run the film backward or forward the large-scale features of the Universe remain unchanged.

It is a simple consequence of all this that the total amount of energy that can be observed at any one time must be equal to the amount observed at any other time. This means that energy is conserved. So continuous creation does not lead to nonconservation of energy as one or two critics have suggested. The reverse is the case for without continuous creation the total energy observed must decrease with time.

We see, therefore, that no large-scale changes in the Universe can be expected to take place in the future. But individual galaxies will change and you may well want to know what is likely to happen to our Galaxy. This issue cannot be decided by observation because none of the galaxies that we observe can be much more than 10,000,000,000 years old as yet, and we need to observe much older ones to find out anything about the ultimate fate of a galaxy. The reason why no observable galaxy is appreciably older than this is that a new galaxy condensing close by our own would move away from us and pass out of the observable region of space in only about 10,000,000,000 years. So we have to decide the ultimate fate of our Galaxy again from theory, and this is what theory predicts. It will become steadily more massive as more and more background material gets pulled into it. After about 10,000,000,000 years it is likely that our Galaxy will have succeeded in gathering quite a cloud of gas and satellite bodies. Where this will ultimately lead is difficult to say with any precision. The distant future of the Galaxy is to some extent bound up with an investigation made about thirty years ago by Schwarzschild, who found that very strange things happen when a body grows particularly massive. It becomes difficult, for instance, for light emitted by the body ever to get out into surrounding space. When this stage is reached, further growth is likely to be strongly inhibited. Just what it would then be like to live in our Galaxy I should very much like to know.

To conclude, I should like to stress that so far as the Universe as a whole is concerned the essential difference made by the idea of continuous creation of matter is this: Without continuous creation the Universe must evolve toward a dead state in which all the matter is condensed into a vast number of dead stars. The details of the way this happens are different in the different theories that have been put forward, but the outcome is always the same. With continuous creation, on the other hand, the Universe has an infinite future in which all its present very large-scale features will be preserved.

Selected Bibliography

First Steps in Cosmological Speculation

Bailey, C., *The Greek Atomists and Epicurus* (Oxford, 1928).

Cornford, F. M., *From Religion to Philosophy* (London, 1912).

Frankfort, H., *et. al., The Intellectual Adventure of Ancient Man* (Chicago, 1946).

Freeman, Kathleen, *Companion to the Pre-Socratic Philosophers* (Oxford, 1949).

Heidel, Alexander, *The Babylonian Genesis* (Chicago, 1951).

Hooke, S. H., *In the Beginning* (Oxford, 1947).

Kramer, Samuel, *From the Tablets of Sumer* (Indian Hills, Colorado, Falcon's Wing Press, 1956).

Leach, Maria, *The Beginning: Creation Myths Around the World* (New York, 1956).

Menon, C. P. S., *Early Astronomy and Cosmology* (London, 1932).

Neugebauer, Otto, *The Exact Sciences in Antiquity* (Princeton, 1952).

O'Neill, John, *The Night of the Gods: An Inquiry into Cosmic and Cosmogonic Mythology,* 2 vols. (London, 1897).

Schiaparelli, Giovanni V., *Astronomy in the Old Testament* (Oxford, 1905).

Smith, E. Baldwin, *The Dome: A Study in the History of Ideas* (Princeton, 1950).

Thompson, Stith, *Motif Index of Folk-Literature* (Bloomington, 1955).

Zimmern, Heinrich, *The Babylonian and Hebrew Genesis* (London, 1901).

The Classic View of a Geocentric-Finite Universe

Clagett, Marshall, *Greek Science in Antiquity* (New York, 1955).

Cornford, F. M., *Plato's Cosmology* (London, 1937).

Crombie, A. C., *Augustine to Galileo* (London, 1952).

Dreyer, J. L. E., *History of the Planetary Systems from Thales to Kepler* (Cambridge, 1906).

Duhem, Pierre, *Le Système du Monde: Histoire des Doctrines Cosmologiques de Platon à Copernic* (6 vols. Paris, 1913, new ed. 1954——).

Heath, Thomas, *Aristarchus of Samos: The Ancient Copernicus* (Oxford, 1913).

Macrobius, *Commentary on the Dream of Scipio,* translated by W. H. Stahl (New York, 1952).

Orr, M. A., *Dante and the Early Astronomers* (London, 1913).

Sambursky, S., *The Physical World of the Greeks* (London, 1956).

Taylor, A. E., *A Commentary on Plato's Timaeus* (Oxford, 1928).

Thorndike, Lynn, *The Sphere of Sacrobosco and Its Commentators* (Chicago, 1949).

Wolfson, Harry A., *Crescas' Critique of Aristotle* (Cambridge, 1929).

The Copernican Revolution and Its Aftermath

Armitage, Angus, *Copernicus* (London, 1939).

Burtt, E. A., *Metaphysical Foundations of Modern Physical Science* (New York, 1925).

Busco, Pierre, *Les Cosmogoniques Modernes et la Théorie de la Connaissance,* (Paris, 1924).

Butterfield, H., *The Origins of Modern Science* (London, 1949).

Clerke, Agnes M., *Modern Cosmogonies* (London, 1905).

Collier, Katherine B., *Cosmogonies of Our Fathers: Some Theories of the 17th and 18th Centuries* (New York, 1934).

Fahie, J. J., *Galileo: His Life and Work* (London, 1903).

Johnson, F. R., *Astronomical Thought in Renaissance England* (Baltimore, 1937).

Macpherson, H., *Modern Cosmologies* (Oxford, 1929).

Rosen, E., *Three Copernican Treatises* (New York, 1939).

Santillana, de Giorgio, *The Crime of Galileo* (Chicago, 1955).

Sidgwick, J., *William Herschel: Explorer of the Heavens* (London, 1953).

Singer, D. W., *Giordano Bruno: His Life and Thought* (New York, 1950.)

Stimson, Dorothy, *The Gradual Acceptance of the Copernican Theory of the Universe* (New York, 1917).

Taylor, F. Sherwood, *Galileo and the Freedom of Thought* (Chicago, 1955).

Modern Theories of the Universe

D'Abro, A., *The Evolution of Scientific Thought from Newton to Einstein* (New York, 1950).

Bondi, H., *Cosmology* (Cambridge, 1952).

Couderc, Paul, *The Expansion of the Universe,* translated by J. B. Sidgwick (London, 1952).

Dingle, H., *The Scientific Adventure* (London, 1952).

Eddington, A. S., *The Expanding Universe* (Cambridge, 1933).

Einstein, A., *The Meaning of Relativity* (Princeton, 1956).

Finlay-Freundlich, E., *Cosmology* (International Encyclopedia of Unified Science, Vol. I, No. 8) (Chicago, 1951).

Gamow, George, *The Creation of the Universe* (New York, 1952).

Heckmann, Otto, *Theorien der Kosmologie* (Berlin, 1942).

Hoyle, Fred, *The Nature of the Universe* (Blackwell, 1950).

Hoyle, Fred, *Frontiers of Astronomy* (London, 1955).

Hubble, E. P., *The Observational Approach to Cosmology* (Oxford, 1937).

Hubble, E. P., *The Realm of the Nebulae* (New Haven, 1936).

Jeans, James H., *The Astronomical Horizon* (Oxford, 1945).

Jeans, James H., *Astronomy and Cosmogony* (Cambridge, 1929).

Johnson, Martin, *Time, Knowledge, and the Nebulae* (London, 1945).

Jones, G. O., Rotblat, J., and Whitrow, G. J., *Atoms and the Universe* (London, 1956).

Jordan, Pascual, *Schwerkraft und Weltall* (Brunswick, 1952).

Lemaître, Georges, *The Primeval Atom* (New York, 1950).

McCrea, W. H., "Cosmology" (in *Reports on Progress in Physics,* of the Physical Society, Vol. XVI) (London, 1953).

McVittie, G. C., *General Relativity and Cosmology* (New York, 1956).

Milne, E. A., *Relativity, World Structure and Gravitation* (Oxford, 1935).

Milne, E. A., *Kinematic Relativity* (Oxford, 1948).

Milne, E. A., *Modern Cosmology and the Christian Idea of God* (Oxford, 1952).

Reichenbach, Hans, *Philosophie der Raum-Zeit Lehre* (Berlin, 1928).

Shapley, Harlow, *Galaxies* (Philadelphia, 1943).

De Sitter, W., *Kosmos* (Cambridge, 1932).

Tolman, Richard C., *Relativity Thermodynamics and Cosmology* (Oxford, 1954).

Tornebohm, H., *A Logical Analysis of the Theory of Relativity* (Stockholm, 1952).

Von Weizsäcker, C. F., *The History of Nature* (Chicago, 1949).

Whitrow, G. J., *The Structure of the Universe* (London, 1949).

Whittaker, E. T., *The Beginning and End of the World* (Oxford, 1942).

Whittaker, E. T., *Space and Spirit* (London, 1946).

Index

Acusilaus, 27
Adam, J., 76n.
Adam of Bremen, 129
Adams, J. C., 302
Aeschylus, 27
Aëtius, 158n.
Æthicus of Istria, 125
Ahmed ben Muhammed ben Ketu al Fagani, *see* Alfraganus
Albattani, *see* Albategni Aratensis
Albategni Aratensis, 134 n., 166
Albertus Magnus, 133
Alexander of Aphrodisias, 154n.
Alexander, the Great, 39, 126, 132
Alfargani, 134, 136, 165n., 166n., 188n.
Alfraganus, *see* Alfargani
Allan, F., 92n., 95n.
Alpetragius, 165
Alpher, Ralph A., 403, 404, 414
Ambrose of Milan, 119, 120, 125
Ambrosius, *see* Ambrose of Milan
Anaxagoras, 22, 83, 156, 185, 209n.
Anaximander, 6, 7, 22, 24, 28, 29, 30, 156
Anaximenes, 6, 7
Apian, Peter, 116
Apollonius, 24
Aquinas, St. Thomas, 132n., 133, 135, 136
Aratus, 209n.
Archimedes, 100n., 160n., 302, 307
Aristarchus of Samos, 7, 40, 173n., 190, 197
Aristophanes, 24, 26
Aristotle, 1, 6, 22, 27, 34n., 38, 40, 61, 62, 63, 64, 65, 77n., 85, 89ff, 101ff, 104, 117, 124, 132, 133, 134, 135, 136, 137, 142, 143, 153n., 155n., 156n., 157n., 158n., 161, 162, 163, 166n., 169, 173n., 175, 176, 177, 180n., 185, 186, 188, 190, 191, 192, 197, 199, 401

Armitage, Angus, 434
Athenagoras, 24
Augustine of Hippo, 1, 20, 132n., 404
Avendeath, *see* Hispalensis, Johannes
Averroes, 166

Baade, Walter, 274, 292, 395, 398, 420
Bacchylides, 26
Bacon, Francis, 354
Bacon, Roger, 134, 135
Bailey, C., 433
Balboa, Vasco Minez de, 155n.
Barankin, E. W., 386n.
Basil, the Great, 117, 118
Beatus, 129
Bede, Pseudo-, 130
Bede, the Venerable, 126, 127, 129, 130, 132n.
Bentley, Richard, 211ff
Berger, H., 39n.
Bessel, F. W., 304
Bethe, H. A., 414
Bible, 6, 7, 29, 30, 117, 118, 119, 120, 122, 123, 131, 133, 135, 199, 209n.
Birkhoff, G. D., 332
Boeckh, A., 34n., 39
Bolyai, J., 378
Bondi, Hermann, 274, 397, 405ff, 426, 434
Boniface, 127, 128n.
Boyle, R., 208
Bradley, J., 235
Brahe, *see* Tycho de Brahe
Bridgman, P. W., 357, 370
Brodetsky, Selig, 149n.
Brunetto Latini, 135, 136
Bruno, Giordano, 143, 174ff, 189, 223
Burtt, E. A., 434
Busco, Pierre, 434

FREE PRESS PAPERBACKS

A Series of Paperbound Books in the Social and Natural Sciences, Philosophy, and the Humanities

These books, chosen for their intellectual importance and editorial excellence, are printed on good quality book paper, from the large and readable type of the cloth-bound edition, and are Smyth-sewn for enduring use. Free Press Paperbacks conform in every significant way to the high editorial and production standards maintained in the higher-priced, case-bound books published by The Free Press of Glencoe.

Many of these books are available in their original cloth bindings.
A complete catalogue of all Free Press titles will be sent on request